D1293767

Planning for cycling

Principles, practice and solutions for urban planners

**Edited by
Hugh McClintock**

CRC Press
Boca Raton Boston New York Washington, DC

WOODHEAD PUBLISHING LIMITED
Cambridge England

Published by Woodhead Publishing Limited, Abington Hall, Abington
Cambridge CB1 6AH, England
www.woodhead-publishing.com

Published in North America by CRC Press LLC, 2000 Corporate Blvd, NW
Boca Raton FL 33431, USA

First published 2002, Woodhead Publishing Ltd and CRC Press LLC
© 2002, Woodhead Publishing Ltd
The authors have asserted their moral rights.

British Library Cataloguing in Publication Data
A catalogue record for this book is available from the British Library.

Library of Congress Cataloging in Publication Data
A catalog record for this book is available from the Library of Congress.

Woodhead Publishing ISBN 1 85573 581 4
CRC Press ISBN 0-8493-1553-0
CRC Press order number: WP1553

Typeset by SNP Best-set Typesetter Ltd., Hong Kong
Printed by TJ International, Cornwall, England

Contents

Preface

Much has happened to develop cycling policy since the publication of the predecessor of this book, *The Bicycle and City Traffic: Principles and Practice*, by Belhaven Press in 1992. I must pay tribute to the strong encouragement to embark on a completely new and greatly expanded version which I received from Iain Stevenson of Woodhead Publishing Limited and the City University.

A great many people have contributed to the evolution of my ideas about planning for cyclists and I must mention specifically a number of these, at different levels, local, national, European and international. They include Howard Boyd, Andy Clarke, Johanna Cleary, David Davies, John Franklin, Lawrence Geary, David Holladay, Don Mathew, Nick Moss, Derek Palmer, Stuart Reid, Tony Russell and Rodney Tolley. I have also learned much from numerous other people in the Cycle Campaigns Network, the CTC, the Local Authorities Cycle Planning Group as well as from many people in other countries, especially the always very diverse and lively range of people at the European Cyclists' Federation Velo-City cycling conferences.

Hugh McClintock
University of Nottingham

Notes on contributors

Colin Ashton-Graham has 11 years' experience in planning and advocacy for balanced transport. He was the Rights and Planning Executive at the Cyclists' Touring Club UK between 1990 and 1996, where he played a leading role in the creation of the UK National Cycling Strategy. He then joined Transport in Western Australia as the Visiting Fellow taking on the Perth Bicycle Network Co-ordinator role between 1997 and 1999, seeing the first stage through from funding application to delivery. Colin is now the Acting Manager of Balanced Transport Strategy with the Department for Planning and Infrastructure Western Australia. He is responsible for development and delivery of the TravelSmart Individualised Marketing programme and for the implementation of pedestrian strategy.

Wolfgang Bohle is a traffic engineer and sociologist, and consultant with Planungsgemeinschaft Verkehr, Hanover, Germany.

Werner Brög founded the Institute of Socialdata in Munich in 1972 and has been the Managing Director since. He specialises in transport modelling and the development and application of public awareness strategies. This pioneering work has seen Werner develop the KONTIV® continuous travel survey tool that is now used in many projects throughout the world. His innovations in the promotion of domestic recycling have expanded to focus on the delivery of travel behaviour change interventions. Werner's company Socialdata own the IndiMark™ methodology that is now used by many public transport providers and governments across Europe and beyond. Werner has authored or co-authored more than 400 publications and papers in his field and has taught at the Massachusetts Institute of Technology and the Technical University of Graz.

Andy Clarke is the Executive Director of the Association of Pedestrian and Bicycle Professionals, a position he has held since June 1999. Andy currently works on-site at the Federal Highway Administration as part of a contract with the University of North Carolina to provide technical assistance for the Pedestrian and Bicycle Information Center. Before becoming APBP's first staff person, Andy was a consultant to the FHWA Bicycle, Pedestrian and Trails Program where he had responsibility for the implementation and explanation of the bicycle and pedestrian-related provisions of the federal transportation programme, including the Transportation Equity Act for the 21st Century (TEA-21).

Prior to his work at FHWA, Clarke spent more than a decade working for a variety of non-profit organisations in the bicycle and pedestrian transportation field. Originally from the UK, Clarke spent three years with Friends of the Earth and the European Cyclists' Federation before coming to the United States. A two year stint as Government Relations Director for the League of American Bicyclists was followed by six years as Deputy Director of the Bicycle Federation of America and two years with the Rails to Trails Conservancy. Throughout this time, Andy has worked to create, interpret and implement the various transportation funding programmes that are available to improve conditions for bicycling and walking.

In addition to his legislative and policy background, Andy has managed a range of bicycle and pedestrian planning projects at the state and metropolitan level and has worked extensively with state and local advocacy groups. He is former Chair of the Transportation Research Board's Committee on Bicycling.

Jo Cleary trained as a town planner during the 1980s in Bristol. On graduation she became national planning adviser for the UK's largest cycling organisation, the CTC, where her role was to liaise with national and local government, and other relevant organisations, to try and ensure the needs of cyclists were safeguarded and promoted in policy and practice. During her time at the CTC she contributed to two working groups of the Parliamentary Advisory Committee for Transport Safety (PACTS), and was a member of the management committee of the European Cyclists' Federation. From the CTC Jo went on to study full time for a PhD in cycle planning at the University of Nottingham, under the supervision of Hugh McClintock. On completion of her studies Jo established a small sustainable transport planning consultancy, based in Nottingham. The consultancy has developed a specialism in school travel, with work ranging from national government research projects to studies with individual schools at the grass root level. Jo lives with her daughter in a car-free household: they walk, cycle and use public transport for the majority of journeys, and hire a car when necessary.

Helen Grey-Smith is the Project Co-ordinator for Socialdata Australia. She has a background in art, teaching, small business management and environment education. Helen was actively involved in environmental issues in Western Australia in the 1970s, and in Canberra from the mid-1980s. She worked with community and school groups on revegetation projects, helped promote community Landcare in the Canberra area, and was a Director of Greening Australia.

Her current role involves managing the field work for travel surveys and IndiMark™ delivery in Western Australia and Queensland.

John Grimshaw, MBE, is Director and Chief Engineer, Sustrans. John studied engineering at Cambridge University before joining contractors Taylor Woodrow; doing a spell with VSO in Uganda; and then consultants Mander, Raikes and Marshall based in Bristol.

In 1980 he left to work full time designing and developing quality cycle routes, often on the alignments of abandoned railways. These immensely popular projects led to the formation of Sustrans Ltd, a registered charity, to carry out practical demonstration projects pointing towards a sustainable transport future. The most significant of these has been the National Cycle Network, partly funded by the Millennium Commission.

Sustrans also launched the Safe Routes to Schools programme and is currently engaged in various initiatives in rural transport; station access; Active Travel – promoting public health; homezones and TravelSmart; all aimed at creating an appetite for change in current transport practice.

James Harrison has been responsible for cycle planning for the City of York Council for over 10 years and is now acting as head of the Transport Strategy team for the city. Prior to that he worked at the University of Southampton researching cycle accidents and remedial measures before joining consultants TPA.

His publications include two TRL reports and occasional magazine articles. He was on the editorial steering group for the production of IHT's *Cycle-Friendly Infrastructure* and was the author of a section of the National Cycling Strategy.

Dave Holladay has had an interest in study and operation of the widest spectrum of transport modes for over 30 years. After obtaining an engineering degree at Bradford and postgraduate qualification in railway engineering at Cranfield he worked for British Rail and Sustrans as an engineer. He has led initiatives to integrate cycling with public transport and enjoys the challenge of innovative problem solving. A member of the Intermodality Working Group of the National Cycling Strategy, he was active in a short term sub-group which produced guidance on bike carriage on buses and coaches, and the use of cycle parking in a fully rounded package of integration.

He has moved from designing bike hanging units for Scotrail's Young Explorer trains to the 1996 Bike on Bus Cycle Challenge and thence to a situation where some 30 vehicles are now equipped to carry bikes on scheduled local services, and many express coach operators are accommodating bikes on inter-urban routes. His study of bikes and trams has grown since a tour of the US in 1990, which provided material for Sustrans' 1993 report on the then proposed Advanced Transport in Avon (ATA) scheme. His work on cycle parking has produced a comprehensive guide for on-street provision, and management and reports for various rail station access and parking proposals in Scotland.

Currently he is active on the Safe Routes to Station project (now known as 'Access to Public Transport'), with a remit to bring in bus, LRT and walking to the core of rail/cycling-based work. A personal travel regime, based around cycling – only as much as necessary – and use of public transport gives him a wide knowledge of the potential and limitations on travels throughout the UK, the rest of Europe and the US/Canada.

Bruce James has over 20 years' experience in community development and programme management in local and state government in Western Australia. He came through a Recreation Officer role to join Transport Western Australia in 1990 as Principal Policy Officer. Bruce pioneered the development of Travel Demand Management and took on the role of Manager, Travel Demand Management, in 1995. Bruce was responsible for the innovation of TravelSmart Individualised Marketing to include cycling and walking information and to apply the technique in the Perth context. Bruce is currently Manager, Balanced Transport, with the Department for Planning and Infrastructure.

Gary John has 20 years' experience in transport, strategic and statutory planning. He joined the Ministry for Planning Western Australia in 1987 after working for a number of years as a town planner in local government. In 1996 he joined the Travel Demand Branch at Transport WA as Senior Policy Officer where he has been involved in research, development and implementation of a range of plans and programmes aimed at reducing car dependence for workplaces, schools, households and major traffic generators (i.e. universities). He is now the Acting Manager of Balanced Transport Implementation with the Department for Planning and Infrastructure in Western Australia where he is responsible for the delivery of marketing and behaviour change projects for cycling, walking and public transport.

Thomas Krag is a civil engineer who has been working in cycling for more than 25 years. Though having the user's point of view as the starting point, he has covered a wide range of issues and approached the problems from many different angles. He was President of the European Cyclists' Federation 1993–96 and has sat on numerous committees and commissions. He started his career as a volunteer in the Danish Cyclist Federation and turned it later into a profession, being head of the secretariat and director of the organisation 1986–2000. Since then Thomas Krag has been working as a consultant dealing with cycling and mobility questions in general.

Marcello Mamoli is an architect who graduated in Venice in 1973. Since 1992 he has been Associate Professor of Urban Planning and Design at the Istituto Universitario di Architettura di Venezia (IUAV). He also practises as an architect and planner mainly in the field of the design of urban spaces and conservation planning of heritage centres and landscapes. As a consultant to local authorities he has since 1984 been developing several cycling and walking routes schemes, some of which have been partly completed; others are still in progress. On behalf of the regional government of Veneto, in 1992 he published a design manual on cycleways and related infrastructure. As co-ordinator of the technical task group of the Associazione Italiana Città Ciclabili (AICC – Italian Cycle Cities Association) he participates in national and international conferences on cycling and mobility.

Hugh McClintock is a Lecturer in the Institute of Urban Planning in the School of the Built Environment at the University of Nottingham. His first involvement in bicycle planning was while working as a planner with Voluntary Service Overseas in Kisumu in Western Kenya in the mid-1970s. Since 1979 he has been extensively involved in bicycle planning in Nottingham through his involvement with Pedals, the Nottingham Cycling Campaign. He has also been involved extensively in researching and writing on bicycle planning, at local, national and European levels.

He was the editor of the predecessor of this book, *The Bicycle and City Traffic: Principles and Practice*, published by Belhaven Press. He also made substantial contributions to the IHT/CTC *Cycle-friendly Infrastructure* report and to one of the National Cycling Strategy technical reports, published in 1996. Since 1998 he has been a member of the DTLR Working Party on Professional Training for Walking and Cycling Professionals, set up under the auspices of the National Cycling Forum. He has also been involved for many years with the Local Authorities Cycle Planning Group and has maintained a database on bicycle planning and related sustainable urban travel and planning topics. This is available on-line at http://www.nottingham.ac.uk/sbe/planbiblios.

Graham Paul Smith is an artist and urban design consultant, lecturing in architecture and urban design at Oxford Brookes University.

He is joint author of the design manual *Responsive Environments* (1985), in its tenth reprint and published in Spanish and Cantonese, and of *Living Streets: A Guide to Cutting Traffic and Reclaiming Street Space* (1999). He wrote 'Traffic calming: the second wave', in *Making Better Places: Urban Design Now* (1993).

He contributes to the 'Streets for People' and 'Homezones' campaigns for Transport 2000; the Children's Play Council; the Road Danger Reduction Forum and Sustrans. He jointly organised and led the 1999 and 2000 Homezones study tours for activists and professionals to the Netherlands and Germany. He plays a central role in the Transport 2000 film *At Home on My Street: Exploring Homezones in the Netherlands and Germany* (1999), director Adrian Sinclair.

Graham's research focuses on the design of movement in public space and the experience of users within it. With Sissons-Joshi he wrote 'Cyclists' perceptions of risk' (1992), *Health Education Journal* and 'Road users' perceptions of risk' (2001), *Health, Risk and Society*. With Freer he wrote *Mixed-Use Main Streets: Managing Traffic within a Sustainable Urban Form* (1996). With McGlynn, Goldberg and Dumonteil he wrote *The Qualitative Impacts of the Oxford Transport Strategy*.

Ton Welleman is a Dutch civil engineer, who studied in Flushing (bachelor, 1968) and Delft (master, 1973).
1974–1976: Ministry of Transport, State Road Laboratory. Fields of interest: aquaplaning, icy road surfaces and weigh axles of moving trucks.
1976–1988: Institute of Road Safety Research SWOV. Fields of interest: cyclists, moped riders, elderly people, infrastructure.
1989–2000: Ministry of Transport, Public Works and Water Management, Directorate-General of Transport. Fields of interest: road safety policy and cycling policy. Project manager of the Bicycle Master Plan from the beginning in 1990 until the end in 1998.
2001: co-ordinator of the Dutch Cycling Council.

Richard Williams is a Chartered Engineer and Chartered Planner. He has worked for national, regional and city governments in Britain. He has been involved in cycling projects in Hampshire, Ireland and in several parts of Scotland, leading Lothian's Cycle Project Team for seven years from 1987 to 1994. He has presented papers on cycling at a number of conferences in Britain, and elsewhere in Europe, North America and Australia.

Michael Yeates is an architect, urban designer, advocate and accredited access consultant. With masters degrees in science (environmental management) and in environmental education, he is currently undertaking research for a PhD in community consultation in transport policy and urban planning. He is Convenor of the Public Transport Alliance and of the Bicycle User Research Group and is National Convenor of the Cyclists Urban Speedlimit Taskforce (an initiative of the Bicycle Federation of Australia Inc). He has attended numerous conferences on planning, transport, environment, equity and ethics and has presented papers addressing walking, cycling, accessibility and public transport both in Australia and overseas. Michael has been a member of various local authority and state government reference groups in south-east Queensland.

Andrzej Zalewski is Assistant Professor at Lodz University of Technology in the Faculty of Building, Architecture and Environment Engineering and Deputy Head of Postgraduate Studies in Urban Planning and Spatial Economy in the Faculty of Architecture in Warsaw University of Technology in Poland. He is an engineer and urban planner specialising in traffic and urban transport planning engineering. Special areas of his professional activity are the problems of bicycle networks planning and traffic calming. The title of his doctoral thesis was 'Influence of transportation and environmental factors on cycling traffic in medium sized towns'. He is the author of many research studies and projects in transportation and traffic engineering and of more than 120 publications in specialised journals and in the proceedings of conferences in Poland and abroad. He is also head of the transportation section of the Society of Polish Urban Planners.

1

The mainstreaming of cycling policy

Hugh McClintock, University of Nottingham

1.1   Introduction

In many developed countries cycling policy has in recent years evolved from being only a peripheral matter with the bicycle regarded as just a marginal means of transport, to be considered after priority has been given to other means, especially the car. Its role has now in many official circles come to be taken much more seriously with the increased importance of links between transport policy and issues such as sustainable development, climate change, health, air quality and social exclusion. This new wider appreciation of the importance of cycling adds new emphasis to the bike's basic importance as a cheap and affordable means of transport, particularly suitable for short trips. It is also seen that it can contribute much to the enjoyment of travel and to the mental as well as physical health of riders (BMA, 1992).

Encouraging use of bikes also offers many benefits to public policy as well as to individuals (Litman, 2001). These include reduced traffic congestion, roadway cost savings, reduced parking problems and savings in the cost of providing and maintaining parking, greater and more equitable transport choice (particularly important for non-drivers including children and the elderly) as well as a variety of social and environmental benefits. These include reduced community severance and increased community interaction, which can result in safer streets. Cycling can help the achievement of more liveable communities particularly when backed by planning policies to reduce sprawl and develop higher density, more compact communities with shorter average trip distances and more facilities within easier cycling (and walking) distance. This also means more efficient use of land.

Public policy makers with an interest in different policy areas are increasingly seeing the connections between the promotion of cycling and a range of wider policy concerns. This is particularly true of those concerned with public and individual health promotion, road safety, environmental policy, improved quality of life, sustainable development, social inclusion and urban design (Health Education Authority, 1999; RCEP, 1994). In some of these there is a new emphasis such as on the links between transport, including transport safety, and health, and between social inclusion and transport. Recognising these links can do much to help the promotion of cycling reach and gain acceptance from a wider audience, both among policy makers and the public and so do more to increase cycling than a narrowly defined 'cycling policy'.

1.2   Health and transport

Transport is having an increasingly significant impact on the environment and health (Health Education Authority, 1999). Road transport is a major contributor to human exposure to air pollution (WHO, 1999). This long term exposure to levels of air pollutants often exceeding air quality guidelines values is associated with a number of adverse health impacts, including effects on cardiovascular diseases and on respiratory diseases in adults and children. This exposure may reduce life expectancy. Furthermore (WHO, 1999), some pollutants, such as benzenes and some types of particles, further increase health risks.

Many people in developed countries are exposed to levels of traffic noise that not only cause serious annoyance and sleep loss but also communication problems, and even learning problems in children (WHO, 1999). As the same report pointed out, 'There is emerging evidence of an association between hypertension and ischaemic heart diseases and high levels of noise. Ambient noise levels continue to grow due to ever-increasing volumes of traffic.' These worsen the quality of life, especially in towns and cities.

People's quality of life is also worsened by other aspects of car dependence such as the way in which heavy road traffic and major transport infrastructure can divide communities and reduce opportunities for social interaction. These disadvantages can be associated with reduced interpersonal networks of support at local level.

Coronary heart disease is a major cause of death in industrialised countries, encouraged by the reduced levels of physical activity in modern lifestyles for many people, spending long periods in their cars, and in front of their computers and their televisions, all sedentary forms of activity. Levels of obesity have increased dramatically. For example a report in England by the National Audit Office (National Audit Office, 2001) said that two-thirds of men and over half of women in England are now overweight or obese, mainly due to inactive lifestyles. During the last 20 years there has been a dramatic increase in obesity in the USA, with the prevalence of overweight among US adults increasing by 61% just between 1991 and 2000 (Center for Disease Control, 2001). Such trends are continuing to accelerate. By 2010, for example, on current trends, it is expected that one in four adults in the UK will be clinically obese (National Audit Office, 2001). Reducing obesity has traditionally focused on lower calorific intake by dieting but there is growing evidence that exercise deprivation is also a major contributor to obesity. With metabolic systems shaped by four million years of highly active hunting and gathering, many people may not be able to maintain a healthy body weight without regular exercise.

The disadvantages of greater car dependence are also very serious for children who have often, increasingly, been driven to school and not allowed to play in the street because of the levels of traffic and associated dangers. Such trends have helped to give rise to a generation of relatively inactive and sedentary children who stay at home and watch television or play computer games, rather than getting exercise and developing social skills obtained from interaction with their peers and other people (Center for Disease Control, 1996).

Although society as a whole suffers from these health risks from transport the adverse effects tend to fall disproportionately on the most vulnerable groups: people with disabilities or hearing or sight impairments; older people; the socially excluded; children and young people; and people living or working in areas of intensified and cumulative air pollution and noise (WHO, 1999).

This increased awareness of the many negative implications of current car dependence patterns has helped to ensure that much more official recognition is now being given to the many health advantages of cycling, in terms of both public and individual health. Only a few years ago these tended to be largely ignored or underrated with cycling regarded as unhealthy just because cyclists were more likely to be involved in traffic accidents.

Numerous reports (BMA, 1992; European Commission, 1998a; WHO, 1999) have now confirmed the advantages of regular cycle use. For example, cycling for about 20 minutes per day prevents cardiovascular diseases. Regular cycling can provide protection against coronary heart disease, strokes, non-insulin dependent diabetes, falls, fractures and injuries (through improved strength and co-ordination), colon cancer, overweight and obesity. Cycling also helps to promote psychological well-being, notably self-esteem (National Cycling Forum, 1999). One of the recent studies that has clearly documented the great health benefits of cycling was based on a study of 91 adult volunteers who did not normally exercise (Boyd et al, 1998). Physical fitness, weight and blood pressure were assessed at the beginning of the experiment, after six weeks and after four months of regular cycling. Most of the subjects showed significant improvements in fitness and mental well-being after relatively small amounts of cycling. The effects were most evident in those who rode 30 kilometres a week or more. Another study (Sharp, 1990) pointed out that 'Regular cyclists have a fitness level equivalent to someone ten years younger.' A study for the Cyclists' Touring Club (CTC) in the UK concluded that 'for every extra 10 per cent of adults who can be induced to cycle, 3 million people will move into the "fit" category' (CTC, 1993).

Table 1.1 Comparison of various transport modes from the ecological viewpoint with a private car for an identical journey with the same number of people (base = 100 (private car without catalytic converter))

	Car	Car + catalyst	Bus	Bike	Plane	Train
Space consumption	100	100	10	8	1	6
Primary energy consumption	100	100	30	0	405	34
CO_2	100	100	29	0	420	30
Nitrogen oxides	100	15	9	0	290	4
Hydrocarbons	100	15	8	0	140	2
CO	100	15	2	0	93	1
Total atmospheric pollution	100	15	9	0	250	3
Risk of accidents	100	100	9	0	12	3

Source: UPI Report, Heidelberg, 1989 and reproduced in European Commission (1999), available at http://www.europa.eu.int/comm/environment/cycling_en.pdf.

1.3   Environmental policy

Concern about worsening air quality has also increased significantly in richer countries in recent years with the undoubted benefits of cleaner engines and fuels often being undermined by sheer growth in the volumes of traffic (RCEP, 1997). The proportion of air pollution attributable to transport is increasing. Bicycles, unlike motor vehicles, do not pollute the air, nor do they consume fossil fuels. Their use helps to reduce CO_2 emissions and other air pollutants (European Commission, 1998b).

One of the most useful analyses of the environmental advantages of the bicycle was the comparison of various transport modes from the ecological viewpoint carried out by the UPI Research Institute in Heidelberg in 1989, quoted in the European Commission report on urban cycling (European Commission, 1999) and reproduced in Table 1.1. This comparison was for an identical journey with the same number of people per kilometre, based on a car as 100.

1.4   Urban quality of life

It is not just in terms of air pollution that high levels of car use have worsened the quality of life, especially in urban areas, and given a new opportunity for the positive contribution of the bicycle to be emphasised. Noise pollution from motor traffic has become more and more pervasive (WHO, 1999) imposing disturbance and discomfort which is often particularly intrusive in residential areas (Litman, 2001). Cars, however clean their fuels and engines, take up large amounts of space, both when parked and, even more, in use. Bikes, on the other hand, take up little space even in use. For example, it is possible to park 10 bikes on an area required for parking one car.

Improving conditions for cyclists can also improve the quality of life for poorer groups, those who are least likely to have access to cars. With the much lower running costs of bikes than cars, improved cycling conditions can help them access job opportunities as well as health and other local facilities over a wider distance. The accessibility of children and older people can also be improved by taming the car and giving an enhanced role to the bicycle, alongside walking. Trends towards reduced independent mobility of children can be reversed, for school and other types of trip. Cycling can enlarge people's activity radius, particularly important as young people develop. As traffic congestion increases, the advantages of cycling, for shorter trips, become more pronounced, as a means of transport offering quick and efficient door-to-door transport and usually offering much more reliable journey times than cars. All this can be achieved by physical provision that is much cheaper than that for motor traffic. More cycling and less car traffic also means lower maintenance costs of highway infrastructure and less community severance.

Even though the advent of cleaner engines and cleaner fuels has reduced some of the problems of car dependence, such as some aspects of pollution, it has not affected others such as the large amount of space required for cars, both when moving and for storage (RCEP, 1997). Dependence on cars means an inefficient use of space and this is likely to remain largely the case, even with further technological changes to manage traffic and to introduce much more radical and cleaner types of engine such as hypercars.

1.5   Road safety

Increased concern with improving urban quality of life has helped to stimulate a new approach to improving road safety, to reduce the dominance of cars in towns and cities and make conditions for vulnerable road users safer and more attractive. Traditional approaches to cycling and road safety have tended to emphasise the dangers of cycling, especially at junctions (DETR, 2000; DETR, 2001) within a context of priority for car users' needs reflected in increased capacity for motor vehicles regardless of the adverse impact on more vulnerable road users such as cyclists and pedestrians. However, there is now much wider recognition (Davis, 1993; RDRF, 2001) that cycling is not dangerous in itself but is made dangerous because of the transport environment. This means that the main dangers stem from high levels and speeds of car use, as well sometimes as lorry use, and it is these that must be tackled rather than relying on the victims protecting themselves, through an undue reliance on physical segregation and safety aids such as reflective kit and helmets (Franklin, 2000; RDRF, 2001).

Moreover, for short urban trips, safety standards improve with an increased share of cycling, provided there is careful planning. This is particularly the case when there are programmes for improving road behaviour both by cyclists themselves and by drivers, including retraining for drivers found guilty of irre-

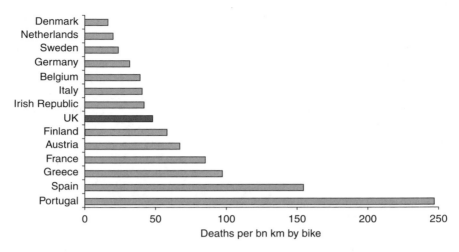

Fig. 1.1   Risk of cyclist death per billion kilometres of cycling in different European countries, 1996. Source: Commission for Integrated Transport (2001), p. 6.

sponsible driving. This can include a failure to notice cyclists at all, as well as a failure to give way to cyclists or a failure to give cyclists adequate space when passing. Efforts to encourage cyclists to ride responsibly and to learn the best way to cope with traffic, for example by careful positioning in traffic lanes, will be of very limited value without fundamental changes to the behaviour of some drivers.

Evidence from various countries (Alrutz *et al*, 1981, 2000; IHT, 1996; Welleman, 1997) suggests that the more people who cycle the safer it gets. For example, between 1980 and 1990, the distance cycled in the Netherlands increased by about 30% while the number of cycle deaths fell by about 30%. Road safety solutions focused on the needs of cyclists and pedestrians have a positive cost benefit ratio.

That high levels of cycling need not be associated with high figures for cycle accidents is shown by international comparisons (CfIT, 2001); see Fig. 1.1. This reveals quite considerable variations among European countries with, for example, cyclists (and pedestrians) in the UK more than twice as likely to be killed than in Sweden or the Netherlands, despite a relatively good overall national accident rate, in terms of deaths per head of population or exposure (deaths per billion passenger kilometres by car) (CfIT, 2001).

The considerable variations among different European countries in cycle accidents are an important indication of differences in safety alongside perceptions of how safe it is to use bikes. Accident rates have often been assessed in isolation from figures for cycle use and the two must be related to give an accurate picture. For example, in the UK, official statistics have often tended to emphasise the relatively low proportion of road deaths involving cyclists. However, as the CfIT survey pointed out, it is in fact high compared to other countries when the modal share of cycling is taken into account (CfIT, 2001).

There are various explanations for these differences in cycle accident rates and in the perceptions of cyclists about road safety (CfIT, 2000a; CfIT, 2001). These include road surface conditions, drivers' attitudes to speeding, drivers' attitudes to vulnerable road users and also several issues of enforcement and acceptance of the law, by drivers and cyclists. In the Netherlands, for example, pedestrians and cyclists have specific legal protection (Environmental Law Foundation, 1998) and the law in recent years has been strengthened so that a driver who hits a pedestrian or a cyclist will only escape liability if they can show that they could not have avoided the crash. This requires the driver to show intention/recklessness on the part of the cyclist or pedestrian. If the driver is found liable, then in terms of compensation they will be automatically deemed 100% liable. Legislation in Germany, similarly, requires drivers to behave towards vulnerable road users in such a way that their safety is not endangered. This sends a clear message that drivers have the primary responsibility for road safety.

As the OECD report on the safety of vulnerable road users (OECD, 1997) pointed out:

Cyclists have a difficult position in traffic. . . . They are sometimes supposed to follow rules for motorists, sometimes rules like those intended for pedestrians. Their needs are similar to those of pedestrians (short routes, smooth surfaces) but they are taken into account in traffic as a last resort. The situation does not encourage homogeneous patterns of behaviour. (II. 3.2, p 29)

There is indeed a great deal of variation in cyclists' behaviour in traffic. The younger cyclists are not yet able to cope with all the traffic signs and rules that apply to them. Young cyclists often like to play and show off, which leads to risk taking. There is also some amount of recklessness among adult cyclists, especially at signalised intersections, where they are often more inclined to act upon their own perceptions of traffic rather than wait for the red light and when performing turning movements (OECD, 1997).

Countries and towns with relatively high levels of cycle use tend also to have substantial numbers of elderly cyclists but their capability to cope with traffic while concentrating on cycling decreases with age; they tend to react more slowly (Van Schagen and Maring, 1991). They are more likely than other age groups to lose control when steering their bikes, even though their average speed is lower (Maring, 1988).

Although it is true that a large number of (less serious) cycle accidents are not reported (BMA, 1992), it is important, even in countries like the UK with a relatively high cycle accident rate, not to exaggerate the dangers of cycling, especially for more experienced and skilled riders (Wardlaw, 2000). Wardlaw maintains that the inherent risks of road cycling are trivial and that it takes at least 8000 years of average cycling to produce one clinically severe head injury and 22,000 years for one death. He quotes a recent study in Glasgow which estimated that 150,000 people are admitted to hospital annually with head injuries in the United Kingdom; road cyclists account for only 1% of this total, yet 6% of the population are regular cyclists and a further 5% are occasional cyclists. He also emphasises the importance of increasing the number of cyclists as a straightforward way of making cycling safer and calls for the positive aspects of cycling to be strongly promoted, i.e. speed, fitness and pride in learning a new skill.

Relying on casualty figures alone can exaggerate the dangers of cycling. Casualty figures need to be related to levels of use per kilometre or per journey to give a more accurate picture (RDRF, 2001). It is also important to try to measure the perception of risk of cyclists to provide some index of their perceptions of the level of dangers in the road environment. A very dangerous junction, for example, may have a very low rate of recorded accidents but be regarded as extremely dangerous by most cyclists, who therefore do not use it, even if it lies in a major desire line for cyclists.

1.6   Sustainable development

New road safety approaches giving much more priority to vulnerable modes are an excellent way of starting to reverse the many unsustainable trends in transport (RCEP, 1994). Indeed, given the major and still increasing contribution of transport trends to climate change it is the area of transport that most of all needs to be addressed in moving towards more sustainable development generally. Ever since the publication in 1987 of the Brundtland Report of the World Commission on Environment and Development (WCED, 1987) and particularly since the first Earth Summit in Rio de Janeiro in 1997, the concept of sustainable development has become much more prominent on the agenda of policy makers. Attempts are being made more and more to apply this concept and to implement the kind of development 'that meets the needs of the present without compromising the ability of future generations to meet their own needs' (WCED, 1987).

When so many current transport trends are so clearly unsustainable, for example in terms of demands on fossil fuels and other non-renewable resources (RCEP, 1994; RCEP, 1997; WHO, 1999), cycling, along with walking, stands out as having many positive sustainability attributes. Cycling places no demands on fossil fuel reserves and also can help to reduce unsustainable air and noise pollution generated by motor traffic. The wide range of sustainability benefits to which cycling can contribute have been particularly stressed by Levett (Levett, 1996). In addition to personal and public health, through exercise and making cycling safer, these include:

- reducing resource depletion and pollution through a modal shift from cars to bikes without offsetting increases in traffic;
- local environmental quality, through safe streets, new public spaces and urban vitality;

- pleasure in both utility and recreational travel;
- fairness in access to amenities;
- job creation, in route construction and maintenance and tourism.

The reference to fairness in access to amenities highlights the growing importance of the social dimensions of sustainable development, which includes the promotion of social inclusion. This not only means recognising the way in which limited access to transport can aggravate social exclusion but also seeing the importance of involving local people in solutions to transport needs and thereby helping to address social inclusion by contributing to skills, community capacity, crime reduction and employment as well as meeting people's transport needs in a way that is less damaging to the environment (Elster, 1999).

1.7   National and regional contrasts in levels of cycle use

Several countries in mainland Europe have been notably successful in encouraging cycling as a realistic and effective alternative to the car for short journeys (CfIT, 2001). The experience of countries like the Netherlands and Denmark has been a constant reminder that the aim of higher levels of cycle use is not just an impossible dream. In the Netherlands a national average of 27% of trips are made by bike (Huwer, 2000) and in Denmark, not all of which is as flat as the Netherlands, 21%. As can be seen from Fig. 1.2, this contrasts with a large number of countries with cycle use of less than 5%.

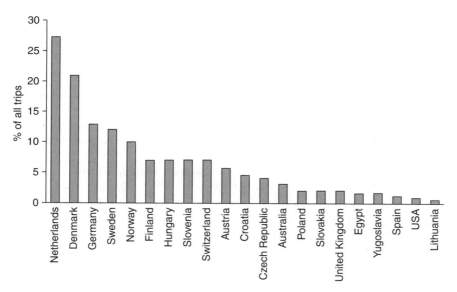

Fig. 1.2   Bicycle share of journeys in different European countries. Source: Huwer (1999), reproduced in Huwer (2000).

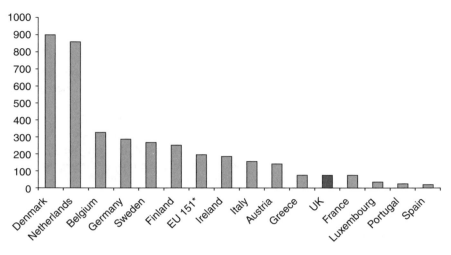

* European average

Fig. 1.3 Cycling distances per km per capita in different European countries. Source: Commission for Integrated Transport (2001), p 20.

It is clear that cycling has a significant role to play, particularly in attracting people out of their cars for short trips. Several European countries have been strikingly successful in encouraging cycling as a realistic and effective alternative to the car for short journeys. Cycle use between countries varies considerably; Denmark and the Netherlands have the highest levels per capita and nearly five times the EU average, while the UK has among the lowest levels with just 77 km per capita per year, as can be seen from Fig. 1.3, which shows the variations in cycling distances per capita in different European countries.

However, even within the Netherlands and Denmark there are clear variations in rates of use between different regions and cities, and there are varying trends of growth and decline, as Krag points out in Chapter 14. Indeed such contrasting trends are evident at European level. As one recent survey showed (CfIT, 2001), with the exception of Sweden and Portugal, cycle use is declining throughout the EU.

Big contrasts are also evident in terms of cycle ownership, as can be seen from Table 1.2. There are currently 294 bicycles per 1000 inhabitants in the UK, in comparison with 980 per 1000 in Denmark, 900 per 1000 in Germany and as many as 1010 per 1000 in the Netherlands (European Commission, 1999). The same source shows that only Ireland, Greece and Portugal rate below the UK. France, Italy, Belgium, Luxembourg, Austria, Finland and Sweden all rate more highly.

In general bikes are used less as a regular daily means of transport in southern European countries though, again, there are regional variations. For example in Italy bikes are much more common in some northern towns and cities such as Ferrara and Padua (see Chapter 16) than in the south (European Commission,

Table 1.2   Bicycle ownership and regularity of use in the European Union

Country	1996 sales	Bicycle stocks	Regular cyclists (at least once or twice a week) %	Occasional cyclists (1–3 times a month) %
Belgium	425,000	5,000,000	28.9	7
Denmark	415,000	5,000,000	50.1	8
Germany	4,600,000	72,000,000	33.2	10.9
Greece	240,000	2,000,000	7.5	1.8
Spain	610,000	9,000,000	4.4	3.9
France	2,257,000	21,000,000	8.1	6.3
Ireland	120,000	1,000,000	17.2	4
Italy	1,555,000	25,000,000	13.9	6.8
Luxembourg	20,000	178,000	4.1	9.7
Netherlands	1,358,000	16,000,000	65.8	7.2
Austria	630,000	3,000,000	–	–
Portugal	380,000	2,500,000	2.6	2.8
Finland	230,000	3,000,000	–	–
Sweden	420,000	4,000,000	–	–
United Kingdom	2,100,000	17,000,000	13.6	0.8

Source: Adapted from European Commission (1999), p 19.

1999). Regional variations are also found in eastern European countries, including Russia, though it is rare for bikes to be used for more than 5% of all journeys.

As Huwer (2000) has pointed out many factors influence these national and regional variations including the availability and price of bikes, the degree of motorisation, relative density and concentration of land use and settlement patterns, as well as income levels. Cycling tends to be more common in urban areas. While 80% of people in the European Union live in towns and cities, there are considerable variations and this too has an effect on actual levels of cycle use.

Within cities too there are wide variations. According to World Health Organisation figures (WHO, 1998), the highest levels are found in the northern Dutch city of Groningen (35%), significantly above other Dutch cities such as Delft (29%) and Amsterdam (23%). Copenhagen also features near the top, with 28%, followed closely by Odense (24%). Bruges in Belgium has 23%, much higher than the average for that country.

What is often forgotten is that in nearly all countries more people own a bike than a car (Huwer, 2000). As Huwer also points out it is particularly noticeable that in wealthier countries bike ownership is higher than already high car ownership rates. This is not necessarily reflected in a high share of journeys by bike, particularly since in many southern European countries, as to some extent in North America, the bike is associated mainly with leisure and/or sport. It is much less widely accepted as a commuter vehicle, although this is certainly changing.

Germany provides a good example of a country where, despite high levels of car ownership, a much greater official acceptance of the role of the bicycle as a daily means of transport since 1980 (CfIT, 2000b) has been reflected in a significantly higher share of trips, with particularly striking results in some places such as Munich (CfIT, 2000b; CfIT, 2001). Between 1970 and 1997, for example, official EU statistics show (European Commission, 1999) bike journeys in Germany rose from 17 to 24 billion kilometres, even more of an increase than in the Netherlands, where it was 11 to 13.5 billion kilometres. This would appear to be in part a reflection of the much greater public commitment to protecting the environment in Germany (as in the Netherlands and Denmark), the greater role of green parties in government and the greater willingness of individuals to make changes in their travel behaviour.

Compared with the UK it also probably reflects, as one survey pointed out (CfIT, 2000b), the role of higher urban densities in reducing the use of the car and in reducing the need to travel altogether. Short distances between homes and shops, leisure facilities and workplaces make cycling, as well as walking, a more viable and attractive alternative to motorised modes for short trips. The UK's predominant patterns, only recently starting to be reversed, for relatively low density housing, means that people are likely to have to travel further to work and for essential services making them more likely to opt for motorised transport as well as making the provision of public transport less viable (Newman and Kenworthy, 1989, 1999).

Weather is sometimes regarded as an explanation for variations in cycle use but its significance can easily be exaggerated. As one benchmarking survey of European integrated transport policies pointed out (CfIT, 2001), 'The British weather need not be a barrier – Denmark, with a similar climate, leads the way in achieving high cycling levels.' The same is true of other mainland north-west European countries such as the Netherlands and Germany. Most important is that the bicycle is accepted fully as a means of daily transport, as it is most emphatically in the Netherlands and much less in the UK, as the same report recognises.

Social attitudes to the use of bikes are also very important, affecting cycle use as a daily utility means of travel as well as for recreation. Despite the advent of the mountain bike in the early 1980s, which certainly encouraged the purchase of bikes by people who would not previously have considered cycling, and despite the much greater publicity for the health and environmental advantage of bikes, there sometimes remains an apparent stigma that cycling is a poor man's means of transport.

Important as these national and regional variations in cycle use and attitudes to cycling are between different parts of Europe, it is also useful to make comparisons with the USA, the most car-oriented country of the developed world. In this perspective even the relatively low amounts of cycling in the UK, by European standards, seem quite high in comparison with the USA, as do rates of walking. Pucher and Dijkstra (2000) have compared walking and cycling levels between the US and several European countries and suggested the following reasons for this disparity:

- the much lower cost of car ownership and use in the US compared with Europe;
- the low cost, and young age for obtaining a driver's licence;
- the lack of appropriate facilities for cycling;
- the American culture and lifestyle, which are almost entirely oriented to the car, and which require extremely high levels of mobility with maximum possible comfort, ease, convenience and speed;
- the real or perceived danger of cycling (and walking) in American cities.

While some of these factors are certainly also relevant for explaining differences in cycling within Europe, others are not such as the much lower costs in the US of owning and running a car and the younger average age for obtaining a licence. And although substantial parts of some European cities have been developed or redeveloped to low density standards more typical of the USA this has never been done to the same extent. Older towns and cities in Europe still mostly have narrower streets, at least in their centres, and this helps people in Europe to some extent to the idea that unrestrained car use there is not appropriate because it damages the urban fabric and quality of life. Much lower average petrol prices in the USA also contribute to these different trends. Cities in Canada and Australia tend to fall in between the European and USA ends of the spectrum, in terms of density and modal share, as shown by the comprehensive analyses of travel patterns and urban form by Newman and Kenworthy (1989, 1999).

1.8 The role of public policy in the development of cycling policy

Variations in cycle use between countries and within countries are particularly significant because they influence not only the social acceptability of cycling but also official willingness to provide for cyclists and give serious recognition to its importance, especially for shorter trips. These official attitudes also affect the way in which infrastructure for cyclists is designed, for example whether or not the bicycle is regarded effectively as a pedestrian with wheels or as a vehicle. This means that, in designing for cyclists, criteria should be used that are analogous to those used for designing for motor traffic, with thought given to sightlines, minimum radii, and so on (City of Edinburgh Council, 1997). This also means catering for speeds of up to 30 kilometres an hour, i.e. considerably higher than those of pedestrians with whom cyclists are frequently expected to mix on equal terms. Bikes need curves and slopes to operate efficiently; sharp corners, dismounts and frequent stops make progress slow and routes unattractive. A failure to recognise that the bicycle is a vehicle and these implications can easily contribute to cyclists' dissatisfaction with the detailed physical provision made with the intention to encourage cycling.

A serious commitment to cycling will not only recognise the significance of the fact that a bicycle is a vehicle but also seek to promote cycling through a range of approaches beyond physical provision whether on the road or off-road and highway planning and traffic management that are sensitive to cyclists' needs. These wider approaches include the co-ordination of cycling policies with other community plans, improved enforcement and taking opportunities to encourage cycling in wider strategies for health, air quality, leisure, recreation and tourism (Litman *et al*, 2001; National Cycling Forum, 2001).

As cycling becomes taken more seriously, cycling policy should evolve from being *ad hoc* and concentrating largely on the easily implementable physical provision of infrastructure. It should become more integrated with a series of other policy activities, and be implemented according to a carefully thought out strategy in partnership with a variety of other agencies, on a regular and systematic basis, including monitoring of the effectiveness of earlier initiatives of different kinds. In doing this it should also increasingly have strong political support since a clear leadership role, by politicians and a range of officials is vital to ensure long term consistency and effectiveness. It is more likely to be effective also with strong community support but at the same time has to ensure that the wider public are both educated and informed on the need for cycling policies and the way they are developed and implemented, in close co-operation with users and other interested groups. Such processes can build on and further develop changes in environmental awareness, for example public awareness of traffic and wider environmental problems such as climate change and resource depletion. Leadership also means the provision of adequate resources and legal powers as well as adequate staff with the right skills and the opportunity to upgrade and develop their skills, including opportunities to exchange examples of good practice.

These general criteria for assessing the value and effectiveness of cycling policy will be referred to throughout the various chapters of the book. Since the provision of infrastructure for cyclists has tended to be the predominant element of cycling policy, the next chapter concentrates on this, with particular reference to experience in the UK.

1.9   References

ALRUTZ, D *et al* (1981), *Dokumentation zur Sicherung des Fahrradverkehrs, Reihe Unfall- und Sicherheitsforschung Strassenverkehr Bd. 74*, Bergisch Gladbach, Germany, Bundesanstalt für Strassenwesen.

ALRUTZ, D *et al* (2000), *Begleitforschung Fahrradfreundliche Städte und Gemeinden NRW. Maßnahmen- und Wirksamkeitsuntersuchung*, Düsseldorf, Ministerium für Wirtschaft und Mittelstand, Energie und Verkehr NRW.

BMA (British Medical Association) (1992), *Cycling: Towards Safety and Health (report of a BMA Working Party edited by Mayer Hillman)*, Oxford, Oxford University Press and London, BMA.

BOYD, H, HILLMAN, M, NEVILL, A, PEARCE, LP and TUXWORTH, B (1998), *Health Related Effects of Regular Cycling on a Sample of Previous Non-exercisers*, Godalming, Surrey, CTC and Bike for Your Life Project.

CENTER FOR DISEASE CONTROL (1996), *Physical Activity and Health: A Report of the Surgeon General*, USA, Center for Disease Control, http://www.cdc.gov/nccdphp/sgr.

CENTER FOR DISEASE CONTROL (2001), *US Obesity Trends 1985 to 2000*, USA, Center for Disease Control, http://www.cdc.gov/nccdphp/dnpa/obesity/trend/maps/index.htm (accessed 15.10.01).

CFIT (Commission for Integrated Transport) (2000a), *European Best Practice in Transport: Benchmarking*, London, CfIT, August, http://www.cfit.gov.uk/ebptbench/index.htm.

CFIT (Commission for Integrated Transport) (2000b), *European Best Practice in Transport: The German Example*, London, CfIT, August, http://www.cfit.gov.uk/ebptgerman/index.htm.

CFIT (Commission for Integrated Transport) (2001), *European Best Practice in Delivering Integrated Transport*, London, CfIT, November, http://www.cfit.gov.uk/research/ebp.

CITY OF EDINBURGH COUNCIL (1997), *Cycle Friendly Design Guide: Guidance on Designing Roads, Traffic Management Schemes and New Developments to Promote Cycle Use*, The City of Edinburgh Council.

CTC (Cyclists' Touring Club) (1993), *Costing the Benefits: The Value of Cycling*, Godalming, CTC.

DAVIS, R (1993), *Death on the Streets: The Mythology of Road Safety*, Hawes, North Yorkshire, Leading Edge Publications.

DETR (2000), *New Directions in Speed Management: A Review of Policy*, London, Department of the Environment, Transport and the Regions.

DETR (2001), *Tomorrow's Roads: Safer for Everyone. The Government's road safety strategy and casualty reduction target for 2010*, London, Department of the Environment, Transport and the Regions, http://www.roads.dtlr.gov.uk/roadsafety/strategy/tomorrow.

ELSTER, J (1999), *Cycling and Social Exclusion*, CASE Report 5, Centre for the Analysis of Social Exclusion, London, London School of Economics, http://sticerd.lse.acuk/case/publications.

ENVIRONMENTAL LAW FOUNDATION (1998), *Options for Civilising Road Traffic*, London, ELF.

EUROPEAN COMMISSION (1998a), *The Common Transport Policy, Sustainable Mobility: Development of Public Health in the European Community, COM (98) 230*, Brussels, European Commission.

EUROPEAN COMMISSION (1998b), *Ambient Air Quality 96/62/CEE: Transport and CO_2, a Community Approach, COM (987) 204 final, 31 March 1998*, Brussels, European Commission.

EUROPEAN COMMISSION (1999), *Cycling: The Way Ahead for Towns and Cities*, Luxembourg, Office for Official Publications of the European Communities, http://www.europa.eu.int/comm/environment/cycling/cycling%5fen.pdf.

FRANKLIN, J (2000), *Cycle Helmets: 25 Years along the Road*, Cycle Campaigns Network and CTC Conference, Cheltenham, April.

HEALTH EDUCATION AUTHORITY (1999), *Making T.H.E. Links: Integrating Sustainable Transport, Health and Environmental Policies – a guide for local authorities and health authorities*, London, Health Education Authority.

HUWER, U (2000), 'Let's bike: the 10 point pedalling action programme to support cycling', *World Transport Policy and Practice* 6 (2).

IHT, Bicycle Association and CTC (Cyclists' Touring Club) (1996), *Cycle-friendly Infrastructure: Guidelines for Planning and Design*, London, IHT/CTC.

LEVETT, R (1996), *Cycling and Sustainable Development: An Overview*, Topic Paper, National Cycling Strategy Appendix, London, Department of Transport.

LITMAN, T (2001), *Quantifying the Benefits of Non-Motorized Transport for Achieving TDM Objectives*, Victoria Transport Policy Institute, Victoria, British Columbia, September, http://www.vtpi.org.

LITMAN, T, BLAIR, R, DEMOPOULOS, W, EDDY, N, FRITZEL, A, LAIDLAW, D, MADDOX, H and FORSTER, K (2001), *Pedestrian and Bicycle Planning: A Guide to Best Practices*, Victoria Transport Policy Institute, Victoria, British Columbia, http://www.vtpi.org.

MARING, W (1988), *Oudere volwassenen also fietser (Older adults as cyclists)*, Report VK 88-14, Groningen, Netherlands, Rijksuniversiteit Groningen, Verkeerskundig Studiecentrum, VSC.

NATIONAL AUDIT OFFICE (2001), *Tackling Obesity in England*, London, The Stationery Office, February, http://www.nao.gov.uk/pn/00-01/0001220.htm.

NATIONAL CYCLING FORUM (1999), *Promoting Cycling: Improving Health*, London, Department of the Environment, Transport and the Regions.

NATIONAL CYCLING FORUM (2001), *Cycling Matters: Advice for Cyclists and Cycling Groups on Cycling Policy*, London, Department of the Environment, Transport and the Regions.

NEWMAN, P and KENWORTHY, J (1989), *Cities and Automobile Dependence: A Sourcebook*, Aldershot, Gower.

NEWMAN, P and KENWORTHY, J (1999), *Sustainability and Cities: Overcoming Automobile Dependence*, Washington DC and Covelo, California, Island Press and London, Earthscan.

OECD (1997), *Safety of Vulnerable Road Users*, Paris, Organisation for Economic Co-operation and Development.

PUCHER, J and DIJKSTRA, L (2000), 'Making walking and cycling safer: lessons from Europe', *Transportation Quarterly*, 54 (3), Summer, and at http://www.vtpi.org.uk.

RCEP (Royal Commission on Environmental Pollution) (1994), 18th Report: Transport and the Environment, Cm 2674, London, HMSO, October.

RCEP (Royal Commission on Environmental Pollution) (1997), *Transport and the Environment: Developments since 1994*, London, The Stationery Office.

RDRF (Road Danger Reduction Forum) (2001), *Safe Roads for All: A Guide to Road Danger Reduction: A Good Practice Guide produced by the Road Danger Reduction Forum*, Keynsham, Bristol, RDRF.

SHARP, I (1990), *On Your Bike*, London, National Forum for Coronary Heart Disease Prevention.

VAN SCHAGEN, INLG and MARING, W (1991), *Probleemanalyse oudere verkeersdeelnemers*, VSC, Report VK 91-09, Groningen, Netherlands, Rijksuniversiteit Groningen, Verkeerskundig Studiecentrum, VSC.

WARDLAW, MJ (2000), 'Three lessons for a better cycling future', *British Medical Journal* 321, 1582–5, http://bmj.com/cgi/content/full/321/7276/1582.

WCED (World Commission on the Environment and Development) (1987), *Our Common Future* (the Brundtland Report), Oxford, Oxford University Press.

WELLEMAN, T (1997), 'The Dutch Bicycle Master Plan', in R TOLLEY (ed) *The Greening of Urban Transport: Planning for Walking and Cycling in Western Cities*, Chichester, Wiley (second edition).

WHO (World Health Organisation) Regional Office for Europe (1998), *Walking and Cycling in the City*, Copenhagen, WHO Regional Office for Europe, http://www.who.dk/environment/pamphlets.

WHO (World Health Organisation) (1999), *Charter on Transport, Environment and Health*, Copenhagen, WHO Regional Office for Europe, http://www.who.dk/london99/transporte.htm.

2

The development of UK cycling policy

Hugh McClintock, University of Nottingham

2.1 Introduction

As mentioned in the previous chapter, levels of cycling in the UK are relatively low by European, especially north-west European, standards, that is only about 2.5% of journeys and an average of only 81 kilometres per person per year ridden (CfIT, 2001; European Commission, 1999). However, there is a particular disparity in the UK between the numbers of bikes owned and bikes actually used which suggests considerable potential, if the conditions were right, for increasing these usage figures. Moreover, the distances to work are basically suitable: two-thirds of all bike journeys to work by men, and two-thirds by women had a 'crow-fly' distance of less than two kilometres, although average distances are greater in metropolitan areas, particularly inner London where most people usually live further from their work.

There are more than 20 million bikes in the UK and more people now own a bike than ever before (National Cycling Forum, 2001). There are also many places with levels of cycling much higher than the national average and this particularly applies to a number of places in the flatter and drier eastern half of England. Table 2.1 shows the top ten areas for bike commuting in 1991.

This chapter discusses UK bicycle planning experience with particular reference to the provision of infrastructure for cyclists, which is where most emphasis has arguably been given. The relationship between this experience and other forms of cycling policy and promotion, for different types of trip and different types of user, are discussed in Chapter 3.

Table 2.1 Districts in England with the highest levels of cycling, 1991

Location	Bike commuting % (not including students)
Cambridge	28.2
York	19.0
Oxford	17.4
Boston	15.5
Gosport	15.1
Hull	14.1
Waveney (Suffolk)	12.5
Crewe	11.9
Scunthorpe	11.7
Peterborough	11.4

Source: UK Government Statistical Service, 'Cycling in Britain', August 1996.

2.2 Landmarks in UK cycling policy development: 1970s to early 1990s

The main resurgence of official interest in cycling in the UK dates from the early 1990s although there had been some signs for about 15 years before that, mainly because the high accident rates for cyclists were seen as a 'problem to be dealt with' rather than for more positive reasons. Some early postwar new towns such as Harlow and Stevenage had introduced substantial segregated provision for cyclists (Dupree, 1987) but there was little interest from older towns and cities. The Labour government in its 1977 Transport White Paper had introduced a special budget to promote 'innovatory cycling schemes' such as special traffic lights and cycle routes and the scope of these was extended by the Conservative government in the early 1980s (DOT, 1982). This encouraged the development in the mid to late 1980s of some partial cycle route networks in places like Nottingham, Stockton, Bedford/Kempston and Cambridge to augment the earlier development of isolated special facilities in some of these towns and in others such as Oxford and Middlesbrough after 1977, with modestly encouraging results (McClintock, 1992; Harland and Gercans, 1993).

Particularly important in changing attitudes in the 1990s was the publication in 1992 of the British Medical Association report on cycling (Hillman, 1992). This directly challenged the key concern about safety in promoting cycling. It pointed out that the life-years lost from cycling are likely to be less than the life-years gained from improvements in health.

The strongly positive tone of the BMA report helped pave the way for important changes in official policy on cycling in the UK, away from a preoccupation with accidents and acknowledging much more strongly health and environmental advantages. This was also the time of the moves, internationally and nationally, towards sustainable development with, for example, the first Earth Summit in Rio de Janeiro in 1992 and the subsequent adoption of national sustainable development strategies including the one for the UK in 1994 (UK Government, 1994). This was the first time that an official policy document had accepted the need to reduce dependence on the car and this aim was also then incorporated in the radically revised national planning guidance on land use planning and transport, Planning Policy Guidance Note 13 (PPG13), issued in March 1994 jointly by the then separate Departments of the Environment and Transport (DOE and DOT, 1994). In this guidance planners were asked to promote other forms of transport including cycling and to foster forms of development that would assist this.

In the mid-1990s there was one very important central government cycle policy initiative which did not focus on special cycling facilities such as cycle routes. This was the Cycle Challenge Project, which announced its Cycle Challenge Programme (Cleary and McClintock, 2000), inviting bids from partnerships of local authorities and other agencies, public, private and voluntary, for matching funding for initiatives to promote cycling other than in terms of providing infrastructure. These particularly supported health-related developments and also schemes such as those in Nottingham (see Chapter 11) and Cambridge to work with employers to promote cycling to work, as well as several rail- and bus-based cycling and public transport initiatives. The successful schemes were implemented in the mid to late 1990s.

2.3   National Cycling Strategy

Even more important in terms of the widening of cycling promotion away from a main preoccupation with cycle facilities was the National Cycling Strategy launched in July 1996 (DOT, 1996) under the auspices of the former Department of Transport but prepared in close consultation with four working parties with a wide range of representatives, from cycling and other environmental groups, from the academic and other research circles, and from the professions and industry. The aims of the NCS were to establish a culture favourable to the increased use of bicycles for all age groups; to develop sound policies and good practice; and to seek out effective and innovative means of fostering accessibility by bike. A central target was adopted to quadruple the number of cycle journeys on 1996 figures by 2016. Relevant local targets were also to be adopted and associated indicators of progress defined. The NCS also defined 24 key outputs to help achieve the doubling of cycle use. These included:

- The development of further government advice and best practice on locations of developments and provision for cycling.
- Partnerships between operators and local authorities to ensure provision for the secure parking and carriage of cycles.
- Initial guidance to local authorities on cycle audit by the end of 1997.
- A commitment to conduct an ongoing review and revision of all design guidance.

- Local highway, planning and public transport authorities to conduct strategic reviews and produce 'Local Strategies for Cycling' by the end of 1999.
- Local authorities to concert a programme of cycle parking provision to be completed before 2002.
- Private sector establishments to review their cycle parking arrangements.
- Local planning authorities to establish cycle parking standards for development plans by the end of 1998.
- Agree a set of graded standards for cycle security devices by the end of 1996.
- Establish working groups on cycle registration in 1996 to report to the National Cycling Forum by the end of 1997.
- All relevant professional institutions to review their training courses and ensure that entry requirements and continuing professional development includes an understanding of cycle issues.
- Establish the collection and dissemination of examples of good practice in cycling provision.
- Extend cycling promotions such as National Bike Week, Green Commuter Plans, TravelWise and other relevant public campaigns.
- Double the number of children cycling to and from school.
- Issue new guidelines to employers for an agreed cycle allowance rate.
- Monitor, maintain and develop the cycling rewards in transport and other funding mechanisms.
- Every local authority to consider responsibilities and staff time for cycling policy and provision.
- Department of Transport to consider the staff resources required for meeting the new policy objectives.
- Central government, its agencies, local authorities, transport providers and large organisations to consider the impact of their activities and expenditure decisions on travel choice.
- Set up a National Cycling Forum chaired by the Minister for Local Transport, to produce an annual progress report.
- All local authorities to liaise with local cycling groups and regularly assess progress towards local cycling targets.
- Review the data sources on cycle use, the range of determinants and indicators of sustainability, and make any improvement by the end of 1997.

2.4   The development of technical guidance for cycle infrastructure

Almost in parallel with the NCS preparation a working party with DOT involvement, but led by the Institution of Highways and Transportation (IHT) and the Cyclists' Touring Club (CTC), had produced completely updated and consolidated advice on the design of cycle infrastructure to replace that previously produced by the IHT in 1983 (IHT, 1983). This guidance (IHT *et al*, 1996) also aimed to consolidate various different sources of advice. It was produced by a working party with wide representation, and tried to discourage too great an emphasis on the provision of special facilities as being the main way of providing for cyclists, relegating facility provision to a lower rank in its preferred hierarchy of:

- traffic reduction;
- traffic calming;
- junction treatment and traffic management;
- redistribution of carriageway space;
- segregated provision (cycle lanes and cycle paths).

The IHT working party was inspired by the comprehensive review of Dutch cycle planning experience published in 1983 (CROW, 1993) and especially its identification of five key characteristics for successful cycle routes:

- attractiveness;
- directness;
- coherence;
- safety;
- comfort.

Reflecting the growing interest in making ordinary streets and roads safer for cyclists and avoiding too much preoccupation with special facilities, the IHT followed up its 1996 guidelines on cycle infrastructure with the publication in 1998 of *Guidelines for Cycle Audit and Review* (IHT, 1998). These gave detailed guidance on how cyclis needs could be thoroughly taken into account in planning any new road or traffic management schemes. The same report also gave guidelines for cycle review, for use in reviewing the cycle-friendliness of existing highway layouts.

2.5   The 1998 Integrated Transport White Paper and Local Transport Plans

The NCS and its targets were subsequently incorporated in the incoming Labour government's Integrated Transport White Paper published in July 1998 (DETR, 1998). This marked a general policy shift in transport in England (with similar documents soon appearing for Scotland and Wales), with substantial extra funding for cycling, walking and public transport and further details of this extra financial commitment were given in the government's Ten Year Transport Plan published in July 2000 (DETR, 2000b). Also of particular importance for cycling in the 1998 White Paper was the introduction of a new system of local transport funding, Local Transport Plans (LTP), not only increasing funding but giving financial commitments over a longer period, five years, than under the previous TPP (Transport, Policies and Programmes) system. The new system was introduced with one year provisional Local Transport Plans in 1999, and following further detailed government guidance (DETR, 2000a), the full LTP system followed in 2000. In terms of cycling the LTP guidance set out the following minimum requirements:

- a discrete strategy for encouraging cycling, which establishes a clear target that contributes to the national targets for increasing cycle use;
- evidence that cyclists have been given a high priority;
- evidence that encouraging cycling is part of all transport policies, including road safety strategies;
- evidence of interaction with local planning authorities to ensure that land use and development planning allow and encourage people to cycle;
- evidence that there has been a review of the road network, to establish where improvements are needed;
- assessment of the quality of existing cycle networks, to identify where improvements are necessary;
- a programme of measures to improve the safety of cyclists and reduce conflicts with other traffic, including pedestrians.

The same document outlined the following characteristics of a good LTP as regards cycling:

- adopts a formal order in which planners consider the needs of different types, placing cyclists near the top;
- partnership for action with health, education, commercial and voluntary bodies;
- plans for improvement of physical provision to be based on methodical application of the *Guidelines for Cycle Audit and Review* (published by IHT);
- cycle audit of all road and traffic schemes;
- aims to improve interchanges, and increase opportunities for combined cycle and public transport journeys;
- encourages cycling through TravelWise, Green Travel Plans, Local Agenda 21 and School Transport Plans;
- minimises conflict between pedestrians and cyclists.

Also of great importance in encouraging cycling in the late 1990s were the plans, announced in 1995, for the 10,000 kilometre National Cycle Network (NCN) by Sustrans, following the successful award of a €70 million (£42 million) grant from the Millennium Commission covering one-fifth of the estimated total cost of €350 million (£210 million). These plans (see Chapter 7) had originated in the work of a local cycle campaign group in the Bristol area in the late 1970s in pressing for the conversion to a shared use pathway of an eight kilometre stretch of disused railway between Bristol and Bath. In subsequent years this had expanded into a national study by Sustrans of the potential for conversion of disused railways throughout England and Wales (and, subsequently, Scotland), as well as the assessment of many other routes, along rivers and canals and quieter rural roads to form longer distance routes, of particular appeal to less confident cyclists, and designed to appeal to commuter as well as leisure cyclists. The 1995 Millennium Commission grant and the plans to complete the first 8000 kilometres of the network by June 2000 added to the momentum of the launch of the NCS and other developments. Many other local authorities were keen to be involved in the NCN and to complement it with their own regional and local routes.

2.6   Revised planning policy guidance on land use and transport planning

The importance of taking full account of National Cycle Network developments was one of the changes in revised government planning guidance on land use planning and transport published by the DETR in March 2001 (DETR, 2001). This advice was similar to that in the LTP guidance and emphasised that effective cycling policies require a wider 'Think Bike' awareness among all planners, traffic engineers and other policy makers and implementers so that improvements for cyclists through specific cycling measures are not undermined by a neglect of cyclists' needs in other schemes and projects. Promoting cycling does not mean that only the designated 'cycling officer' should take cyclists' needs seriously. To help develop this wider awareness the revised PPG13 reminded local authorities, as part of their LTP strategy, that they should also promote cycling through measures such as:

- reducing traffic volumes, including, where relevant, restricting or diverting heavy goods vehicles;
- traffic calming – reducing speeds, particularly in residential areas and close to schools;
- reallocation of carriageway, to provide more space for cyclists, such as cycle lanes or bus lanes where cyclists are to be permitted;
- improvement of facilities off the carriageway, such as cycle tracks or paths;
- encouraging health and education providers and employers to promote cycling to and from schools, hospitals and places of work, ideally in the context of site-specific travel plans;
- encouraging more use of public rights of way for local journeys and helping to promote links in rights of way networks;
- carefully considering the shared use of space with pedestrians when alternative options are impractical. Unsegregated shared use should be avoided, particularly in well-used urban contexts.

In addition to this specific advice on cycling this revision of PPG13 also strongly emphasised in its general principles the importance of influencing the design, location and access arrangements of development, including restrictions on parking, to ensure it promotes cycling, and also of seeking the provision of convenient, safe and secure cycle parking and changing facilities in town centres and developments including the provision of cycle storage facilities at transport interchanges, including park-and-ride sites. It also emphasised the importance of locating day-to-day facilities which need to be near their clients in local centres

so that they are accessible by public transport, walking and cycling. Also important, it stressed, is to ensure that development comprising jobs, shopping, leisure
and services offers a realistic choice of access by cycling, walking and public
transport, particularly in urban areas. Another important strand of advice in the
revised guidance, the principle of giving people priority over ease of traffic movement and providing more road space to alternative modes is also significant for
cyclists, especially in town centres, local neighbourhoods and other areas with
a mixture of land uses. The revised guidance was also important for reinforcing recent government guidance advocating the development of higher density
housing, both to help reduce the extent of development on new greenfield sites
and also, through the resulting more compact development patterns, to make it
easier for housing to be within more convenient cycling (and walking) distance
of local facilities.

2.7   Assessment of UK cycle policy experience: some challenges for the future

While it is easy to criticise the situation for cyclists in the UK in comparison with
many continental countries, it is important to bear in mind that the situation is
now very different from that which prevailed until the early 1990s. Even leaving
aside the special cases of some of the early postwar new towns with attempts at
special physical provisions such as Stevenage, Harlow and Milton Keynes (from
the 1970s), there are quite a number of older towns and cities with a record
of some provision for cyclists that goes back to the early 1980s or even late
1970s. These include York, Cambridge, Peterborough, Oxford, Middlesbrough,
Nottingham (see Chapter 11), Bedford, Bristol, Southampton (McClintock, 1992)
and Edinburgh (see Chapter 10). In addition there are a number of London
boroughs with a sustained commitment to cycling which tried to maintain the
momentum for cycling in London generally established for a few years by the
former Greater London Council in the early 1980s, with its dedicated cycling
project team.

Although the results of these efforts, in terms of generating new cyclists rather
than just providing better conditions for existing cyclists, were often only modest
(Harland and Gercans, 1993) there have been several cases of impressive local
increases in cycling, as for example in York (see Chapter 9) and Nottingham (see
Chapter 11).

Whilst the national cycling policy framework remained relatively cautious, if
not at times distinctly discouraging, until the late 1990s, much work was done
for cyclists, of varying quality and value to users (DTLR, 2001; Jones, 2001).
Some of the poorer schemes were marked by a greater readiness to take space
from pedestrians than from drivers, for example narrow shared paths converted
from pavements and often with no adjustment to intruding features such as lamp
columns and other street furniture. Other awkward features such as uneven kerbs

and even dangerous ones such as sharp bends and poor inter-visibility were also quite common. Route continuity often proved a problem; many schemes were fragmented, suddenly abandoning cyclists at points where they most needed protection, and signs and markings were also often poor, even before vandalism and poor maintenance made them worse. As Jones has commented, quoting the evidence of a series of seminars held during 2001 to discuss National Cycling Strategy implementation (Jones, 2001): 'Problems occur at many levels, for example cycling left out at early planning stages of new development, as well as in the detailed design and implementation of schemes.'

A recurring theme of criticisms by cycle campaigners and other users has tended to be what they perceive to be the abuse of the principle of flexibility. Clearly, with common constraints of space and finance, ideal standards of provision are often hard to achieve, especially in many urban areas, but at the same time being too flexible and achieving distinctly substandard solutions can sometimes put cyclists more at risk than if no provision was made at all. And, it seemed, this was not just a matter of interpreting guidance too liberally but, in some cases, not even to be using common sources of guidance such as the IHT/CTC *Cycle-friendly Infrastructure* report (IHT, 1996) at all.

A great deal of useful research work has been done on cycling over the years, particularly in the 1990s by the Transport Research Laboratory (TRL) (Reid, 2001a). Much of this recent work has looked at the newer forms of cycle facility such as contraflow cycle lanes, toucan crossings and advanced stop lines, forms of on-road provision which have tended to be more widely appreciated than some cycle paths and shared paths. In addition to evaluations of specific cycle facilities the TRL has done much more general work on traffic calming and the impact of particular types of feature on cyclists' safety, comfort and convenience. Advice on monitoring of cycle flows has also been included, to help with the increased requirements for monitoring in Local Transport Plans. However, there remain many important challenges in the provision for cyclists and these will have to be tackled effectively if the revised government targets are to be met. They include the following areas.

2.7.1   Land use planning

Land use planning's role is fundamental, especially in the longer term. Since 1994 government planning advice has encouraged planners to make provision for other forms of transport than the car and to reduce car dependence. This advice was strengthened in the 2001 revision of central government's Planning Policy Guidance Note 13 (DETR, 2001). Following this advice has become all the more important with the projected major new housing development in many different parts of the country. Higher density developments, steering most development to areas well served by public transport and local facilities, and detailed layouts designed to promote safe, convenient and attractive direct routes are all essential to encouraging cycling and walking for short trips.

2.7.2   Making general roads and streets safer for cyclists

Provision for cyclists in older urban areas is always going to be more difficult because of constraints of space and existing development. However, the way forward must be to rely more on making the general road layouts safer for cyclists to use, with less reliance on special cycling facilities and, above all, shared paths. It is evident that cyclists can benefit greatly from the reallocation of carriageway space away from drivers, from well-designed traffic calming and from tighter geometry at roundabouts, to reduce high entry and exit speeds, especially at roundabouts with more than one approach lane.

Roundabouts are the most dangerous type of junction for cyclists and the most common incident at a roundabout involves a driver entering the roundabout and colliding with a circulating cyclist. This problem is particularly acute at small roundabouts whose layout encourages a 'racing line' to be taken (Reid, 2001a). However, substantial research has now been done which has shown that where motor vehicle inflows are below 2500 per hour 'continental' roundabout geometry is likely to reduce accidents (Reid, 2001a). This geometry features radial, not tangential arms, single-lane entry and exit, a larger island and a tighter circulatory carriageway.

Although lower speeds will not automatically result in increased cycling (Babtie Group, 2001), there is plenty of experience, especially in places in continental Europe where extensive area-wide 30 km/h zones have been introduced, to suggest that it is very probable. Another recent report, for example, with particular reference to Munich, commented that such measures:

> have been fundamental in prompting both strong growth in walking and
> cycling and in the ability of public transport to compete with the private
> car. The balance has been shifted away from 'movement space' to
> 'exchange space' where the focus is on personal interaction in quality
> urban space rather than on mobility in car dominated streets. (CfIT,
> 2001, Chapter 4, p 2)

Speed reduction through physical measures, i.e. traffic calming, also has a vital role in improving cyclists' safety but sensitivity to cyclists' needs is also very important in the detailed design of traffic calming schemes, for example ensuring a minimum distance of four metres at roadway narrowings to allow motor vehicles to pass cyclists safely (Reid, 2001a). Such details of traffic calming schemes have often made conditions worse for cyclists, not better (DTLR, 2001). The same has often been true of other features such as pavement built-outs and pinch points.

Full use of cycle audit procedures and those for cycle review (covering existing roads and streets) will help in this regard as will a much greater emphasis on lowering speed limits and enforcing them. Volumes and speed of traffic, width, gradients, lighting, surveillance and surface quality can all affect the willingness of cyclists to use different roads as well as cyclists' desire lines and the locations of trip attractors and trip generators. Methods of assessing the cyclability of roads have been suggested (TRL, 2001). This report found that cyclists cited safety, the attractiveness of the route and the smoothness of the road surface as the most important determinants in shaping their perceptions. It is important to take account of:

- motor vehicle flows (including the volumes of heavy goods vehicles and concentrations in the peak period);
- motor vehicle speeds;
- number and types of junction per kilometre (minor and major);
- the roughness of the surface and the materials used;
- social safety (including levels of lighting and surveillance).

Although dedicated cycle facilities can reduce actual and perceived dangers, wider traffic management and reduction measures can also do much and quite possibly sometimes more to benefit cyclists. These include wider traffic management to reallocate road space away from cars, road closures, and area access restrictions including the development of traffic cells from which most motor vehicles are excluded, at least when making through journeys. These can give cyclists a series of quiet on-road short cuts compared to cars. Cycle use will always tend to benefit from wider policies to restrain traffic (Mathew, 1995; Wardman *et al*, 1997), especially where constraints of space preclude the development of good standard segregated provision.

2.7.3   Reducing reliance on shared path provision

Converting footways to shared use by cyclists and pedestrians has often seemed a simple way to begin making improvements for cyclists but experience shows that this can often be very controversial and increase the risks to other vulnerable road users, especially the elderly, the visually impaired and those with hearing impediments (CTC, 2000). Perceptions of the risk of conflict may be significantly higher than actual incidents and this can deter some people from using such shared paths at all (Countryside Agency, 2001). Surface and width improvements can attract more cyclists but also encourage them to go faster, increasing the risks to other users and, in some cases, provoking criticisms about the much more intrusive character of the 'improved' path.

However, this does not mean that shared use paths should never be used, especially where there are smaller numbers of likely cyclists and pedestrians. Experience suggests that such paths, though often unpopular with both sets of users because of the increased possibility of conflict, can encourage both cycling and walking because of the reduced risk of conflict with motor traffic.

New shared paths tend to be more acceptable than paths converted from sole pedestrian use and clearly signed and marked segregation can make sharing more attractive, along with careful attention to social safety and reducing the risk of crime. It is also important (IHT, 2000) that the differing needs of pedestrians and cyclists are taken into account in the design and selection of shared routes.

Both in government and professional guidance (DETR, 2000a; IHT, 2000) it now seems to be accepted that shared use should be considered only where all other options have been rejected and this is consistent with the IHT hierarchy of design solutions mentioned above (IHT *et al*, 1996) with on-road solutions re-allocating road space from motor vehicles to cyclists being given priority over off-road solutions. Careful attention to detail in the design of shared path solutions is also important, taking full account of the local circumstances, including the level and type of use (IHT, 2000). Such local assessment also means taking full account of the likely levels of use of shared paths by various groups of impaired people and, for example, not only providing clear markings to segregate users, including tactile markings, but also ensuring that the area for cyclists is closest to the carriageway.

2.7.4   Improving the quality of cycle networks

There is evidence that, overall, cycling schemes have a good safety record in terms of achieving an overall reduction of 58% in injury accidents, according to data from the TRL's Molasses database (DTLR, 2001). Cycle contraflow systems have a good safety record and are well liked by cyclists, the same report stated. It also commented that advanced stop lines have been found to be a useful facility for cyclists at signalised junctions. However, cycle paths and cycle lanes often have a less clear safety benefit, at least for cyclists. For motorists on the other hand cycle paths may be liked because they are seen as 'helping to get cyclists out of the way'. It is particularly irritating for cyclists when they are abused by drivers for not using a cycle path that they know to be distinctly substandard, in terms of route attractiveness, distance, surface or safety – very important perceptions for cyclists (TRL Ltd, 2001) – of which passing drivers are likely to be wholly unaware. Cyclists want safe, convenient, attractive and direct access and cycle paths often fail on at least one of these criteria if not several.

The quality of cycle networks should be reviewed, to ensure that all existing facilities do play a net positive role, for less confident cyclists at least. Quality as well as quantity of cycling provision is important and the former needs to be given more emphasis, in initial design and construction and in maintenance. Cycle lanes are often too narrow, encouraging drivers to pass cyclists dangerously (Jones, 2001). Cycle routes are often fragmented, sometimes leaving cyclists feeling 'dumped' at dangerous locations where they most need protection.

Attention to detail is so important, avoiding sharp upstands, barriers and posts, and poor lighting, signing and lining, and ensuring route continuity and coherence, for example overcoming barriers such as rivers, main roads and railways and not being truncated at difficult junctions.

Suggested Canadian guidelines for auditing the safety of bikeway systems (Matwie and Morrall, 2001) suggest paying particular attention to these design elements, each of which has the potential either to increase or decrease the safety of cyclists, depending on the situation:

• lateral and vertical clearance;
• sight distance;

- grades (gradients);
- pathway/street furniture;
- lighting;
- directional signing;
- pathway/roadway width;
- ride quality.

The need to review and upgrade older provision must be explicitly recognised, using detailed feedback from both regular and occasional, and more and less confident cyclists.

Indeed it could be said that there are three basic categories of cyclist; the fast commuter who will tend to ride much more on roads, the utility cyclist and the particularly vulnerable and probably less confident and slower leisure cyclist. Less confident cyclists will tend to give more weight to safety in their choice of route and be more willing to accept detours and longer slower routes while more confident cyclists emphasise directness and speed and generally being able to maintain their momentum. Child cyclists are unable to cope with heavy traffic which more confident commuter cyclists might tolerate for speed and directness. Child and recreational cyclists are more likely to put safety and enjoyment first, rather than directness and speed of journey.

Making due allowance for the differing weight placed on some criteria by differing cyclists is one reason why it can be difficult to give an unqualified answer to the question of whether dedicated facilities do on balance help promote cyclists' safety. Another reason is the differences that often exist in local conditions, including topography, space available and so on. With tight constraints on space it is certainly more difficult to achieve quality provision but an answer to the question as to whether off-road or on-road cycling is safer also depends on other local factors. These include the number of pedestrians, the frequency of side roads, the amount of traffic on the side roads and the main carriageway and the extent of inter-visibility at junctions. There is always a risk that the potential safety benefits on the link sections of off-road paths, between junctions, will be undermined by extra dangers at junctions but this balance needs to be looked at very carefully in the light of detailed local conditions.

It is also important to ensure that the routes in a cycle network go where cyclists want. Davies et al (1998) suggest giving priority to routes which serve schools, railway stations, large employers and town centres, with additional priority to routes which serve leisure and utility purposes, for example linking town centres and countryside. Creating short cuts and new direct links for cyclists tends to be appreciated, provided that social safety is not overlooked in some cases such as routes across parks it may be necessary to sign alternative routes for use after dark. Some authorities have now begun to develop dual networks, to suit the differing needs of more and less confident cyclists, with the former concentrating on more direct main roads routes and the latter making more use of cycle paths, shared paths and back streets, even if these are substantially longer.

2.7.5 Maintenance

This is particularly important for cyclists, both on special facilities and on the highway and this also is a vital dimension of quality. Basically good facilities can soon deteriorate without adequate maintenance and, in some cases, acquire such a poor reputation that many cyclists will avoid them altogether. To achieve adequate maintenance there need to be clear performance standards, and adequate staffing and revenue funding covering the maintenance of both on- and off-highway routes, with reference to surface quality, signing, markings and cutting back intrusive vegetation. Regular inspection is vital as well as clear and well-publicised mechanisms for reporting defects. Maintenance issues for cyclists also need to be thought through in the case of general roadworks, for example in ensuring that any necessary diversion routes are well signed.

2.7.6 Enforcement

Just as poor maintenance can give the impression for cyclists that they are still regarded as second-class road users so too can poor maintenance. If cycle facilities, or shared bus and cycle lanes, are blocked by parked cars or drivers allowed to abuse areas banned to through motor traffic, cyclists will be discouraged. The same is very much also true of drivers who are allowed to ignore speed limits with impunity. A failure to enforce speed limits will in turn make it harder to tackle the problem of poor riding behaviour including cyclists taking to riding on pavements because of their fears of riding on the road being exacerbated by regular speeding by motor traffic (RDRF, 2001). Adequate enforcement is essential in encouraging mutual respect among different road users (Litman *et al*, 2001).

2.7.7 Expanding the integration of cycling with other modes and policies

Cycling policy has now started to be much more integrated with wider planning and traffic management decisions, in line with the principles of the 1998 Integrated Transport White Paper (DETR, 1998), and this must continue as well as the further integration with other policies and strategies, such as those on health, leisure, tourism, air quality, climate change, urban regeneration and social inclusion.

In terms of integration with other modes, cycle use of bus lanes has become widespread but, as in the case of traffic calming, the great potential benefits for cyclists of such provision can be undermined by poorly thought-through design details such as insufficient width or car parking in bus lanes (Reid, 2001b). Permitting motorcycle use of shared bus and cycle lanes, as in Bristol, can be controversial and most other places in the UK have, at least so far, been reluctant to follow this example. The reintroduction of trams in British cities has so far had only a mixed record in the provision for cyclists in their detailed arrangements (MVA, 1998). Integration with public transport, especially rail services, is of great importance for longer trips and we have much to learn from Dutch experience

(McClintock, 2000). The integration of cycling and public transport is discussed in more detail by David Holladay in Chapter 8.

2.7.8   Expanding the promotion of cycling

Cycling still often has a poor image in the UK and still tends to be regarded mainly as a dangerous means of transport, of appeal only to low income people who cannot afford cars (Jones, 2001). Promotion of cycling has started to receive more emphasis and this greater emphasis should continue, in partnership with wider travel awareness and health campaigns and in partnership with other public bodies and also commercial and voluntary organisations. This can be considered of major importance given the continuing relatively low status of the bicycle in the UK, in comparison with the Netherlands (CfIT, 2001).

Indeed, some cycle campaigners feel that this 'softer' side of cycling promotion should be given far more prominence than infrastructure particularly as 'cycle infrastructure' tends to be closely associated with the dangers of cycling which are often, some feel, given exaggerated importance (Franklin, 2001).

There will be increased opportunities to promote cycling through other very important sustainable transport schemes such as homezones, Safer Routes to School and safe routes to leisure projects.

Partnerships with local employers are vital as also are partnerships with the police, particularly in giving a higher profile to reducing cycle theft. The National Cycling Strategy had a series of recommendations on bike security, including a comprehensive assessment of the adequacy of local bike parking, but these do not appear to have been widely taken up.

Cycling needs to be promoted for purposes other than leisure and work, for example for shopping and other utility trips. There is also a special challenge to promote cycling to women and, in particular, school age children (Jones, 2001). Good cycle parking is also an important part of cycling promotion but is still often overlooked, despite the specific recommendations of the National Cycling Strategy. For all types of trip secure, well-sited and convenient bike parking is important with the location of such facilities well signed and publicised.

Finally, the importance of monitoring of cycle use, by different groups, for different trip purposes, should also be emphasised, to help assess the impacts of various forms of promotion of cycling and to help provide evidence on specific questions such as the extent to which increased leisure cycling is helping to encourage use of bikes for other 'utility' purposes.

The importance of using a variety of 'soft measures' to promote cycling is discussed in detail in Chapter 3.

2.7.9   Training of cyclists

An important part of Safer Routes to School projects is the training of young cyclists, as well as providing safer access routes and bike parking at schools, and including relevant curriculum content. At the same time there is increased recognition, for example by the DETR and the CTC, of the importance of high quality on-road adult cycle training, to give confidence to adults returning to cycling, often after many non-cycling years (RDRF, 2001). A number of local initiatives have already started in some areas but these need to be expanded and co-ordinated nationally, including the development of nationally recognised accreditation standards for providers of training. Expanding adult training is also one way in which the problem of irresponsible cycling, which attracts regular local press coverage, can be addressed, although there is a difficult challenge in ensuring that those who most need to benefit from training do in fact receive it.

2.7.10   Staffing, improving professional training and current awareness

Staffing and skill upgrading have become critical issues including the need to raise the profile of staff working on cycling and the need to ensure that there are enough staff with relevant skills and with knowledge of the most important current sources of technical guidance. Many local authorities do now have cycling officers, or cycling and walking officers, to help ensure that cyclists' needs are specifically attended to but this awareness has often been slow to permeate more widely, it would appear (McClintock, 2001). This lack of awareness of the importance of catering for the mode and the benefits associated with it, among officials and local politicians, as well as the public, is still quite common in the UK (Gaffron, 2001) and aggravates the still often common lack of senior and/or political support for cycling. This in turn increases the likelihood of inadequate funding for good cycling provision and increases the risk of adverse criticisms from users. Lack of staff with enough time and the relevant skills has often compounded the problem, as well as inadequate communication between individuals and departments, and these difficulties have in some cases been aggravated when there are difficulties in obtaining land for new schemes (Gaffron, 2001).

It is clear that many of those responsible for implementing cycling policies have often lacked the support of senior management (Gaffron, 2001; Jones, 2001) and indeed still encountered negative attitudes from some of their colleagues. This makes it that much harder to implement in practice the cycle-friendly policies that have increasingly been adopted. Wider awareness and positive support for cycling are crucial for ensuring that cyclists are given quality provision, based on a real understanding of their needs and reflected in getting the details right, for different types of user. Related to this is the challenge of ensuring that those responsible for decisions on the planning, delivery and promotion of cycling have access to the latest technical guidance and relevant research findings, using electronic as well as written sources and other methods such as conferences, seminars and workshops. This should include opportunities for accessing and discussing information on best practice solutions from elsewhere that could be implemented locally. Site visits and discussions with colleagues from elsewhere may be much more effective as a means for inculcating good practice than having to wade through mountains of research reports and technical guidance documents even when these are easily accessible and digestible, it would seem!

The CTC, which over the last 20 years has greatly increased its interest in the plight of daily cyclists as well as cycle touring, embarked in 2001 on a 'benchmarking' project to help raise standards of good practice, working closely with a group of local authorities (Russell, 2000). This initiative should also be a great help in tackling the challenges for the future and in helping to raise further the respect that cyclists have started to be given in local transport policy. The experiences of several places involved in this project, i.e. York, Edinburgh and Nottingham, are discussed in three later chapters (Chapters 9, 10 and 11).

2.8   References

BABTIE GROUP (2001), *Urban Street Activity in 20mph Zones*, Manchester, Babtie Group.

CFIT (Commission for Integrated Transport) (2001), *European Best Practice in Delivering Integrated Transport*, London, CfIT, November, http://www.cfit.gov.uk/research/ebp.

CLEARY, J and MCCLINTOCK, H (2000), 'Evaluation of the Cycle Challenge Project: a case study of the Nottingham cycle-friendly employers project', *Transport Policy* 8, 117–25.

COUNTRYSIDE AGENCY (2001), *How People Interact on Off-road Routes*, Research Note CRN 32, Cheltenham, Countryside Agency.

CROW (1993), *Sign up for the Bike: Design Manual for a Cycle-friendly Infrastructure*, Ede, Netherlands, CROW.

CTC (Cyclists' Touring Club) (2000), *Cyclists and Pedestrians: Attitudes Towards Shared Use*, Godalming, CTC.

DAVIES, DG (2001), 'From guidelines to practice: cycle audit and review', *IHT Journal*, March/April.

DAVIES, DG, EMMERSON, P, and GARDNER, G (1998), *Achieving the Aims of the National Cycling Strategy: Summary of TRL Research*, Crowthorne, Berkshire, TRL.

DAVIS, R (1993), *Death on the Streets: The Mythology of Road Safety*, Hawes, North Yorkshire, Leading Edge Publications.

DETR (Department of the Environment and Department of Transport) (1994), *Planning Policy Guidance Note 13: Transport*, London, HMSO.

DETR (Department of the Environment, Transport and the Regions) (1998), *A New Deal for Transport: Better for Everyone – The Government's White Paper on Integrated Transport Policy*, Cm 3950, London, The Stationery Office, June.

DETR (Department of the Environment, Transport and the Regions) (2000a), *Guidance on Full Local Transport Plans*, March, London, Department of the Environment, Transport and the Regions, http://www.local-transport.detr.gov.uk/fulltp/index.htm.

DETR (Department of the Environment, Transport and the Regions) (2000b), *Transport 2010: The 10 Year Plan*, London, Department of the Environment, Transport and the Regions and The Stationery Office, http://www.detr.gov.uk/trans2010/index.htm.

DETR (Department of the Environment, Transport and the Regions) (2001), *Planning Policy Guidance Note 13: Transport*, London, The Stationery Office, http://www.planning.detr.gov.uk/ppg/ppg13/index.htm.

DOE (Department of the Environment) and DOT (Department of Transport) (1994), *Planning Policy Guidance Note 13: Transport*, HMSO.

DOT (Department of Transport) (1982), *Statement of Cycling Policy*, London, Department of Transport.

DOT (Department of Transport) (1996), *National Cycling Strategy*, London, Department of Transport.

DTLR (Department for Transport, Local Government and the Regions) (2001), *A Road Safety Good Practice Guide*, London, Department for Transport, Local Government and the Regions, http://www.roads.dtlr.gov.uk/roadsafety/goodpractice/18.htm.

DUPREE, H (1987), *Urban Transportation: The New Town Solution*, Aldershot, Gower.

EUROPEAN COMMISSION (1999), *Cycling: The Way Ahead for Towns and Cities*, Brussels, DGXI, and Luxembourg, Office for Official Publications of the European Communities, http://europa.eu.int/comm/environment/cycling/cycling%5Fen.pdf.

FRANKLIN, J (2001), *The Cycle Campaign Network's Campaign for High Standards*, paper presented to the Cycle Campaigns Network/CTC Cycle Planning Conference, Ryde, Isle of Wight, 5 May, http://www.cyclenetwork.org.uk/conferences/spring2001/ryde2.htm.

GAFFRON, P (2001), *Successful Implementation of Local Cycling Policies in Great Britain: What does it Depend on; How Could it be Improved?*, paper presented to the Velo-City 01 International Cycling Conference, Edinburgh, September.

GARDNER, G and GUTHRIE, N (1998), *A Study of Selected Cycle Challenge Cycle Centres*, TRL Report 340, Crowthorne, TRL.

HARLAND DG and GERCANS, R (1993), *Cycle Routes*, Project Report PR42, Crowthorne, TRL.

HILLMAN, M (ed) (1992), *Cycling: Towards Health and Safety*, London, British Medical Association and Oxford, Oxford University Press.

IHT (Institution of Highways and Transportation) (1983), *Guidelines for Providing for the Cyclist*, London, IHT.

IHT (Institution of Highways and Transportation) (1998), *Guidelines for Cycle Audit and Review*, London, IHT.

IHT (Institution of Highways and Transportation) (2000), *Guidelines for Providing for Journeys on Foot*, London, IHT.

IHT (Institution of Highways and Transportation), Bicycle Association and Cyclists' Touring Club (CTC) (1996), *Cycle-friendly Infrastructure: Guidelines for Planning and Design*, London, IHT and Godalming, CTC.

JONES, M (2001), 'Promoting cycling in the UK: Problems experienced by the practitioners', *World Transport Policy and Practice* 7 (3), 7–12, http://www.ecoplan.org/wtpp.

LITMAN, T, BLAIR, R, DEMOPOULOS, W, EDDY, N, FRITZEL, A, LAIDLAW, D, MADDOX, H, and FORSTER, K (2001), *Pedestrian and Bicycle Planning: A Guide to Best Practices*, Victoria Transport Policy Institute, Victoria, British Columbia, September, http://www.vtpi.org.

MCCLINTOCK, H (1992), *The Bicycle and City Traffic: Principles and Practice*, London, Belhaven Press.

MCCLINTOCK, H (2000), 'When in the UK, cycle like the Dutch?', *Town and Country Planning* 69 (12), December, 356–7.

MCCLINTOCK, H (2001), 'Practitioners' take-up of professional guidance and research findings: Planning for cycling and walking in the UK', *Planning Practice and Research*, 16 (2), May, 193–203.

MATHEW, D (1995), *More Bikes: Policy into Best Practice*, Godalming, CTC.

MATWIE, CT and MORRALL, JF (2001), 'Guidelines for a safety audit of bikeway systems', *World Transport Policy and Practice* 7 (3), 28–37, http://www.ecoplan.org/wtpp.

MVA CONSULTANCY (1998), *The Interaction of Cyclists and Rapid Transit Systems*, Woking, MVA.

NATIONAL CYCLING FORUM (2001), *Cycling Solutions: A Guide to Innovative Projects*, London, NCF/Department for Transport, Local Government and the Regions.

RDRF (Road Danger Reduction Forum) (2001), *Safe Roads for All: A Guide to Road Danger Reduction: A Good Practice Guide produced by the Road Danger Reduction Forum*, Keynsham, Bristol, RDRF.

REID, S (2001a), 'Pushing bikes', *Surveyor*, 21 June, 18–20.

REID, S (2001b), *Bicycles in Bus Lanes: Should they really be there?*, Paper presented to the Velo-City 01 International Cycling Conference, Edinburgh, September, and available from the author at TRL Ltd, Crowthorne.

RUSSELL, A (2000), 'Selling the cycle habit', *Surveyor*, 19 October.

TRL LTD (2001), *Cyclists' Assessments of Road and Traffic Conditions: The Development of a Cyclability Index*, TRL Report 490, Crowthorne, TRL.

UK GOVERNMENT (1994), *Sustainability Strategy*, London, HMSO.

WARDMAN, M, HATFIELD, R and PAGE, M (1997), 'The UK National Cycling Strategy: Can improved facilities meet the targets?', *Transport Policy* 4 (2), 123–33.

3

Promoting cycling through 'soft' (non-infrastructural) measures

Hugh McClintock, University of Nottingham

3.1   Introduction

The previous chapter discussed the changes in official policies on cycling, with particular reference to the provision of physical infrastructure and, especially, special facilities for cyclists. However, infrastructure and, in particular, cycle routes, are:

> neither a necessary nor a sufficient condition for high levels of cycle use. Attention must be given to other factors, like the availability of car parking at employment sites, distances between where people live, shop, work and go to school, and the urban environment, traffic speed on roads, congestion problems for motorists, and so on. In particular, attention must be given to cultural attitudes. (Jones, 2001)

It has become increasingly clear in recent years that serious promotion of cycling must embrace a much wider range of approaches than infrastructure. This is especially the case in a country like the UK with relatively low levels of cycling, where the bike still often has a poor image as a means of transport primarily for those who cannot afford cars, as well as often continuing fairly lukewarm official attitudes to its role. The very fact that most bikes are cheap to buy and use can reinforce this image. Without a much wider range of approaches the provision of cycle infrastructure on its own, even of good quality, will have limited impact. Moreover, without these kinds of changes many of the people that could benefit from taking up cycling, or cycling more, will continue to be reluctant to do so. While recognising that well-designed and well-maintained infrastructure itself can help to promote cycling this chapter therefore concentrates on the non-physical or so-called 'soft' elements of cycling policy and discusses these within the broader context of promoting more sustainable transport in general and the links to environmental and health policy in trying to change travel behaviour, by individuals and organisations, away from high levels of car dependence.

Various surveys have shown that many people would be willing to cycle, given the right conditions. For example 44% said they would cycle more on short journeys if the roads were safer, in one report on public attitudes to transport (CfIT, 2001a). In the same survey more than half the population identified issues which, if addressed, would encourage them to cycle more. The factors most often cited as particularly likely to encourage cycling were better/safer and more cycle routes, better facilities for parking bikes and a more considerate attitude by drivers. At present drivers' attitudes to cyclists are often very poor, sometimes aggravated by poor riding behaviour and the use of unroadworthy bikes (Scottish Executive, 2001). Such findings underline the importance of a wider approach than just the provision of cycle routes.

'Softer' cycling policies concentrate on the positive advantages of cycling whereas infrastructure, even if introduced for positive reasons, can be seen as having negative connotations, being implemented primarily to deal with the negative attributes of cycling in terms of safety, an aspect that some believe is greatly exaggerated, with very harmful consequences for increasing cycle use (Franklin, 2001; Wardlaw, 2000).

Other barriers than perceptions of dangers, real or imaginary, about cycling exist and these too need to be addressed. These include the weather, hilliness and distance. As Ryley comments,

> Some reasons people give for not cycling, such as the weather and the gradient, are difficult to change through policy measures. Others such as safety and distance could be improved through policy measures. For example, the installation of a new cycle path may make journeys along a route more attractive to cyclists by increasing safety and/or convenience. (Ryley, 2001)

Perceptions about difficulties of taking parcels and even small amounts of luggage on a bike can also deter people, and it has been suggested that this may be a particular problem in Britain because of the type of bike mostly sold (Guthrie, 2001).

Some of the reasons given for not cycling are exaggerated because of poor perceptions such as people not realising how few days it actually rains in the year, during peak commuting time, compared with their impressions of the number of rainy days. Indeed some surveys have shown how many of non-cyclists' perceptions are exaggerated, when comparing their attitudes with those of a similar group who had had some exposure to cycling (Bracher et al, 1991; CTC, 1999; Davies and Hartley, 1998).

The role of public policy can do more to tackle some perceived obstacles to cycling than others, for example the weather. However, even in these cases it can, through education and communication, help to correct common misconceptions. Policies will be more effective if they clearly have a good understanding of such misconceptions and also of various dimensions of environmental awareness. They can build on this increased awareness, at least among some groups, and in turn it will help policies to be more effective if they are building a wider political consensus (Maddox, 2001). Building political support for well-attuned and consistently pursued cycling policies, which are likely to need a long period to become fully effective, is therefore also very important.

3.2   Cycling to work and to education

Cycling to work in the UK accounts for nearly 40% of journeys, the single most important journey purpose (DTLR, 2001a; Ryley, 2001). Experience of Travel Plans in the UK, which are particularly for trips to work, and often known as Green Commuter Plans, dates from the mid-1990s, pioneered in Nottingham but very much reflecting experience with similar initiatives in other parts of the world, especially Southern California in the USA, where improving poor air quality was a prime motivation, and the Netherlands (McClintock and Shacklock, 1996), strongly government-led. In the Californian approach, particularly concerned to reduce single occupant car commuting, encouraging cycling has played a fairly insignificant role, compared with the emphasis on ride sharing and the use of minibuses. However, cycling has been very important in Dutch initiatives to work with employers to encourage alternative travel behaviour by their employees (Touwen, 1997).

In the late 1990s, following the 1998 Integrated Transport White Paper (DETR, 1998a) the use of Travel Plans received strong encouragement from the government and this emphasis was strengthened in the most recent revision of government policy guidance on land use planning and transport in Planning Policy Guidance Note 13 (DETR, 2001). Travel Plan initiatives have mainly focused on encouraging alternative travel by staff in their trips to and from work with varying attention to travel by customers, by staff in the course of their work and their general fleet management. Increasingly also, initiatives have been linked to other work, Local Agenda 21 and other awareness-raising exercises, road safety and Local Transport Plan preparation. Travel Plans are designed to be a systematic attempt, covering all modes, to assist staff in managing journeys to and from work in an environmentally sustainable way. Reducing the number and proportion of drive-alone car journeys made by staff from home to work is important as well as increasing the number and proportion of journeys to work by car sharing, public transport, cycling and walking.

Employers, especially major employers, have often now become more sympathetic to cycling (TRL, 2000). They have come to realise that Travel Plans can mean significant advantages for them such as reduced areas needed for car parking, more space for expansion, a fitter workforce with less staff absenteeism and improved workforce performance, reduced peak time traffic congestion and car parking pressure in their vicinity and generally a greener image (DETR, 2000c; National Cycling Forum, 2001a; Rye and McGuigan, 2000; Transport 2000 Trust, 1998). Staff are more likely to arrive at work on time, to be more alert when they arrive and to take fewer sick days (BMA, 1992). To be effective a Travel Plan must be based on a detailed understanding of current staff travel patterns including the distances they now travel to and from work, the modes they use and the reasons for their current choice of mode, and what incentives are needed to persuade them to change, on an occasional or regular basis. This means that regular communication with staff, not just in terms of an initial travel survey, is vital (BRE and ETSU, 2000). To be successful other elements are also important such as the employment of a dedicated staff member, demonstrable support from senior management, the provision of an emergency ride home facility, sustained marketing of the Plan, a Plan that really does attract employees to participate, and dynamic monitoring of the effectiveness of the Plan with provision for adjustments if necessary.

In terms of cycling important incentives for staff travel have proved to be expanded and much more secure cycle parking, preferably undercover, together with convenient showering and changing facilities. These must be backed by the provision of extensive information on cycling, particularly local cycling maps, deals to give discounts at local cycle shops and mileage allowances for cycle use during the course of work. Cycle shops deals can include discounts on accessories and the purchase of bikes and it also helps if there is provision for cycle loans for staff to enable them to buy bikes and related equipment such as a sturdy lock.

Even better is to introduce a Bicycle Use Group (BUG) for cyclists, whether regular or occasional, to give their views on further improvements needed including improvements to the local road network as well as on-site facilities. New cyclists can be encouraged by special events such as Bike to Work Days and Cyclists' Breakfasts and the provision of adult cycle training and bike buddy schemes where cyclists are matched up with experienced cyclists with detailed knowledge of particular routes in their areas. Providing company pool bikes can lessen the need for people to feel they need to bring their cars into work so that they can use them for work trips during the day. The exact package of measures needs to be tailored to the local situation and be clearly based on the views of current cyclists, and their needs and constraints, and those who are willing to take up cycling, at least on an occasional basis. Regular newsletters, forums and email discussion groups can help exchange views on what is needed and test reactions to new initiatives. As generally in the case of Travel Plans at the workplace a demonstration of commitment from senior management for the project's aims and objectives is vital, as well as an ability to identify and overcome onerous disincentives to cycle use for work journeys which arise and to provide flexible solutions.

Some of the schemes promoted under the government's Cycle Challenge Project in the late 1990s focused on working with employers to encourage cycling to and from work, especially in Nottingham, Sefton and Cambridge, and these produced some substantial and interesting results (Cleary and McClintock, 2000). For example, monitoring of the Nottingham Cycle Challenge facilities (Cleary Hughes Associates, 1999, and see Chapter 11) showed that the provision of shower facilities not only provided a facility of use to lunchtime joggers as well as cyclists but that it also resulted in a smaller season decline in cycling in winter. Another particular lesson from this project was the need to promote the accessibility, versatility and benefits of cycling, rather than focusing exclusively on what some might perceive as complications, such as showering facilities. Also clear from this survey is that cyclists value high quality and well thought out facilities and that they appreciate having their needs considered and being 'taken seriously'. Good standard provision appears to help improve the image and appeal of cycling. It is also clear from the 'after' survey that cyclists are rather less inclined than non-cyclists to express a demand for 'more cycle facilities'. By contrast, when asked what further *en route* improvements they would like to see, cyclists are relatively more likely than non-cyclists to say that they want 'better cycle facilities' (Cleary Hughes Associates, 1999, p 59), suggesting that experience of facilities is likely to make them more discriminating about quality. This is an important point.

Employers can also encourage cycling by taking an active part in local partnerships, for example on tackling cycle theft, with the police and others, or with health agencies, to promote the health benefits to their employees and to reinforce the effect of their own in-house efforts.

To help translate general local targets for increased cycling into action it is helpful to have precise targets for different kinds of trip. Trips to and from work are, most of the time, the most obvious contributors to traffic congestion but school travel is also very important, exacerbated by the general trend in the UK in recent years of sharply increased car traffic by parents escorting their children to and from school (Hillman *et al*, 1992). As well as further aggravating traffic congestion, especially during school terms, this has also had the serious effect of undermining children's independent mobility development and contributing to their increasing levels of unfitness, with substantial long term health implications. Higher levels of use of bikes by children going to and from school, as well as walking, are common in several continental European countries, even ones like Germany with higher levels of car ownership. However, the strongest incentive for the Safe Routes to School projects that have become much more common in the UK since the late 1990s has been the experience of Denmark where this has received special emphasis and it was this experience that the pilot projects in the UK, developed by Sustrans with a pioneer group of local authorities, sought to build on and develop (DETR, 2000a; DETR, 2000b; RDRF, 2001) (discussed in more detail by Johanna Cleary in Chapter 6).

Focusing on increased cycling for particular purposes such as work or school trips can seem a more practical way of achieving cycling than keeping to a general target for overall cycle use. It can also be easier to measure the results of such specific promotion efforts.

3.3   Cycle use for shopping trips

In promoting the use of bikes it is always important to link cycling to wider initiatives for these can help to make cycling encouragement more effective and offer cycling as one of a range of options for people to try, occasionally or full time. This is also true in the case of retail trips where use of cars has become more and more dominant, closely linked to the growing domination of superstores, especially at out of town or edge of town sites. This pattern, so particularly common in the UK in the 1980s and early 1990s, is now beginning to be slowed down but it has helped to engrain very strong car dependent habits that will take time to reverse.

One means of reducing car dependence for shopping trips is to encourage expansion of home delivery services, eroding the common view that there is no alternative to the use of the car for bulk shopping trips because of the need to carry large loads. This can help facilitate use of bikes for shopping trips, at least over short distances, especially if bikes are fitted with panniers or baskets to permit relatively small and light items to be transported without waiting for home delivery.

Cyclists, for obvious reasons, are less attracted to out of town sites. In addition to expanding the market for health and fitness-related goods, cyclists are more likely to shop regularly so that the total volume of their purchases over a week may not differ so much from that of car-borne customers, making a large weekly shop.

It should also be remembered (National Cycling Forum, 1998) that the different shopping patterns of cyclists can also be beneficial, particularly to the smaller or independent retailer. Cyclists have a tendency to shop locally. Just because they are more likely to make more frequent visits means there is also more scope for related activities including window shopping, browsing and socialising. In addition, as the NCF points out (National Cycling Forum, 1998), 'the tendency of cyclists to make a variety of different purchases in a centre means higher levels of cycling can contribute to a healthier and more diverse retail mix.' This in turn suggests that there may be significant benefit to town centres from implementing traffic management to encourage cycling in that the consequent broad retail mix can encourage a distinctive local character. Less motor traffic in a centre helps to make it more attractive and encourages people to browse and wander around, less inhibited by traffic.

3.4   Cycle parking

As in the case of work trips the provision of secure, convenient and attractive cycle parking facilities is of vital importance for encouraging use of bikes for retail trips, along with measures to provide safer access from areas within convenient cycling distance of shopping centres. This can include exemptions for cyclists from streets closed to through motor traffic which will be especially appreciated if they offer more direct as well as safer routes. Shopping trips will of course usually be much shorter than work trips, meaning that the provision of convenient as well as secure bike parking is all the more important, helping to offer much closer cycle parking than car parking facilities. Cycle parking must therefore be well distributed throughout a centre to make it easier for cyclists to visit easily the shops they wish to visit (National Cycling Forum, 1998). To minimise the risk of extra clutter, cycle parking can be included as an integral part of wider street enhancement schemes when such opportunities arise, with allowance for further provision if necessary. Allowance also needs to be made for bikes with trailers which may be encouraged by favourable traffic conditions. In some cases the provision of cycle centres in shopping centres may be convenient for shoppers as well as commuters, as with the very successful Leicester Bike Park (DETR, 1998b; Gardner and Guthrie, 1998). Indeed such facilities, offering more security as well as shelter, can be of particular attraction to the employees of nearby shops and it should not be forgotten that there is much that shops can do to encourage cycle use by their staff as well as by their customers. In decisions about the location of the provision of cycle parking security must always be a major consideration as well as making convenient provision for cyclists (Gardner and Ryley, 1997; National Cycling Forum, 2001b), while avoiding inconvenience for others, especially the visually impaired.

The National Cycling Strategy (DOT, 1996) had a series of recommendations on bike security, including a comprehensive assessment of the adequacy of local bike parking, but these do not appear to have been widely taken up. Good cycle parking is, arguably, a basic but most important form of cycling provision. For all types of trip secure, well-sited and convenient bike parking is important with the location of such facilities well signed and publicised.

3.5   Leisure cycling

Many people who, for various reasons, do not wish to cycle for utility trips such as work and shopping can still be persuaded to cycle for leisure, if the conditions are right (Gardner, 1998), and particularly as it may be easier to provide conditions for such trips where cyclists can get right away from traffic or have much less traffic to cope with than in many busy urban situations. A great advantage of cycling as a leisure activity is that people can take it at different levels, depending on their fitness and other motivating factors. They can choose their pace, their distance and their companions or cycle on their own. Ideally leisure cycling should not be thought of as just a rural activity or one confined to large areas of open space such as parks in urban areas. Traffic restraint and traffic calming measures should aim to make it easier for people who live in towns and cities to cycle to and from their front door so that they do not feel the need to load their bikes on cars to drive out into the countryside before they start to ride. This will also of course be affected by the distances from where they live to the countryside and whether there is a convenient and attractive traffic-free route near their home, such as on a disused railway or in a river valley or on a canal towpath. For longer distances it is also important to offer attractive and accessible opportunities for taking bikes on public transport, including trains and buses (as discussed in Chapter 8).

The provision of attractive, up to date and reliable information is arguably of particular importance in the promotion of leisure cycling, as are well signed, safe and attractive routes. Directness of routes may be of less importance than in utility trips and smooth surfaces may also be of rather less importance, provided that they are well maintained and drained. Routes need to be linked to a series of attractions and viewpoints, as well as safely and easily accessible refreshment facilities. People may also be encouraged to cycle for leisure if there are organised group rides where they can just turn up if they feel like it and the weather is fine. These opportunities, including rides of a variety of lengths and types, also need to be widely publicised, not forgetting that for many people without much cycling experience to complete even a short ride, of less than 10 kilometres, can seem a great achievement. Some local authorities such as Nottinghamshire now have many years' experience of this kind of promotion, augmented by the publication of a series of maps and guides to encourage people with a bit more confidence to go off and try out routes on their own (Hughes, 2000). This may in turn encourage some to take up cycling for other trips and means that leisure cycling should not be promoted in isolation from the use of bikes for other purposes. However, this progression is not automatic. One survey into the attitude of cycling motorists (Lawson and Morris, 1999) noted that there are differences between utility and leisure-only cyclists and that someone who cycles only for leisure is not necessarily more likely than a non-cyclist to be encouraged to become a utility cyclist.

3.6   Other cycling promotion initiatives

An increased interest by people in doing more to improve their own fitness is clearly an important opportunity to interest more people in cycling and leisure cycling. Organised guided rides can be a particularly suitable way to do this, seeming much less stressful than plunging into urban traffic. Furthermore, people can be encouraged by events such as Dr Bike Clinics offering basic maintenance checks, and also sessions to teach basic riding skills and basic bike maintenance. It should not be forgotten that novice cyclists also need help with choice of bikes and accessories. At the same time the needs of existing bike users should not be overlooked as they need encouragement to remain regular cyclists, particularly by improving the convenience of cycling and the security of parking, as well as introducing incentives and rewards (Davies et al, 1998).

Some former regular cyclists will have become only occasional cyclists or indeed may not have cycled at all for a long time. It is clear that many adults who have not cycled for many years have the motivation to learn again but want some guidance, either on a one to one basis, such as via 'bike buddies', or via the kinds of adult cycle training now being developed to give such riders more confidence. This is a vital part of learning to cope with traffic dangers, in addition to introducing more effective measures to reduce the dangers at source, especially excessive driver speed. As Davies *et al* commented:

> Much of the demand for cycle routes, on or off road, is a response to danger and unpleasantness from motor vehicles. If these deterrents to cycling were reduced by other means, such as enforcement of speed limits, improved motor vehicle engineering leading to reduced noise and pollution from vehicles, and traffic reduction measures, the demand for cycle facilities may reduce. (Davies et al, 1998, p 24)

As Davies also commented in the same report, while understanding the impact that danger has on levels of cycle use:

> It is equally important to reiterate that danger is not the only significant deterrent to cycling. Reducing danger is a necessary but not significant condition to increasing cycling. The research into attitudes to cycling showed that there are many factors involved and that there are 'layers' of motivation, reasoning and behaviour change involved. . . . These other factors include the practical convenience of cycling, the image and status of cycling, fear of cycle theft, fear of personal attack, the physical effort involved, and enjoyment of cycling. (p 25)

As the hitherto separate fields of transport and health come to work more closely together it seems that this will offer great potential for the further encouragement of cycling, especially since the opportunity to become fitter is of particular importance for the young (George Street Research, 2000). Another report has looked in detail at the lessons from various health promotion campaigns for promoting cycling (Davis, 1999) and emphasised the need for targeting particular sectors such as 'empty nesters' whose children have grown up and left home.

Clearly, marketing is very important (Davies *et al*, 1997; IHT, 2000), with different approaches targeting different market segments such as teenage girls. Davies *et al* segmented the population according to current cycling habits, and developed the following four categories:

- those who cycle already;
- those who wouldn't take much persuasion to change;
- those who would take a lot of persuasion;
- those who would always try to stay in their cars whatever measures are introduced.

Within the general population there appears to be a group of people in favour of non-motorised modes (cycling and walking) (Cullinane, 1992; DTLR, 2001b). A large household survey of attitudes to car ownership and use (Cullinane, 1992) identified them as people who represent the vanguard of a movement wanting to help the environment, the community and/or their children (Ryley, 2001).

The successful experience in different countries over some years in understanding motivations for changing travel behaviour by the German firm, Socialdata, also shows that it is very important, if time and labour intensive, to differentiate groups in terms of their willingness to make changes in their travel behaviour and to concentrate then on a range of incentives for those who appear more receptive (see Chapter 18, by Colin Ashton-Graham, for a discussion on the application of this approach in Western Australia).

One attempt has been made to draw out lessons from wider health promotion for cycling policy (Davis, 1999; IEEP and Adrian Davis Associates, 2000). This stressed that campaigns to alter travel behaviour must focus on people's needs and perceptions and not just on trip types, in order to achieve fundamental changes in attitudes. It suggested targeting specific groups and encouraging them to change their travel behaviour, for example matching people willing to try cycling to work with experienced colleagues through a bike-mate matching scheme. Understanding the effect of major life events on personal behaviour such as moving house or being involved in a major accident can also have the potential to alter travel choices, the report also said, especially when combined with better and more widespread information on local amenities and services.

Others are more cautious about deciding who may and may not cycle in the future because of the fact that most analyses of likely groups are one-dimensional and take no account of the fact that many people fall into more than one of the likely target groups, i.e. the groups are not mutually exclusive (Lawson and Morris, 1999).

There is also some evidence that more success can be gained by giving people opportunities to try cycling (Bunde, 1997; Reid, 2001). Even if people only cycle occasionally, on a couple of days a week or during the periods of better weather, that can be a worthwhile gain in terms of their health, reduced traffic and pollution, and so on.

3.7   Partnership, leadership and the importance of a co-ordinated approach to cycling promotion

Another crucial lesson for successful cycling promotion is the development and maintenance of partnerships for action with health, education, commercial and voluntary bodies. Cycling conditions are affected by so many agencies that this much wider support is crucial (HEA, 2000). This was one of the lessons from the experience of the various Cycle Challenge Projects where there was often close co-operation between organisations such as employers, local authorities, health authorities, schools and local communities. Indeed there is often scope to extend this further, for example in developing partnerships with leisure and tourist organisations, public transport operators and others in the public and private sectors (Cleary and McClintock, 2000; Sustrans, 1999).

Close partnerships with the police and with local cycle traders, as well as other public, commercial and voluntary bodies, can also help to reduce the incidence of cycle theft and vandalism. This needs to be given a higher profile.

Good leadership, by politicians and senior officials, is also crucial in helping to develop successful partnerships for the promotion of cycling and the development of more positive official attitudes across a range of departments and agencies (Russell, 2001). This can help to implement effectively the clear strategies to promote cycling that are also a prerequisite of success and also complementary strategies such as those for improving air quality, for public transport and reducing traffic dangers generally. It will also help to co-ordinate actions on a wide range of policies, from infrastructure and environmental and health promotion to a more responsible standard of road behaviour by all road users, including drivers and cyclists. At the same time special emphasis should be given to promoting the many positive benefits of cycling (Jackson, 2001; Wardlaw, 2000) and the further growth of environmental awareness should be both responded to and encouraged (Maddox, 2001).

3.8   Monitoring of cycle use and learning from experience

Finally, the importance of monitoring of cycle use, by different groups, for different trip purposes, should also be recognised, to help assess the impacts of various forms of promotion of cycling and to help provide evidence on specific questions such as the extent to which increased leisure cycling is helping to encourage use of bikes for other 'utility' purposes. Monitoring should also survey the attitudes of cyclists, both regular and occasional, to help establish the value both of specific cycling policy initiatives, and of the impact on cyclists of complementary projects such as travel awareness, Travel Plans, Home zones and Safe Routes to School schemes, and the best ways to develop these in future. The views of other interested parties are also important, to assist a more comprehensive evaluation of what works and what does not work, in various local circumstances, and what might be considered as best practice for others to learn from and apply elsewhere.

3.9   References

AUTOMOBILE ASSOCIATION (1993), *Cycling Motorists: How to Encourage Them*, Basingstoke, AA Public Policy Group.

BMA (British Medical Association) (1992), *Cycling: Towards Safety and Health (report of a BMA Working Party edited by Mayer Hillman)*, Oxford, Oxford University Press and London, BMA.

BRACHER, T, LUDA, H and THIEMANN, H-J (1991), *Fahrradfahren in der Stadt: Zusammenfassende Auswertung von Forschungsarbeiten in der Stadt*, Bergisch Gladbach/Berlin/Bonn, Bundesministerium für Verkehr.

BRE (Building Research Establishment) and ETSU (2000), *Travel Plans: The Role of Human Resources Staff and Trades Union Representatives in Supporting Travel Plans*, General Information Report 81, Watford, BRE and Harwell, ETSU.

BUNDE, J (1997), 'The BikeBus'ters from Arhus, Denmark: We'll park our cars for 200 years', in R Tolley (ed) *The Greening of Urban Transport: Planning for Walking and Cycling in Western Cities* (second edition), Chichester, Wiley.

CFIT (Commission for Integrated Transport) (2001a), *Public Attitudes to Transport in England*, London, CfIT.

CFIT (Commission for Integrated Transport) (2001b), *European Best Practice in Delivering Integrated Transport*, London, CfIT, November. http://www.cfit.gov.uk/research/ebp.

CLEARY, J and MCCLINTOCK, H (2000), 'Evaluation of the Cycle Challenge Project: A case study of the Nottingham cycle-friendly employers' project', *Transport Policy* 8, 117–25.

CLEARY HUGHES ASSOCIATES (1999), *Nottingham Cycle Challenge Project: Final Report*, Hucknall, Nottingham, Cleary Hughes Associates.

CPAG (Cyclists' Public Affairs Group) (2001), *BikeFrame: A Model Cycling Policy*, CPAG, CTC and the Bicycle Association, Godalming, CTC/CPAG.

CTC (Cyclists' Touring Club) (1999), *New Cycle Owners: Expectations and Experience*, Godalming, CTC.

CULLINANE, S (1992), 'Attitudes towards the car in the UK: some implications for policies on congestion and the environment', *Transportation Review* 26A, 291–301.

DAVIES, D, HALLIDAY, ME, MAYES, M and POCOCK, RL (1997), *Attitudes to Cycling: A Quantitative Study and Conceptual Framework*, TRL Report 266, Crowthorne, Transport Research Laboratory (TRL).

DAVIES, DG, EMMERSON, P and GARDNER, G (1998), *Achieving the Aims of the National Cycling Strategy: Summary of TRL Research*, Crowthorne, TRL.

DAVIES, DG and HARTLEY, E (1998), *New Cycle Owners: Expectations and Experiences*, TRL Report 369, Crowthorne, TRL.

DAVIS, A (1999), *Active Transport: A Guide*, London, Health Education Authority.

DETR (Department of the Environment, Transport and the Regions) (1997), *Traffic Advisory Leaflet 11/97: Cycling to Work*, London, DETR.

DETR (Department of the Environment, Transport and the Regions) (1998a), *New Deal for Transport: Better for Everyone – The Government's White Paper on Integrated Transport*, Cm 3950, London, The Stationery Office.

DETR (Department of the Environment, Transport and the Regions) (1998b), *Traffic Advisory Leaflet 5/98: Cycle Centres,* London, DETR.

DETR (Department of the Environment, Transport and the Regions) (2000a), *The School Travel Resource Pack,* researched by Sustrans for the DETR, London, DETR in conjunction with the Department of Health and the Department for Education and Employment.

DETR (Department of the Environment, Transport and the Regions) (2000b), *School Travel Strategies and Plans: Case Studies Report*, London, DETR.

DETR (Department of the Environment, Transport and the Regions) (2000c), *A Travel Plan Resource Pack for Employers (An Essential Guide to Developing, Implementing and Monitoring a Travel Management Strategy for your Organisation)*, February, http://www.local-transport.detr.gov.uk/travelplans/resource/index.htm.

DETR (Department of the Environment, Transport and the Regions) (2000d), *Guidance on Full Local Transport Plans*, London, DETR, http://www.local-transport.detr.gov.uk/fulltp/index.htm.

DETR (Department of the Environment, Transport and the Regions) (2000e), *New Directions in Speed Management: A Review of Policy*, London, DETR.

DETR (Department of the Environment, Transport and the Regions) (2001), *Planning Policy Guidance Note 13: Transport*, London, The Stationery Office, http://www.planning.detr.gov.uk/ppg/ppg13/index.htm.

DOT (Department of Transport) (1996), *National Cycling Strategy,* London, Department of Transport.

DTLR (Department for Transport, Local Government and the Regions) (2001a), *Transport Statistics Bulletin: National Travel Survey 1998/2000 Update*, London, DTLR Transport Statistics.

DTLR (Department for Transport, Local Government and the Regions) (2001b), *Focus on Personal Travel*, London, The Stationery Office.

EUROPEAN COMMISSION (1999), *Cycling: The Way Ahead for Towns and Cities*, Luxembourg, Office for Official Publications of the European Communities, http://europa.eu.int/comm/environment/cycling/cycling%5Fen.pdf.

FRANKLIN, J (2001), 'Quo Vadis?', *Cycle Campaign Network News* 55, November, http://www.cyclenetwork.org.uk/.

GARDNER, G (1998), *Transport Implications of Leisure Cycling*, TRL Report 347, Crowthorne, TRL.

GARDNER, G and GUTHRIE, N (1998), *A Study of Selected Cycle Challenge Cycle Centres*, TRL Report 340, Crowthorne, TRL.

GARDNER, G and RYLEY, TJ (1997), *Trip End Facilities for Cyclists*, TRL Report 309, Crowthorne, TRL.

GEORGE STREET RESEARCH (2000), *Why People don't Drive Cars*, a study for the Scottish Executive, Edinburgh, The Stationery Office.

GUTHRIE, N (2001), 'Better by bike?', *Surveyor*, 6 December, 18.

HEA (Health Education Authority) (2000), *Making the Links*, London, Health Education Authority.

HILLMAN, M, ADAMS, J, and WHITELEGG, J (1992), *One False Move: A Study of Children's Independent Mobility*, London, Policy Studies Institute.

HUGHES, T (2000), 'Exploring Nottinghamshire by bike', *Countryside Recreation* (University of Wales, Cardiff) 8 (3), August.

IEEP (Institute for European Environmental Policy) and Adrian Davis Associates (2000), *Delivering Changes in Travel Behaviour: Lessons from Health Promotion*, London, IEEP.

IHT (Institution of Highways and Transportation) (2000), *Guidelines for Providing for Journeys on Foot*, London, IHT.

JACKSON, ME (2001), *Promoting Cycling as a Normal Part of a Healthy Lifestyle . . . Strategies for Behavior Change*, Velo-City 2001 Falco Lecture Prize Paper, Velo-City Conference, Edinburgh and Glasgow, September, http://www.cyclenetwork.org.ukpapers/falco011.html.

JONES, M (2001) 'Promoting cycling in the UK: Problems experienced by the practitioners', *World Transport Policy and Practice* 7 (3), 7–12, http://www.ecoplan.org/wtpp.

LAWSON, S and MORRIS, B (1999), 'Out of cars and onto bikes: what chance?', *Traffic Engineering and Control*, May.

MCCLINTOCK, H and SHACKLOCK, V (1996), 'Alternative transport plans: encouraging the role of employers in changing staff commuter travel modes', *Town Planning Review* 67 (4), October.

MADDOX, H (2001), 'Another look at Germany's bicycle boom: implications for local transportation policy and planning strategy in the USA', *World Transport Policy and Practice* 7 (3), 44–8, http://www.ecoplan.org/wtpp.

NATIONAL CYCLING FORUM (1998), *Cycling in Urban Areas: Issues in Retailing*, London, Department of the Environment, Transport and the Regions.

NATIONAL CYCLING FORUM (2001a), *Cycling Works! How Employers can Benefit from Increased Cycling*, London, Department of the Environment, Transport and the Regions.

NATIONAL CYCLING FORUM (2001b), *Cycle Security*, London, Department of the Environment, Transport and the Regions.

RDRF (Road Danger Reduction Forum) (2001) *Safe Roads for All: A Guide to Road Danger Reduction: A Good Practice Guide produced by the Road Danger Reduction Forum*, Keynsham, Bristol, RDRF.

REID, S (2001), 'Pushing bikes', *Surveyor*, 21 June, 18–20.

RUSSELL, A (2001), *Benchmarking of Local Cycling Policy*, paper presented to the Velo-City 2001 Conference, Edinburgh and Glasgow, September.

RYE, T and MCGUIGAN, D (2000), *Green Commuter Plans: Do they Work?*, London, The Stationery Office.

RYLEY, T (2001), 'Translating cycling policy into cycling practice', *World Transport Policy and Practice* 3, 38–43, http://www.ecoplan.org/wtpp.

SCOTTISH EXECUTIVE (2001), *Sharing Road Space: Drivers and Cyclists as Equal Road Users*, Edinburgh, Scottish Executive.

SUSTRANS (1999), *Cycle Tourism*, Bristol, Sustrans, http://www.sustrans.org.uk.

TOUWEN, M (1997), 'Stimulating bicycle use by companies in the Netherlands', in R Tolley (ed) *The Greening of Urban Transport: Planning for Walking and Cycling in Western Cities* (second edition), Chichester, Wiley.

TRANSPORT 2000 TRUST (1998), *Health on the Move: Policies for Health Promoting Transport*, London, Transport 2000 Trust.

TRL (Transport Research Laboratory) (2000), *Institutional and Organisational Attitudes to Cycling*, Crowthorne, TRL.

WARDLAW, MJ (2000), 'Three lessons for a better cycling future', *British Medical Journal* 321, 1582–5, http://bmj.com/cgi/content/full/321/7276/1582.

4

Making space for cyclists – a matter of speed?

Michael Yeates, Architect and Urbanist, Cyclists' Urban Speedlimit Taskforce, Brisbane, Australia

This chapter is dedicated to the memory of Kerry Fien who as Brisbane City Council Bikeways Planning Officer and later, Program Manager Bicycles, undertook much of the research and negotiation necessary for the early trials and later development and implementation of the concept. Kerry was tragically killed on September 10, 1999.

4.1   Introduction

The Cyclists' Urban Speedlimit Taskforce (CUST) was initiated by the Bicycle Federation of Australia (BFA) in 1995 to develop support for reducing the national General Urban Speed Limit (GUSL) due to recognition that the GUSL of 60 km/h, although apparently generally accepted by road authorities as safe, was and remains much too fast for the safety of people with access and mobility disabilities in urban areas, i.e. the young, the elderly and those with specific, degenerative, temporary or permanent disability irrespective of whether their mode is walking, cycling or use of a mobility aid. The fact was and remains that urban areas in Australia are designed and managed primarily for the safety and convenience of motorised traffic and transport. Despite the needs of non-motorised modes being increasingly obvious (e.g. in road tolls, transport, health and safety strategies and policies) and these views being increasingly presented by advocates for these modes, traffic and transport expertise continues to promote a perspective which favours the priority for and hence dominance by the motorised modes.

In Australia, the urban speed limit is 60 km/h (approx. 37 mph), which is very high compared to Europe and parts of North America. Some Australian state governments are introducing 50 km/h, primarily in local streets and some lower volume roads, and one state, South Australia, has been the location of an extensive trial of 40 km/h in all local streets in one large precinct of approximately 6 square km.

For advocates of walking and cycling and public transport, this causes two related problems. The first is that road authorities implicitly promote the roads as 'safe' but in many cases are reluctant to provide facilities, whether crossings for pedestrians or bike lanes for cyclists, on those roads. The other is how to increase walking and cycling activity including their use as a means of access to public transport in urban areas which, having a relatively low density, require somewhat longer journeys to local services. This is exacerbated by the fact that, despite a relatively high cycling usage before the Second World War, cycling usage is very low in most places.

To establish the case for lower traffic speed and the traffic and urban planning and environmental benefits that flow from reducing speed and dominance of traffic, CUST researched and produced a report *Towards a Safe Urban Speed Limit* (CUST, 1996) which was officially launched at the Velo-Australis conference in Fremantle in 1996. At that conference, Staysafe Committee Deputy Chairman, Mr Jeff Hunter MP, presented Staysafe report No 34 supporting reducing the GUSL in NSW from 60 km/h to 50 km/h. However, international cycling engineering, mainly based on extensive research in the Netherlands, accepts that 50 km/h is too fast for cyclists and that speeds of 30 km/h are regarded as 'safe' as Staysafe Chairman, Mr Paul Gibson MP, confirmed during his visit to Europe. *Towards a Safe Urban Speed Limit* provides an extensive examination of the issues together with a considerable reference list and examples of how lower speed can, and has been shown to, produce a 'safe' urban traffic environment.

CUST continues both an advocacy and a monitoring role. CUST has been represented at local, regional and national levels in Australia and internationally, including attending the international Velo-Australis conference in Fremantle in 1996, international Velo-City conferences in 1995, 1997 and 1999 (in Graz, Austria, the first city in the world with a 30 km/h GUSL), and 2001 and in June 2000, as a panellist, at Velo-Mondial 2000 in Amsterdam.

Despite the technical knowledge being available in Australia, CUST is very concerned that the interests of cyclists continue to be excluded by 'experts' or perhaps more accurately 'expertise', whose concerns for road safety appear to reduce, rather than endorse, opportunities for safer urban road environments for all road users by choosing not to improve road safety for cyclists. By so doing, current traffic and transport engineering views of road use and road safety are thus effectively endorsed rather than encouraging new ways of traffic management which can achieve more walking and cycling in support of the many and various policy goals by integration of the many seemingly conflicting goals.

The concept which is the focus of this chapter offers many and varied opportunities for integrating cycling into 'mainstream' traffic management and urban design practice both safely and at low cost. However, before addressing the possibility of 'solutions', the policy setting and reasons for ensuring cycling is included and no longer excluded, i.e. the 'problems' or 'concerns', must first be examined and accepted.

4.2   The policy position regarding cycling

Not so very long ago, cars, trams and trains required a person waving a red flag and ringing a bell to accompany such vehicles when they travelled along or crossed roads. This of course was in an era when concern for pedestrian safety and convenience provided some priority over trains and trams, and cars, being a rarity, had little priority. While the bell and flag have in most cases been super-seded, for example by traffic lights, it is the relative priority of people walking and cycling which has subsequently been substantially reduced relative to the priority of motorised traffic. In relatively recent times, perhaps in the last 30 years at most, the excesses of too much traffic were increasingly recognised as a blighting effect on urban living leading to various protests against major roadways, primarily urban freeways.

It was the rural activists, perhaps most notably in the UK, who raised the issue of the impact of town bypasses. Bypasses, while ingeniously designed to improve the urban condition, encouraged even more traffic, which while dividing the rural scenes also continued to add to the impacts in urban areas. By the 1990s, it had become clear that the priority of traffic over urban and rural settings is not an essential or inevitable outcome as has been increasingly well demonstrated by its reversibility, for example in extensive traffic calmed or low speed urban areas and pedestrian precincts in Europe, North America and some examples in Australia, each of which could be considered as home to high levels of car use. The issue of traffic priority in urban areas was finally able to be seen as a policy if not political issue, an issue of design and management, and of considered and therefore deliberate policy, and where and when to apply it.

4.3   Urban planning and traffic in towns

Without doubt, the seminal work on planning and traffic in recent times must be *Traffic in Towns*, the *Report of the Steering Group of the Study of the Long Term Problems of Traffic in Towns* which includes the report of the Working Group (The Steering Group, 1963) also known as the Buchanan Report. However, being published during the great modernist enthusiasm for reconstruction, much of the valuable analysis was lost in the dramatic reconstruction proposals for inner London for example. By comparison, the scheme for Norwich has, perhaps due to the location of Norwich resulting in limited need for the reconstruction approach in the period since, been reflected in practice. In the Norwich scheme, traffic was constrained by capacity within the town, with the town being modified by managing the road use and functions, rather than reconstruction.

It is the Steering Group Report itself which provides relevant clues as to the fate of the Buchanan Report recommendations which is best illustrated by three main conclusions, namely that 'there is no one easy and complete solution', that each of the available approaches 'reacts immediately on the others . . . [necessitating they be] applied in a carefully co-ordinated way', the forerunner of the concept of integrated transport planning, and that 'most importantly, any such organised attempt to solve the problem will necessarily involve very large-scale redevelopment of our cities and towns on a significantly different pattern. If we are to have any chance of living at peace with the motor car, we shall need a different sort of city' (The Steering Group, 1963, para 35).

The Steering Group thus provided a choice which suggested that cities and towns could be developed while 'living at peace with the motor car' yet this option was never tested, it was only assumed. It has taken nearly 40 years to discover that this suggestion simply reflected the optimism of the postwar modernist approach expressed by huge urban renewal schemes usually supported by massive freeways. Few people considered where the cars went after and before they used the freeways. So like the bypasses seen as important by Buchanan, the solution simply added to or relocated the problem.

The concepts in *Traffic in Towns*, although now somewhat dated by scientific advances, confirm that there must be limits to the growth and impact of traffic in towns if health, safety, character and amenity are to be preserved for the visitors, occupants and residents. Issues such as noise and air pollution, traffic congestion and efficiency were addressed in both the analysis and the various schemes. The concept of the 'environmental capacity of streets' for example considered the threshold of nuisance and amenity to be far lower than the maximum or 'crude' capacity for through traffic and parking commonly utilised by traffic engineering. As the Buchanan Report states,

> from this idea that an environmental capacity could be defined for a
> *street* and the adjacent buildings it serves, it is only a step to the
> possibility that both environmental and crude capacities could be
> evaluated for complexes of streets or for whole environmental areas.
> (para 129)

Where did *Traffic in Towns* fail? Did it fail? Arguably, the one failure was the assumption that traffic would, indeed could, continue to grow and that, when it reached the limits for the town, bypasses could be constructed to ensure that capacity was not exceeded. The immediate result, in a period when freeways were beginning to be viewed as symbols of progress, was a schism between urban and town planning on the one hand and traffic planning on the other, a schism which suggested the need for more and bigger roads through and, where necessary, bypassing towns and cities. It is this continuing schism between town planning and traffic planning which results in the continuing unresolved conflicts between land use, transport and traffic. It resulted in people being forgotten . . . people walking or cycling, people who are ubiquitous in urban areas, people who were vulnerable and therefore a danger to themselves and the increasingly fast and greater volumes of motorised traffic. It is these conflicts which primarily cause the problem of lack of mobility and safety in urban areas for people walking or cycling or using mobility aids.

4.4   The Australian context

Being a 'young' country, it is easy to overlook that most 'older' parts of Australian urban areas were 'designed' well before the dominance of cars and hence their structure reflects the needs of people who walk or cycle or use public transport much as do European 'walking' cities. It is only the more recent areas that, like most other cities in other countries, have been designed to suit motorised transport, being either older areas retrofitted for increased traffic or designed as newer 'car-based' areas. Despite cycling being a major mode of transport previously, and, unlike for example the Netherlands where efforts have been made to maintain cycling, Australian urban policy failed to include cycling which, as it increasingly became both less valued and more dangerous, soon reduced. Cycling trips such as those to local work, school, shopping or public transport all but disappeared. It is these short trips to these destinations that must be made sufficiently safe that people can again choose to undertake them by walking or cycling.

Only recently has this begun to be recognised. Arguably, nowhere has it begun to be reversed, despite increasing provision of cycling facilities. The prime reason it can be confidently stated no reversal is or has been occurring (with the possible exception of some very important local examples) is that at national, state and regional level, few if any authorities have accepted that, rather than continue allowing motor traffic to dominate urban areas, non-motorised accessibility, mobility and permeability should be more dominant thus reflecting the ubiquity of walking and cycling in urban areas (Brindle, 1984).

Thus despite a second national cycling strategy, *Australia Cycling: The National Strategy 1999–2004* (Austroads, 1999a), despite Australian design standards and guidelines (Austroads, 1999b), despite concerns about urban road injury and fatality costs as expressed in the *National Road Safety Strategy* (Austroads, 1999c), and despite obligations to reduce greenhouse emissions and to improve health, well-being and equity, it is the road authorities, the local government authorities and the road safety authorities that continue to promote current practices and alternatives while rarely including provision for cycling even when opportunities occur. Interestingly, it would appear that it is the perceived risk of cycling in current traffic conditions which dissuades many 'experts' from supporting cycling. While this conflicts with the evidence and the policy position, it succinctly confirms that the urban road network should be safe enough that people who would like to walk or cycle actually feel encouraged to do so.

The fact that the 'experts' generally do not choose to include and promote cycling, arguably because it is considered or felt that cycling is not safe, raises many questions of an ethical nature, not the least being the position of those promoting cycling as a transport policy on a road and street system which they believe is not safe. By adopting a concern about the unsafety of cycling on current roads and streets, many experts then choose not to include cycling. However, for credibility, and given the policy position outlined above, cycling should, if not must, always be included, even if that requires a reduction in the dominance or priority given to high speed motorised traffic.

Of particular relevance, most current policy goals can be more easily achieved by integration of the many issues which arise in urban planning and by reducing the current almost total dominance given unquestioned to motor traffic. Arguably, cycling must be included, if necessary by compulsion, and not arbitrarily marginalised or excluded, to enable 'inclusive' urban and traffic design to support current goals seeking improved urban safety and amenity for all (Yeates, 1999, 2000a). Exclusion or non-inclusion of cycling leaves most current 'model' urban projects open to both critique and failure.

4.5   Safety

Safety is a relative concept. While cycling or walking might feel unsafe, in practice, the environment in which such activities take place, whether on roads or in rural settings, often feels more threatening than it is. As recent projects promoting bike paths have shown worldwide, when designed with safety and security of the users in mind, the facilities are used. The more users, the better the environment feels. It is vitally important, however, to gain the support of numerous users, and to do this facilities must feel and be sufficiently safe. In the urban context, there are few variables. Space is usually occupied or owned. Problems exist but, usually being permanent, can be difficult to change. Dutch cycling design aims for combinations of coherence, directness, attractiveness, safety and comfort (CROW, 1994), a concept summarised as 'safety+convenience' (Yeates, 2000a), a summation of the assessment by user audit of the combination of both safety and convenience.

The concept of safety+convenience should apply to everyone, equitably. Accordingly, in much the same way that freeways are designed for the safety+ convenience of high speed motorised traffic, crossings of freeways by people walking or cycling should be safe+convenient for them, necessitating overpasses or underpasses designed accordingly and located at frequencies to suit the users. In urban areas, this model or concept is perhaps best exemplified in postwar new towns such as Stevenage in the UK or, somewhat earlier, Canberra in Australia, where the higher speed motorised traffic is separated from the walking and cycling traffic. Increasingly it has been recognised that separation is extremely expensive in that, in many areas, it requires expensive duplication which may be unnecessary if the road system is managed for a more equitable safety+convenience outcome for all modes. In addition, off-road paths, overpasses and underpasses are not necessarily safe from a personal safety perspective even if they are safe from traffic.

From a safety perspective, it is therefore possible to argue that in urban areas generally, the road, street and movement systems should be designed for safe and convenient movement for people with a mobility disability or vulnerability, rather than for motorised traffic, and that with appropriate allowances for operational space for the various modes, this will also be an urban environment safe for people walking or cycling as a priority (Yeates, 1999). The concept of 'sharing the road' then makes sense, not only from a safety perspective but also from Buchanan's concept of the 'environmental capacity' as an amenity indicator of a street, road or precinct. Using these concepts, it becomes apparent that the measure of the safety+convenience of the various users is a powerful indicator not only for safety, mobility and accessibility but also for amenity and for environmental outcomes such as noise and air pollution.

Thus while road safety authorities praise themselves for reducing the road toll, rarely do they assess whether, and if so how and why, exposure has changed. It is not surprising that safety+convenience audits can predict both high risk areas and areas where exposure has reduced, especially when all modes, that is, all potential rather than existing users' needs are taken into account by the audit. By adopting the idea of safe roads and streets rather than celebrating the reduction in injuries and fatalities, Sweden's *Vision Zero* aims for a zero road toll requiring reconsideration of the whole road system design and management (Wramborg, 1999).

Now that the idea of a zero road toll has been adopted in Sweden, it becomes more difficult for other road safety agencies to ethically defend their annual targets. This allows others to seek a similarly equitable and people-friendly road system management from their road safety authorities (Yeates, 2001a), arguing not just for safe walking and cycling facilities but for a system safe+convenient for all users (Yeates, 2000a, 2001b). Increasingly, once it becomes accepted that separate facilities for the different modes are arguably impractical with obvious exceptions, the idea of integration, again identified by Buchanan, can be applied.

4.6   Integration – or separation?

It is important to acknowledge that there are many differences in approach to road safety but those grounded in equity and equality appear increasingly to be not only more successful but also to have the capacity, if not the imperative, to result in outcomes which address a multitude of problems. It is therefore equally important to confirm that these diverse ways of addressing urban problems are essential to development of long term better outcomes. Arguably then, the use of bike lanes can probably be seen to reflect the apparent need for separation. But why the need for separation? As the Dutch design manual makes clear, there is a need for separation as motor vehicle speeds exceed 30 km/h and volumes increase (CROW, 1994, pp 80–1), a need reflected by studies in India in dense, mixed, congested traffic which showed that traffic mixes relatively safely until traffic speeds increase above 20–30 km/h and the modes separate into lower and higher speed 'lanes' (Tiwari et al, 1995). Arguably, however, there is still a need to provide space on the road for cyclists even when the traffic is stationary.

Acknowledging that there is a need to provide space for the bicycle on the road and to reduce conflict between cyclists and pedestrians on footpaths and shared paths wherever possible (CTC, 2000), the concept of sharing the road rather than segregating users requires a means of identifying where cyclists are expected to travel on the road (Mackay, 1995). In Australia, a bicycle symbol has been used as the standard graphic, e.g. in bike lanes. The standard symbol is typically 1.15 m wide and thus ideal for identifying where cyclists are expected to travel on the road. By use of a series of symbols along a road, with location selected to suit the traffic speed, speed limit, parking and other road requirements, a zone or bicycle lane without edges is created. Unlike Europe, advisory lanes are not part of the accepted standards in Australia thereby ensuring that, in principle at least, substandard or extremely narrow advisory lanes are not provided in lieu of sharing the road concepts.

The concept of a bike lane without edges may initially appear controversial. However, if the intention is to share the road while creating space for the bicycle, it is the edge line of bike lanes which creates the division and the potential for territoriality and the resulting claims of ownership and trespass whether it is cyclists moving out of the bike lane or motorists driving into it. In principle, the idea is to create a road space with sufficient space that sharing the road is feasible and safe, and as in the research from India mentioned above, the various modes can separate in higher speed areas. In Australia, the nearest comparable facility is described as a wide (usually) kerbside lane (Austroads, 1993). However, if the symbol is located on the road according to protocols which reflect the situation and the intended uses, there is no need for the facility to necessarily be in the kerbside location which might for example most often be used or allocated for parking (see Figs. 4.1–4.3).

In Brisbane, the state capital of Queensland, the Brisbane City Council has utilised this concept widely to include improved safety and amenity for both new and existing road projects (see also section 4.12). To best describe the concept as a design and management tool, BCC has combined the concept of cyclist-friendliness and the idea of the 'lane' being a zone, hence the concept was identified as a 'Bicycle-Friendly Zone' or BFZ (Yeates, 2000b). Use of the BFZ concept has emerged as an economical means of endorsing cyclists' likely route of travel on roads and streets throughout a city the size of Brisbane with some 7000 km of roads and streets by not only identifying bike routes but also making motorists and cyclists more aware of the location where cyclists would be expected to travel on the road. However, as BCC urban design teams have developed the concept, another unexpected benefit has emerged, the opportunity to design the road space to encourage lower traffic speed. This occurs because, unlike bike lanes where the segregated lanes each have minimum lane widths to ensure larger vehicles can travel in the adjoining lane without entering the cycle lane, the BFZ operates quite differently in that it acts as an advice or warning to vehicle drivers that cyclists are likely to be ahead and sharing the road.

Fig. 4.1 Shows the BFZ symbol which in this case is accompanied by a white statutory (regulatory) edge line which prohibits motorists from driving to its left. However, cyclists are exempt from that restriction and may cycle either side of the line. The symbol placement therefore has multiple meanings but the effect as shown here is to encourage motorists to travel further to the right thus adding to the space for the cyclist. Also, because there are no lane restrictions as there are with bike lanes, cycling two abreast is feasible whereas in a minimum width bike lane it is not. In some locations with car parking and without edge lines, the symbols are used in the same location but more frequently.

The location of the bicycle symbols can be used to narrow the appearance of the available road space for the motorised traffic while preserving the space for cyclists. In shopping areas, for example, the space for the motorised traffic could be as narrow as 2.4 m allowing cars to proceed but giving the impression of a tight profile thus encouraging lower speed. With traffic space as narrow as this, buses and trucks will most likely run over the symbols, but of course the intention is to provide increased priority for the cyclists when the width is so narrow. In some very narrow locations, the symbol could be in the middle of the lane.

However, from a cycling advocacy perspective, the BFZ has yet another role. Many road authorities appear to forget cyclists when new works and modifications are being carried out. Typical are approaches to traffic lights where extra lanes are created or at build-outs and other squeeze points such as at pedestrian crossings and the entry and exit roads on roundabouts. Here the BFZ functions more as a reminder to designers to maintain cycling space while assisting with provision of a tighter, lower speed profile yet still allowing wider vehicles the ability to travel through the tight radius bends needed for speed reduction.

Fig. 4.2 Shows the BFZ symbol in a tighter road profile, here with a median island with the road narrowed to minimise the crossing distance at a pedestrian crossing. However, while tight, there is ample space for the cyclist given the BFZ symbol is always at least 1.2 m wide. It is the narrow tight profile which encourages lower traffic speed, yet the likely presence and location of cyclists is clearly and repetitively provided.

Fig. 4.3 Shows the BFZ in a local shopping strip with the kerb built out on the departure side of a bus stop. While this would normally create a squeeze point, made more so by the left turn, the clear marking and the fact that space is not only provided but shown to be for cyclists, not only warns motorists but endorses the presence of cyclists and, while very much a concept based on sharing the road, is also a strong political statement endorsing cycling.

This experience draws on the European development of lower speed environments, notably 30 km/h, as a 'safe' urban speed limit for all road users. This has revolutionised traffic and travel demand management, not only in the 30 km/h speed zones and Dutch 'living streets' but now with the idea that a whole city might have a 30 km/h speed limit with higher speed limits posted by signs as in the Austrian city of Graz. In conditions such as this, the need for traffic calming and for separation is substantially reduced and mobility by walking and cycling demonstrably prioritised, again at relatively little cost but with wide-ranging benefits in 'environmental capacity' at precinct, city and regional scale.

The question that increasingly arises therefore is whether a cycling-friendly city or town requires any separate facilities for the safety of cyclists if the traffic speed is safe. The answer appears increasingly to be that separate facilities or sharing the road facilities are needed to provide space for the cyclist but that safety and increased usage are much more dependent on traffic speed. This reflects current practice as proposed for example in Sweden's *Vision Zero* where the local streets are either designed as 'walking speed streets', or for 30 km/h or 50 km/h with 30 km/h intersections (Wramborg, 1999).

4.7   Speed management in Australia

To share the road safely, speed management is therefore a major issue. In Australia, the urban speed limit is 60 km/h (approx. 37 mph) which, as mentioned before, is very high given much of Europe and North America has 50 km/h (30 mph) and some of the USA has 25 mph (40 km/h). As the first step in the process, some Australian state governments are introducing 50 km/h, primarily in local streets and in some states, on lower volume roads. One state, South Australia, has been the location of an extensive trial of 40 km/h in all local streets in the city of Unley, one of the local authorities in the state capital, Adelaide. Unley promotes both its 40 km/h speed limit and itself as a local authority area as 'safe' using a variety of promotional materials.

While the various state road authorities are apparently reluctant or unable to reduce the urban speed limit to 50 km/h and local street speed limits to 40 km/h or less, two major problems occur. As in Victoria's campaign 'Think safe. Think 50', the first problem emerges because Australian road authorities promote 50 km/h in residential streets as 'safe' with a likely 50% fatality rate for children hit by a car travelling at 50 km/h despite, for example, Sweden's *Vision Zero* material suggesting an 80% fatality rate for pedestrians. Arguably therefore with high traffic speeds and relatively low levels of walking and cycling, one of the problems in assessing road safety in Australia is exposure (Yeates, 2001a).

Related directly to this, the other problem is how to increase walking and cycling activity including as a means of access to public transport in urban areas which are relatively low density, requiring somewhat longer journeys to local services yet on roads where traffic is permitted to travel at speeds of 50 km/h in local streets and generally, on most urban roads, at speeds of 60–70 km/h, especially now that, despite a relatively high cycling usage pre-war, cycling usage is very low in most places.

4.8   Where to start first?

As much urban congestion results from local trips and these are induced by the relative perceived safety and convenience of using a car, trips to local common destinations must be made sufficiently safe that people again choose to undertake them by walking or cycling. Whether to local work, school, shopping or public transport and various combinations thereof, these trips are potentially what, in the Netherlands, are described as links in various individual 'trip chains'. All provide opportunities for starting cycling.

However, while many of these areas are being considered for, and in some cases provided with, improved cycling facilities, the measure of success lies with the number of people who actually feel sufficiently inclined to change modes that they actually do change modes, if not always, at least regularly enough to provide evidence of a shift. This is necessary not only to satisfy research or policy needs but to provide a critical mass sufficient to encourage others. It has been well argued that the perception of social support at both individual and community levels is needed to encourage and endorse changes in behaviour that might otherwise be considered 'new' or radical, e.g. cycling to work. Hence both comparative safety and comparative convenience compared to other modes must be assessed and arguably, if the 'new' modes are to be encouraged, substantial changes must be made to ensure that the perception of the safety+convenience is not only sufficiently improved to encourage changed behaviour, but when the behaviour change is made by individuals, the reality meets or exceeds their perceived needs. If not, changes will not take place or if they do, they will be either temporary or very occasional.

Current strategies imply that the idea of road safety is supported. However, review of specific measures suggests that rarely if ever are both safety and convenience i.e. safety+convenience, taken into account. For example, extremely localised school speed limit zones clearly do not encourage parents to allow children to walk or cycle to school as is practised increasingly under the Safe Routes to Schools rubric. Elsewhere, barrier fences try to prevent natural pedestrian crossing of roads where traffic conditions dominate pedestrian movement thus preventing preferred crossing provisions. Reduced speed limits are limited only to 50 km/h and only in local streets, although it is known that 50 km/h is far too fast for safe cycling or pedestrian movement for all potential road users. Being very narrowly defined in Queensland, most if not all main roads are specifically excluded from being reduced to 50 km/h or less, seemingly ignoring safety, amenity and convenience reasons and local use by people walking or cycling.

A current NSW review by the Roads and Traffic Authority (RTA, 2000) of *Sharing the Main Street* (FORS, 1993) by the Roads and Traffic Authority (RTA), and a recent conference, *Street Design for Road Safety*, are thus of particular relevance and concern. It appears public consultation on the content of the RTA review may have been extremely limited and/or constrained by both the review process and current perceptions that cycling on Australian urban roads and streets is not safe enough to promote. By not addressing the safety issue sufficiently, lessons learned both in Australia (e.g. at Campsie, Marrickville, Hurstville in Sydney, NSW) and overseas will again be ignored, resulting in a second generation of examples which fail to adequately address the inclusion of cycling (e.g. Taree, NSW – see section 4.14). The fact that no expert cycling presentation was included in the *Street Design for Road Safety* conference seems to confirm that the benefits of including, i.e. integrating, cycling and the danger of current 'expert' perceptions of cycling are not sufficiently understood or remain excluded.

4.9   The local town centre as just one of the first steps?

The failure of current and previous practice and guidance to include if not mandate cycling arguably has its most serious consequences in local urban centres and strip shopping areas. This arises from advocacy for separation rather than integration and expertise that seeks to address concerns by and needs of a number of differing and apparently conflicting interests including those of motorists, through traffic, local traffic, people walking, people catching public transport, people with access disabilities, people who choose not to (or do not) own or have access to cars, people cycling, and not least, property owners and business owners and the experts' own perceptions.

To date, these interests have not been integrated by policy or design. Rather, one or more dominant views have ensured either marginalisation or exclusion of the others. For example, it seems that through traffic must be entitled to travel at 60 km/h (in NSW increasingly at 50 km/h) on main roads through urban centres which, with little real change, can be 40 km/h or less if through traffic is calmed, managed or relocated. Arguably, this reflects an orthodoxy that allows through traffic to dominate local walking and cycling. Increasingly, however, it is obvious this orthodoxy conflicts with purported policy concerns about road safety, equity, health, etc.

It is therefore much more useful and pragmatic to consider such difficult questions and their solutions as trials of solutions, rather than solutions in themselves. Arguably, all solutions should then be considered as 'integrated resolution' by integration of all the interests and not a response primarily to the dominant modes. By assessing the comparative safety+convenience of all modes and then ensuring their accessibility and mobility is equitable, where appropriate (e.g. in town centres and strip shopping centres), people places can be made more people-friendly in the same way as freeways are high speed motor vehicle-friendly and very people-unfriendly; by establishing and legitimising the relative priority. Arguably, therefore, residential areas, shopping areas, school and public transport precincts should clearly be much more people-friendly and should exhibit a much higher comparative safety+convenience, if not priority, than for motor traffic.

One of the most effective ways of doing this is by speed management that clearly reflects these goals. Most Australian road authorities support 40 km/h as 'safe' for local school zones and thus imply sufficient safety+convenience to encourage children to walk or cycle to school. Arguably the 25 km/h speed limit supported by a supplementary plate 'when children present' used in South Australia provides a much higher safety+convenience for children and a much higher safety and much lower convenience (thus resulting in a lowered safety+ convenience and priority) for motorists. It is only a small step for much wider application of such zones to be supported, e.g. as 'pedestrian priority' or 'cyclist priority', as is increasingly achieved elsewhere.

Local town centres and shopping strips thus provide a very useful and arguably most appropriate location for increasing the legitimacy of cyclists by increasing their safety+convenience either by bike lanes or by other means. Cyclists can and should be included rather than excluded by non-provision or relegating them to back lanes or back streets as suggested by one of the 'experts' during the *Street Design for Road Safety* conference. By utilising experience from elsewhere to help solve problems, sharing the main street can be moved from excluding cyclists to including them to make main streets safer for all.

4.10   Sharing a safer main street?

When *Towards a Safe Urban Speed Limit* (CUST, 1996) was being developed, Bicycle Victoria was developing *It can be done . . . A bicycle network on arterial roads* (Bicycle Victoria, 1996). It provides a design basis for completing Melbourne's Principal Bicycle Network which identifies some 2000 km of roads in metropolitan Melbourne where the state roads authority of Victoria, VicRoads, and local government will make provision for cycling. It was clear, however, that in Sydney and Brisbane in particular, but also in many other places throughout Australia, 20 m wide road reserves with standard footpaths provided a traffic surface which was very safe for 60 km/h traffic operating in four lanes each slightly above 3 m wide. However, as 60 km/h is known to be too fast, and 3 m lanes too narrow to allow overtaking, only a reduction in speed could improve cycling conditions. Thus *Towards a Safe Urban Speed Limit* suggests that speed should be reduced if there is no space and thus, cities can easily and economically make space for cyclists . . . by slowing the traffic, while achieving a range of other policy goals.

Slower traffic in urban areas provides major benefits for road safety as well as for amenity, environment, mode shift, etc. Narrow lanes assist in reducing traffic speed based on speed environment principles. Hence, positive endorsement of cyclists in narrow traffic lanes can provide both legitimacy for the cyclists and the opportunity for cyclists to provide a 'mobile speed management' role. In practice, this is exactly what happens in many of the cities which boast high volumes of cyclists using old roads and streets in historic cities where there is no space for separated bike lanes, footpaths or bike paths.

Thus rather than the traffic safety issue arising from conflict between pedestrians and cyclists as occurs on footpaths and bike paths (e.g. CTC, 2000), the issue is actually a conflict between potentially high speed traffic (allowed to travel at speeds of 50–60 km/h or more on urban roads and streets) and cyclists (whose normal comfortable speed varies between 15 and 30 km/h). Arguably, provision of facilities or conditions that are sufficiently safe+convenient for cyclists on roads would encourage many more, if not most, cyclists to use such roads. As elsewhere, if footpath use is banned, failing to provide such facilities or conditions effectively bans cycling, contrary to equity, road safety and national cycling strategy goals. Where cyclists are banned from footpaths, safe+convenient facilities and/or conditions on roads should be mandatory.

4.11   Sharing a safer main street? A concept from Denver, Colorado

During the Velo-City conference at Basle, Switzerland in 1995 (and again in a similar paper presented at Velo-Australis in 1996 at Fremantle), James Mackay (Bicycle and Pedestrian Planner for the City of Denver, Colorado) presented a paper on Bicycle-friendly infrastructure (Mackay, 1995). He identified the success of roadway markings in Denver which indicate 'the likely travel corridor for bicyclists' using symbols and 'arrows in "shared use" lanes where there is not sufficient width to provide bicycle lanes or sufficient bicycle usage to justify adding bike lanes'.

While recognising that 'many traffic engineers and beginner cyclists want all cycling to take place off-street', Mackay notes that:

> (s)eparate bikeways within street right-of-ways pose special problems at intersecting streets and driveways (and doors and gateways). The physical separation creates the *illusion* that motorists and bicyclists don't need to keep track of each other. Motorists frequently forget that there is two-way bicycle traffic when turning, and bicyclists focus on continuing straight ahead when they should be scanning for turning automobiles.

Identifying 'the likely travel corridor for bicyclists' is potentially a major road safety benefit as well as an important facility for endorsing the presence of cyclists on routes frequently used by cyclists.

4.12   Bicycle-friendly streets in Brisbane

After many years of inability to introduce bike lanes in Brisbane due to road width and parking constraints and following the implementation of a major permanent trial project using bike lanes which resulted in severely reduced parking and access for residents and businesses, a series of trial projects commenced.

These initially included several variations of wide shared parking lanes. However, as with other bike lane combinations, the typical restricted road space resulted in the bike lane being very narrow. This potentially increases the need for cyclists to leave the bike lane suddenly and enter the adjoining traffic lane to avoid sudden or unforeseen events, e.g. a car door being suddenly opened or broken glass on the 'unswept' area. The edge line dividing the cyclists from the traffic created false safety and territorial expectations thus increasing the likelihood of crashes in circumstances unable to be anticipated by drivers or cyclists. The minimum width traffic lane also created a speed environment encouraging higher rather than lower speed. To overcome these problems, Brisbane City Council agreed Mackay's Denver concept (as discussed previously in section 4.6) might be useful to endorse cycling without bike lanes by use of standard size bike symbols to identify 'the likely travel corridor for bicyclists'. Following development of an initial protocol, the council agreed to trial use of the symbols, later described as 'Bicycle-Friendly Zones' or BFZ.

The construction of a major 'green' (cycling and walking only) bridge at Indooroopilly (an inner middle ring Brisbane suburb with a well-used bicycle route to a nearby university and the Brisbane Central Business District) created the need for quality infrastructure support on both approaches. The southern approach has little regular parking so combinations of bike lanes with and without parking restrictions were implemented. The northern approach is quite different, as it emerges into a congested area with shops and a major suburban railway station and multiple land uses and road users – a classic and complex 'main street' problem area. By retaining kerbside car parking and providing endorsement for cyclists by a system of symbols, the concept developed by Mackay could potentially assist in providing interim space for cyclists by slowing the traffic and avoiding the problems associated with bike lanes.

Crucial to the implementation of the trial, research by the council's Bikeways Planning Officer showed support for the idea as encouraging traffic to travel closer to the centre line had been shown to encourage lower traffic speed and decrease, not increase, crash risk and severity. For the drivers of larger vehicles such as buses and trucks, the presence of cyclists and the indication of the space required for them provides a guide to the actual space even allowing for overruns at corners, bends and intersections. In effect, the concept of the symbols acting as a system to create a zone is similar to but more adaptable and flexible in use than the advisory bike lanes increasingly commonly used in Europe because it has no edges.

Providing such symbols (see Figs 4.1–4.3) shows the likely travel corridor for bicyclists. Where it is desirable to reduce traffic speed and maintain a narrow lane, legitimising cyclists' presence is important to motorists, warning them to expect cyclists and advising them that cyclists are legitimate users of the road space. In Brisbane, the initial trial involved use of symbols in support of bike route signs (Austroads, 1999b). Identifying space for cyclists has led to increasingly broadened use of the symbols in other ways including in particular to support 50km/h on roads and through shopping centres and at squeeze points.

Further application could also include in or along bus lanes and in narrow slow speed precincts.

Although the concept of bicycle-friendly streets was developed in Brisbane by Brisbane City Council and CUST from experience in Denver, in fact, symbols have been used in a similar way by local authorities on streets in Adelaide in South Australia for many years (see Austroads, 1999b). The fundamental issue is simple in principle. As cyclists are entitled to use the road system, road authorities should include, not exclude, cyclists' safety and legitimacy in road and traffic design on all road corridors by maximising safety+convenience for all road users, i.e. by including for cyclists.

4.13   Sharing a safer main street in practice

The BFZ concept was initially applied along identified bike routes (Austroads, 1999b). However, with increasing recognition of both the practicality and the adaptability of the concept, the use of systems of symbols has spread, resulting in many examples of their use both in Brisbane and also on the Gold Coast and more recently on the Sunshine Coast, both areas with substantial tourist as well as sport, commuter, and utility or 'useful' cycling.

Although by no means final and always subject to specific site conditions, the protocol developed has the following characteristics where traffic speeds are less than 70 km/h. To reduce the speed of through traffic and to encourage it to travel close to the right of the kerbside lane or close to the centre line, the through lane width should not exceed 3.0 metres (preferred 2.7 m in 50 km/h and 40 km/h areas). The symbols should always be nominal minimum 1.2 m wide (preferred 1.5 m). Parking space, which can be line marked, should not exceed 1.8–2.0 m. Thus a typical urban 12 m wide road in a 40–50–60 km/h mixed traffic environment would have 2.7 m (traffic), 1.5 m (symbols) and 1.8–2.0 m (for parking, build-outs, etc.) while a through road situation would have 3.0 m (traffic), 1.2 m–1.5 m (symbols) and 1.8–2.0 m (parking and build-outs). The BFZ clearance for build-outs and intersections (e.g. in kerbside lanes) is essential to avoid squeeze points where cyclists are required to take up a position in front of faster flowing traffic on the approach to the squeeze point, e.g. at refuges, build-outs, intersections and, notoriously, roundabouts.

Thus while related to issues regarding speed limits, the issue of endorsing both cyclists' presence and space for cyclists requires clearly marking the space or zone for cyclists.

4.14   Bicycle-friendly streets in New South Wales?

Coincidentally and prior to the *Street Design for Road Safety* conference, the main street of Taree (a high profile demonstration project in rural NSW) was inspected and subjected to an 'audit' which identified the complete lack of provision for cyclists and a lack of facilities to legitimise the presence of cyclists in the traffic. Although the through lane did not appear sufficiently narrow to clearly identify it as a 40 km/h or lower speed environment, the design speed of around 30 km/h has been successfully achieved. However, while the through road appears to be too wide, it also appears suitable for use of a symbol system similar to the BFZ used in Brisbane to identify the precinct as bicycle-friendly. The questions which must then be asked are first why the road is so wide given the acknowledged benefits of narrow roads in speed management, then whether cycling could and should be identified and legitimised as described above, and finally, if that would be permitted. As it is permitted (Austroads, 1999b) would it not make Taree a better example of 'street design for road safety' if design of its new main street included safety for cyclists of all ages and ability in much the same way as was sought and arguably achieved for other road users? Adding the BFZ symbol systems to Taree's main street would surely not reduce safety.

In practice, Taree has achieved slower speeds as sought. Thus, not only is the 40 km/h speed environment suitable, it and the precinct amenity might be further improved and speeds reduced even lower if the presence of cyclists was endorsed by use of the symbols.

The crucial issue emerges as to why it is that both the conference on *Sharing the Main Street* and the case studies at the conference failed to consider endorsing adequate safety+convenience for cyclists when the presence of cyclists is inevitable and can assist in achieving goals of a safer road environment for all road users including cyclists.

Many discussions and consultation processes before, since and during the *Street Design for Road Safety* conference confirm both the interest in and necessity of including cycling. Hence, it is now clear there is a very urgent need to ensure adequate inclusion of space for cyclists in all projects by providing sufficient guidance (e.g. in the new RTA guidance) to ensure if not mandate implementation, which *Sharing the Main Street* and other guidance (e.g. Austroads, 1993, 1999b) has, to date, failed to provide. In the current policy situation described above, it is clear current projects fail to meet policy expectations. All the new guidelines must ensure these goals are achieved or supplements must be urgently issued to provide and ensure the endorsement of cyclists in order to achieve the policy goals.

4.15   Where to from here?

In summary, this chapter questions why it is that conferences such as *Street Design for Road Safety* and guidelines such as *Sharing the Main Street* and *Bicycles: Guide to Traffic Engineering Practice* (Austroads, 1993, 1999b) continue to fail to ensure cycling is adequately included in urban road projects both by policy and by design when a range of suitable options have been identified.

Arguably, cycling remains excluded by experts who choose not to include it as part of urban traffic management and urban design. Despite the perceived danger of cycling (and walking) expressed through current road safety strategies and despite consultation processes where cycling advocates seek inclusion of cycling, exclusion is still occurring despite well-known ways of including cycling. These include:

- speed management and safety and equity goals, e.g. of main street or school trip projects;
- improving road safety for all as sought by road safety strategies; and
- achieving or exceeding current transport, traffic, health, environment, urban amenity, equity, equality, social justice and other policy or strategy goals.

This chapter therefore confirms the merit of endorsing both the presence of, and space for, cyclists based on the use of the system of symbols outlined herein together with appropriate speed management. This is essential if increased 'useful' cycling is to be supported by government and local authorities. Only when both endorsement and safe space are provided and meet with community perceptions will goals seeking more walking and more cycling including to schools, shopping, work and to public transport begin to result in positive outcomes. The system of symbols concept provides a low cost means of integrating road safety, speed management and urban design over wide areas quickly, easily, consistently and at very low cost. The experience with use of similar systems both in Brisbane and other parts of Australia provides the necessary endorsement for authorities to implement cycling.

This chapter also confirms the need for urgent attention to, and if necessary review of, any new guidelines by a review process, e.g. by parliamentary committees such as the Staysafe Committee (NSW) or the Travelsafe Committee (Queensland). This will ensure that new guidance prepared by authorities, e.g. the RTA in NSW, is reviewed by the relevant parliamentary process, in NSW the Staysafe Committee, advised by cycling 'experts', and thus best meets users' needs in order to better ensure that:

- Safe+convenient conditions and/or facilities for cycling are required to be included in all main street and all other road and road safety policies, guidelines and projects (such as school road safety and local non-motorised access) because it is a legal, legitimate and, with walking, vulnerable mode of on-road transport that is strongly supported by transport, health and community safety policy; and
- If this is not required by the new guidelines, then the parliamentary committee can request the relevant road authority to amend the new guidelines to better meet the users' needs.

In conclusion, CUST has been pleased to provide the initial idea from Denver, to develop supportive material and to participate in the development and trial of the necessary protocols for use of symbol systems which identify bicycle-friendly zones as described above. These trial outcomes must be relevant to leadership at both state and then national level as an ongoing extension of the national leadership exhibited by several state parliamentary committees and finally at international level such as the Velo-City conferences where the BFZ concept was discovered and transferred to Brisbane.

In Australia, this process has been exemplified by the interest of the Staysafe Committee, the NSW government and the RTA, the NSW road authority, not only in the problems of and solutions for main streets, but also for example in (i) the provision of awareness and warning signs, crossing facilities and bike routes on urban freeways and on the interstate Pacific Highway between NSW and Queensland and (ii) in the process for implementing the range of excellent outcomes allowed by the NSW protocols for reducing the urban speed limit. For similar reasons, Safe Routes to School and the ability for local authorities to implement local street speed limits of 30 and 40 km/h should also be included as a high priority.

However, reviews should ensure that issues beyond state boundaries are also addressed. Hence, the failure of the current main street projects in NSW to include advanced implementation by projects elsewhere, e.g. the system of BFZ symbols in use in Brisbane, must and should be overcome in order to further benefit road safety and transport strategies, goals and experience nationally.

CUST strongly recommends that a number of recent and proposed main street projects including some best practice examples sought from interstate should be reviewed for cycling adequacy in further and more specific detail beyond that provided during the *Street Design for Road Safety* conference, to establish measures for current and potential cycling-friendliness (i.e. safety+convenience). CUST also suggests that these and any other current or future main street projects be assisted by the RTA and other state road and local authorities (if necessary by financial assistance and access to adequate cycling expertise) as trials to implement bicycle-friendly main streets based on the inclusion of cycling, including the use of the system of BFZ symbols as developed in Brisbane. The fact is that space is needed for cyclists' safety, and slowing traffic can provide that space.

4.16   The future

If the current trends continue, it seems inevitable that integrated solutions will finally emerge that include walking and cycling safely as essential requirements and therefore set thresholds for safety+convenience for these modes. While provision of space on the road for cyclists will be required, the reason will not be for safety but for equity thus ensuring that the advantages of cycling are not reduced by congested traffic but rather advantaged by provision of adjoining road space. In general, however, to ensure cycling and walking are safe+convenient for the full range of potential users of these modes, urban areas will have 30 km/h local street speeds with perhaps 40 km/h in the less dense urban areas of the USA and Australia. There will be few roads with speeds of 50 km/h without bike lanes in urban areas and the few urban roads with speed limits higher than 50 km/h will have separate cycling facilities rather than bike lanes.

While some separation will therefore be required on the higher speed roads, most roads and streets in urban areas will be safe due to the lower speed limits, encouraged in part by the sharing the road concept resulting in increasing cycling and walking and in part by use of facilities like the BFZ to assist in showing how to share the road safely. The result is a more equitable urban environment where no modes are prohibited and all modes have appropriate priority. In reality, many places already reflect many of these concepts and many others provide examples. The challenge for designers and advocates is to support integration of road safety and speed management rather than separation to ensure that speed limits are reduced thereby improving both the safety and the competitiveness of walking and of cycling by providing endorsed space on the road for cycling.

4.17   References

AUSTROADS (1993), *Guide to Traffic Engineering Practice: Part 14: Bicycles*, Sydney, Austroads.

AUSTROADS (1999a), *Australia Cycling 1999–2004: The National Strategy*, Sydney, Austroads.

AUSTROADS (1999b), *Bicycles: Guide to Traffic Engineering Practice. Standards Australia, Part 14* (2nd edition), Sydney, Austroads.

AUSTROADS (1999c), *National Road Safety Strategy*, Sydney, Austroads.

BICYCLE VICTORIA (1996), *It can be done . . . A bicycle network on arterial roads*, Melbourne, Bicycle Victoria.

BRINDLE, R (1984), *Town Planning and Road Safety: A Review of Literature and Practice*, Office of Road Safety Report CR33, Melbourne, Australian Road Research Board.

CROW (1994), *Sign up for the Bike: Design Manual for a Cycle-friendly Infrastructure*, the Netherlands, Centre for Research and Contract Standardization in Civil and Traffic Engineering.

CTC (Cyclists' Touring Club) (2000), *Cyclists and Pedestrians: Attitudes towards shared use*, Godalming, CTC.

CUST (Cyclists' Urban Speedlimit Taskforce) (1996), *Towards a Safe Urban Speed Limit*, Adelaide, BFA (Bicycle Federation of Australia).

FORS (1993), *Sharing the Main Street*, Canberra, Federal Office of Road Safety.

MACKAY, J (1995), *Bicycle Facility Signs and Pavement Markings in the USA*, Proceedings of the 8th Velo-City Conference, Basle, Switzerland, September.

RTA (2000), *Sharing the Main Street* (2nd edition) Sydney, Roads and Traffic Authority.

TIWARI, G et al (1995), *Lessons from Heterogeneous Traffic Flows for Planning Integrated Facilities*, Proceedings of the 8th Velo-City Conference, Basle, Switzerland, September.

THE STEERING GROUP (1963) *Report of the Steering Group of the Study of the Long Term Problems of Traffic in Towns* (which includes the report of the Working Group), London, HMSO.

WRAMBORG, P (1999), *On a New Approach to Urban Planning, Traffic Network and Street Design with a Special Focus on Bicycling*, Proceedings of Velo-City Conference 1999, Graz.

YEATES, M (1999) *Integrating Urban Design: Meeting the Needs of People with Access Disabilities . . . and Cyclists*, Proceedings of Velo-City Conference 1999, Graz.

YEATES, M (2000a), *Road Safety: For all Road Users?* Proceedings of Road Safety, Policing and Education Conference, Brisbane.

YEATES, M (2000b), *Making Space for Cyclists by Sharing the Road . . . Brisbane City Council's 'Bicycle Friendly Zones'*, Safe Cycling Symposium, Brisbane.

YEATES, M (2001a), *Zero Road Toll . . . a Dream, a Realistic Vision . . . or a Challenge?* Proceedings of 24th Australasian Transport Research Forum, Hobart, www.transport.tas.gov.au/atrf.

YEATES, M (2001b), *Can Urban Roads be Safe for Motorists . . . If They are not Safe for People Walking or Cycling?* Proceedings of The National Speed and Road Safety Conference, Adelaide, www.transport.sa.gov.au/nsrsc/pdfs/Speedco2.pdf.

5

Homezones and traffic calming: implications for cyclists

Graham Paul Smith, Joint Centre for Urban Design, Oxford Brookes University, Oxford

5.1 The homezone idea

The idea of the homezone has been taken directly from the Dutch creation of the *woonerf*. The Dutch word refers to a shared surface street and translates as a 'living yard'. The 'yard' refers to the kind of yard found in a farm, a place where many things happen, and the 'living' refers to a form of sociability, of vitality. The *woonerf*, and later simply the *erf*, which is more widely applicable, is legally defined and requires the implementation of physical measures to achieve appropriate behaviour from road users (CROW, 1989).

5.1.1 What purpose does the homezone serve?

Early progress in the Netherlands in creating the homezone seems to have been sparked by a 'revolutionary' response. Towards the end of the sixties residents, objecting to commuters filling their (mostly narrow) streets with cars, began putting out flower boxes and taking tables and chairs out onto the street. Their idea was to occupy for residents the space that was being used as parking space by non-residents.[1] The aim of the subsequent policy of reconstruction and reclassification of residential streets is to restrict the strain placed on residential areas by increasing levels of motorised traffic.

5.1.2 The street as social space

The orthodox view had grown, in thinking about traditional areas in the Netherlands, that the street was only for passage and, by default, for motor vehicle movement and increasingly for motor vehicle parking. The Dutch have over-turned this perception with the homezone and have re-created the street as a social space, as a place for play and a place where walking and cycling could again be safely enjoyed. The important lesson for cycle policy is that these social streets are attractive for cyclists and their arrangement, their layout, has been reproduced in modern Dutch planning to a great extent. The Dutch often quote Buchanan's *Traffic in Towns* (Ministry of Transport, 1963) as a stimulus for their develop-ment of environmental areas separate from the main traffic streets. The contrast with British policy is dramatic, however, in that the functional hierarchy pursued in Britain has effectively focused the residential areas into tightly bound 'islands' whose separation from each other and from other facilities has encouraged car usage and curtailed walking, cycling and public transport.

5.2   Who is the homezone/*woonerf* for?

The homezone is for everyone and everywhere. The essential elements of the homezone can be established in most residential streets and areas. The shared surface street operates most successfully in lower traffic speed and low traffic volume streets (CROW, 1989). In urban areas most streets a little distance from the centre are predominantly residential. It is a characteristic of the majority of urban areas in the Netherlands, Germany and elsewhere that most residential streets are now planned to be reconfigured so that 'residents come before cars'.

5.2.1   The safety benefits
The safety benefits of emphasising this multiplicity of use for the street have been dramatic; in the Dutch experimental areas in Rijswijk and Eindhoven traffic acci-dents, mainly incidents involving cars, fell by 46%. This fall was recorded for whole areas including access roads. The safety effect seems to have influenced the accident rate for the main roads associated with the experimental areas (a reduction of 35.5%) but this cannot be confirmed with statistical certainty. The rate of accidents involving slow vehicles, mainly mopeds, has not fallen, as their speed was barely constrained by the measures (Mathijssen, 1985). In British experience considerably greater falls in accident rate have been reported, perhaps resulting from the lower presence of mopeds and cyclists.

5.3   Popularity of the homezone

5.3.1   Growth of the idea
The implementation of the homezone has been rapid in the Netherlands. From 1976 when the legal definition was established 150 authorities had, within a year, set up 362 *woonerven* covering 855 streets. By 1985 some 4000 *woonerven* covered 7400 streets.[2] The benefits to be gained are widely appreciated and

together with the less costly to implement 30 kph (20 mph) zones the current aim, the Dutch 'Sustainable Safety' policy, is for all residential areas to be reconstructed to the 30 kph standard by 2010 (Ministry of Transport, 1997).

5.3.2   The image of the homezone

The impetus for establishing a kind of social equity in the street led to some local municipalities rolling the homezone idea into urban renewal schemes in the 1970s. Whilst the reasons for such a policy may be laudable, given the greater exposure to traffic injury among children from poorer families, the policy had an unforeseen consequence. By the late seventies the homezone was seen as having an unhappy association with the urban renewal concept which was seen as an indicator of the poverty of the area concerned. Another observation was that the homezone was not being used any more. One potential reason for this was the materials being used. The Dutch railways were re-laying track and replacing wooden sleepers with concrete ones. The wooden sleepers were used to make planters and other features. By the beginning of the 1980s many of the devices and planters in the homezone areas were in poor condition and looking tatty. However, at the same time in a range of new development areas, e.g. Tanthof in Delft, De Heeg in Maastricht, the use of the homezone concept, applied to new, mixed developments of public and private housing, helped establish the idea as applicable to all kinds of residential area. As in many walks of life the influence of fashion has an impact even in the apparently objective world of highway design. For visitors today, to residential areas in the Netherlands, a common experience is that the tool of 'environmental traffic management' is being freely used.

5.4   National policy

5.4.1   Changing transport priorities

The homezone idea for residential areas shadows the changed priorities in Dutch thinking about transport priorities. In *Cities Make Room for Cyclists* the authors describe how in the 1960s the

> rapid spread of car ownership and use . . . brought a disturbing rise in accident rates. Cycleways were developed and road junctions made safer [but] even so, the main emphasis in transport and planning policy was on accommodating rather than controlling car-use, which was increasing at a spectacular rate and at the expense of cycling. . . . In the mid 1970s the Dutch government began rethinking the role of cars and bicycles in urban transport. Increased road safety was one aim; another was the relief of congestion in urban areas. (Phillips, 1995, p 7)

5.4.2   The homezone, permeability and the bicycle

An essential feature of the homezone areas is that the network in which they are located is connected and permeable. Cycling is promoted by physical conditions offering convenience and safety and the connected, traffic calmed and mixed use area delivers these qualities. This arrangement, typical of historic and nineteenth century developments, is to be found in many new areas in the Netherlands.[3] The potential barrier to the convenience of cycling, of unnecessary distance, is minimised by streets being connected and permeable. This permeability is an important component of the 'fine network' of routes for cyclists described by Dirk ten Grotenhuis in the Delft Cycle Plan: 'For the city level a pattern of network of ±600 by 600 metres . . . for the district level 300 by 300 metres, for the quarter level 150 by 150 metres' (Grotenhuis, 1978). In the *West Kwartier* of Delft, where the homezone concept is said to have originated, a grid of 150 metres would include several of the narrow streets. In the nineteenth century area where Cornelis Trompstraat, the best known original reconstructed homezone is built, south-east of Delft centre, a 150 metre grid could accommodate a couple of streets. So homezone qualities will help promote cycling by offering a fine local network, the basic level in a cycle-friendly infrastructure.

5.4.3   Reducing perceived risk

By reducing the impact of motorised traffic (which is concentrated onto the main arteries where speeds are less constrained), the advantages of the traffic calmed network can be better enjoyed by pedestrians as well as cyclists. Perceptions of risk are much higher among pedestrians and cyclists than among all other road users surveyed in a recent British study (Sissons *et al*, 2001). Dutch commentators are clear that risk, the threat of an accident, is a major disincentive to cycling. Reducing speed is a key to accident severity reduction and thus to perceptions of safe streets. The homezone can be seen to be perhaps the most important building block in constructing significant levels of cycling. Making the environment outside the front door into an environment safe for cycling is the first step in making the choice of cycling a plausible one.

5.4.4   Location policy

Differences in the planning and land use regime inevitably have an impact on the attractiveness of pedestrian and cycling journeys. Much has been written elsewhere on the Dutch 'ABC location policy', which locates traffic generating developments at transit nodes where non-car modes are most likely to be chosen (Netherlands Second Chamber of the States-General, 1990). The British planning and development system may be described as being at the threshold of learning how to deliver a denser, mixed use development and one that reduces the need to travel (DETR, 1998). Origins and destinations need to be perceived as being reasonably close for walking and cycling to be seen as legitimate modal choices.

5.5   Inherent disadvantages of the hierarchical system

In Britain the hierarchical layout system which is designed to segregate roads from places has given rise to the ubiquitous cul de sac, with through flowing traffic excluded. With housing located at the end of the hierarchy, many of the qualities of the homezone can easily be achieved but with some important differences:

- The hierarchical system inevitably leads to the maximisation of journey length and consequently the probability of a greater proportion of journeys being made by motorised means.
- Crime such as burglary seems to be focused in the quieter areas of segregated layouts.
- Social safety is compromised in the relatively unpopulated routes that proliferate in traffic segregated cul de sac layouts because vitality and social exchange are reduced.
- The very intelligibility of places that are segregated can result in people simply fearing being lost.

5.5.1   Journeys

In a detailed study by Chellman, for the White Mountain Survey Co (1991) of journeys in the relatively dense, built-up and inter-connected street layout of the traditional town of Portsmouth, New Hampshire he found that the daily trip generation rate for vehicles was one-half of what would be expected. The American Institute of Transport Engineers' projection methods, which are normally reliable, give figures for segregated 'modern' developments:

> Very important from a traffic management perspective, we also found morning and evening peak hour reductions in traffic of sixty to seventy per cent. Such peak hour reductions are phenomenal and engineers I have discussed this with know of no other way to effect such a reduction.[4]

The pattern of the traditional layout, in a desirable place like Portsmouth, means that people can make access to multiple destinations by non-car means. This is a result both of the more convenient proximity of facilities and also because of the relative inconvenience of returning to a parked car and moving somewhere else.

5.5.2   Crime

Hillier and Hanson (1984) have studied the qualities of and the relationships between urban space in different kinds of layout, traditional and hierarchically segregated, an enquiry method they call Space Syntax. The connected, grid-like arrangements of traditional Barnsbury, London, when compared with the isolating qualities of the hierarchical, segregated layout of the Marquess Road estate,

Canonbury, London, reveal with a statistical significance the conclusion that the more segregated a layout, the more rates of burglary rise (Hillier, 1987).

5.5.3   Vitality

The chance of meeting another person passing through public space is similarly affected by the layout. Encounter is a factor in defining the social safety, the vitality of the public realm. From many observations 'in street-based traditional residential street areas' in London, Hillier records an encounter rate:

> of about 2.6 people per 100 m/minute of walking time [contrasted] to somewhere between 0.4 and 0.7 inside estates. The effect is that whereas in streets you are in contact with other people most of the time, on estates you are on your own most of the time. Put another way, the daytime encounter field in the estates turns out to be like night-time in ordinary urban streets. In terms of their naturally available encounter field, people on estates live in a kind of perpetual night. The sense of isolation is not imaginary, it is an objective fact. (Hillier, 1987, p 45)

5.5.4   Intelligibility

The difference between traditional and modern hierarchical layouts is that the latter are designed with traffic segregation in mind. Hillier (1987, p 41) describes how the modern design has a kind of recognisable pattern that, from the plan or from the air seems recognisable.

> But if we try to move around them we quickly lose all sense of where we are. The similarity of the parts . . . guarantees that on the ground they lack intelligibility. The old town plan has the opposite properties. From the air it seems disordered . . . But on the ground it has a degree of natural intelligibility which means that we do not need signs to tell us where to go.

5.6   Induced traffic and reducing traffic

In the Dutch context the inter-connected street grid, in minimising trips and journey length, tends not to result in unacceptable traffic travelling through the residential area because of the effective traffic calming which is applied to the homezone network. To the British mind the very thought of a short journey through a residential area, known as 'rat-running', is just unacceptable. Making journeys shorter is, however, a part of a more sustainable movement system. If those shorter journeys can be attractively and conveniently made by bicycle this can act as a powerful 'pull factor'. What is needed is a 'push factor' and this can be achieved by skilfully employed traffic management measures that reduce the attractiveness of making the journey by car without prohibiting car accessibility.

5.6.1   Capacity

British thinking about layout design effectively caters for motorisation and thus induces more use of the car. In 1994 SACTRA (the government's Standing Committee on Trunk Road Assessment), recognised that building roads can generate traffic. Conversely, a study by Cairns *et al* (1998) has provided evidence that road capacity reductions can lead to less traffic. It would appear to be the case in the Dutch homezones that in layouts where car space is reduced and traffic management measures used, a kind of friction is induced on the convenience of car use, resulting in less use of the car. Whereas the number of cars in the UK continues to grow (RAC, 2000) the number in the Netherlands shows a gentle fall in the most recent years (Mitchell and Lawson, 1999). The Dutch are not necessarily travelling less and there seems little evidence that the Dutch standard of living is lower than in Britain, in fact national car ownership rates in the two countries are now roughly similar.

5.6.2   Modal split

The Dutch Bicycle Master Plan sought to increase by 30% the number of kilometres travelled by bike between 1986 and 2010. Ton Welleman points out (Director General for Passenger Transport, 1999) that it is not yet safe to claim that the Bicycle Master Plan itself has affected behaviour, because of the inevitable lag between policy and outcome, but bicycle use has grown. The context that enabled the Master Plan to be established existed in the 1980s and since then the increase in kilometres travelled by bike is clear, around +10% by 1997: 'Considering the slight increase in the number of kilometres travelled by bicycle annually since 1986, the ample growth potential of bicycle traffic for trips shorter than 7.5 kilometres . . . the current state of affairs appears to make the . . . target feasible.' So the growth potential is based largely on local streets, and the achievement of a safer and supportive local environment is therefore important for the goals to be achieved.

5.7   Case study: De Strijp

Having considered aspects of the creation, definition, popularity and policy development opportunities presented by the homezone the following section considers a modern development area and aspects of its design, which contribute to achieving policy objectives. De Strijp (The Strip) is a new housing development comprising 1360 homes completed in 1998 on the western flank of Rijswijk, near The Hague.

5.7.1   The VINEX programme

The housing extension of De Strijp is the municipality of Rijswijk's part of the massive government promoted VINEX housing expansion programme to build

634,800 homes from 1993 to 2005. VINEX is an acronym signifying the fourth Report on Town and Country Planning. The programme is 'to provide a good living environment [with] varied and "pleasant" new city areas of the future . . . the original objectives for the VINEX sites lay in "expansion close to the city; the compact city policy".'[5] The hope behind the programme is to prevent urban sprawl and a Dutch suburbia. Government support under the VINEX programme has enabled private developers to purchase a range of industrial and agricultural sites, with the support of local authorities, in the 10 areas around the country's big conurbations and numerous other smaller initiatives. Joint development companies have been formed to pursue the policy objectives. The policy to expand near to the city is designed to make best use of existing public transport lines and to enable sustainable transport modes. The site for De Strijp was formerly a market garden site and was the only available site for expansion in the municipality.

5.7.2   Location of De Strijp
The site is 3 km west of the historic centre of Rijswijk, 1.25 km from a major in-town shopping centre, In De Bogaard, and 1.5 km from the main line rail station. Public transport runs past the north and south edges of the scheme. Cycle connections from De Strijp are segregated with priority measures across the dual carriageways to the north and east making a convenient journey to existing local facilities.

5.7.3   Development history
The development is a joint project between three housebuilders – Dura (50%), Wilma (25%) and Bouwfonds (25%) – and the municipality of Rijswijk. The development is 85% family homes and 15% flats. The building ownership is 70% market price and 30% social rent. The agreement was signed in December 1993, the first earth was turned in December 1994 and the thousandth home completed in November 1996.

5.7.4   Description
The house styles in this development are astonishingly mixed to British eyes. With large, high quality, terraced private housing, long (Georgian style) terraces, a range of family, elderly and single person housing types and including a semi-permanent Romany housing section, the mix is greater than found on most British schemes of a similar size. There are a variety of architectural styles and a broad palette of colour and materials. A number of opportunities for children's play are built into the development with a basketball pitch, grassed areas and a number of smaller play areas located on- and off-street. The layout has two long and straight roads built as 30 kph zones (c 20 mph) with road humps at approximately 60 metre frequencies. The major north/south link is Strijp Laan. Parking provi-

sion along this road is located in the tree-lined road centre of the dual single lane carriageway; it is shared, at right angles and double banked.

Cycling within the scheme is mostly unsegregated, on-street with the majority of roads being faced by dwellings. The environment feels entirely calm and unthreatening. Children are to be found using the public space with freedom since *drempels*, sinusoidal road humps, keep speeds low. A segregated length of cycle road with footpaths strikes out east-south-east towards Rijswijk centre, opposite the circular footprint building on the western edge of the scheme (see plan, Fig. 5.1). The other provisions for cycling, two shared cycle/footbridges, link the south-west corner with the north-east corner of the area south of the cycle road.

De Dijk homezone
In Fig. 5.1 the De Dijk area occupies much of the centre of the plan and comprises about 15% of the housing in the De Strijp development. This area has high quality terraced private housing with 203 wide frontage, three and four bedroom houses built more or less on the edge of a curving, shared surface highway that is indicated by a 'dotted line' of white setts (see Fig. 5.2). The setback varies but is rarely more than 2 metres. The housing is a mix of two and three storeys. Streets are relatively narrow with a minimum carriageway of about 3 metres. There are sometimes as little as about 7 metres between facing houses and around 12–14 metres generally when the carriageway is split into two. There is a minimum of street furniture and the street lighting is fixed to the housing with access established by covenant. At the rear of all the houses, a short private garden steps down to a man-made dike. There is no rear access. In the plan these dikes show as curving, amorphous shapes. The rear separation of the houses is at least 20 and often 30 metres or more. In the Netherlands a location next to water is much sought after and guarantees a premium value for the housing. The wide private space at the rear forms a balance for the relatively narrow public space on the street.

Parking friction
The parking provision for De Dijk is built at 160%. Each house has a built-in garage giving 100% with the balance of 60% parking provision in publicly shared, road centre spaces or spaces at the end of the street at the edge of the area. Within a short time of the housing being occupied about half of the garages were being used as storage, so the on-street parking was under pressure.[6] A small amount of planting around the edge of the estate, separating the right angled parking into small groups, has been converted to give additional parking spaces. Where a car is not in the garage the municipality accepts that the car can sometimes be manoeuvred into a parallel position, in the 2 metre or so space between the building front and the highway, adjacent to the garage door. Within the homezone (*woonerf*) the Dutch regulations demand that cars may only park where a place is marked so this acceptance is important.

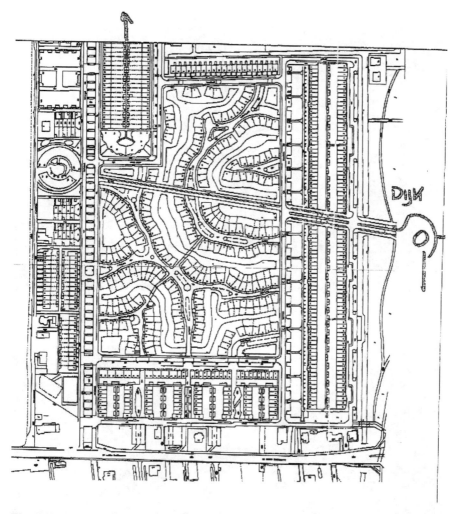

Fig. 5.1 Part of the plan of De Strijp, focusing on the De Dijk housing to the southern end of the development. A segregated length of cycle road with footpaths strikes out east-south-east towards Rijswijk centre, opposite the circular footprint building on the western edge of the scheme.

In Fig. 5.3, Liseiland, there are 7 metres between facing houses at the narrowest part. The two way single carriageway opens into a single lane 'dual carriageway' area with road centre play space, planting and parking, around 12–14 metres generally. These dimensions mean that it is never easy for cars to pull into the garages, sometimes requiring more than a three-point turn in the scant 5 metres available. There is not room for a car to pull forward within the property boundary

Fig. 5.2   Rieteiland, De Dijk, homezone.

Fig. 5.3   Liseiland, De Dijk: accommodating the car.

when exiting a garage. One observable consequence of this design is that vehicles are constantly constrained by the layout and traffic speeds are gentle. Some garage doors have planting that indicates the garage is definitely not used for a car.

The estimated current parking provision is about 130–140%. This is a surprisingly low amount for a high quality, modern, for sale development. With many journeys made by bicycle in the Netherlands, around 30–40% in all sectors of society, national policy seeks a continuation of high and higher levels of bicycle use. The relative constraint here, of parking friction, and the decisions made by residents themselves to exchange garage space for other uses, helps to make travel by non-car modes such as cycling a legitimate activity.

Vitality
With a garage on one side of the front door, on the other side a section of the through living room looks onto the street. This helps to ensure that the public realm is under surveillance and not just mutely faced by garage doors.

Play on the street
The traffic environment and the layout encourage the use of the street as a social space. As well as informal play, as evidenced by coloured chalk marks in numerous locations, more formal play is encouraged with play furniture for climbing, spinning and bouncing, set into the highway and next to road centre parking. The ground surface in this play area is rubberised. On six or seven visits to the estate the author has not failed to see children playing on these streets with total freedom, maybe from as young as 3 or 4 years old. In fact the private gardens facing the canals are clearly more for the use of adults, most being luxuriously planted and offering minimal space for children's play! One of the contentious features of the Dutch homezones is the use of play to define traffic behaviour. Of course physical constraints help define acceptable behaviour but the presence of children playing obliges care among drivers (see Fig. 5.4).

One of the design difficulties of terraced housing is where to locate the refuse bins. In this development there are recycling and refuse containers set into the road and the parking areas, with highly finished and relatively unobtrusive chutes projecting above the ground surface. (One of the chutes can be discerned behind the cone-shaped roundabout in Fig. 5.4.) No bins to be carried through the house and more efficient removal for the refuse and recycling service, which uses one-man operation refuse trucks with a hydraulic lifting arm.

5.8   The homezone: a conclusion

Whilst the homezone concept is fashionably new in the UK, its progenitor, the Dutch *woonerf* (Vahl, 1985) is over 30 years old. It has been a great success and is achievable in new developments too. Will the current excitement over homezones (Children's Play Council, 1998) get anywhere? Will the DETR invitation for pilot projects[7] to create new pedestrian and child-friendly residential streets

Fig. 5.4   Vilgeneiland, De Dijk: play on the street.

open the floodgates and result in changed design norms? Will layout guidance change to capitalise on the potential of achieving a greater permeability? Will cycling benefit? The reply might be no – because current orthodoxies may envelop the emerging 'new' idea and ingest it, excluding the possibility of approaching road layout design in new ways. For the response to be a positive yes, the experimental projects will have to show that a different way of living with the car, and travelling without it, are indeed possible. The UK's approach to highway design and layout has been significantly different from our continental neighbours but, in the government's homezones pilot projects invitation, the ability to experiment was positively made. Let us hope the opportunity will be seized.

5.9   Notes

1 Conversation with Marion van Caspel, 29 April 1985, ROVU (Ruimtelijke Ordening van de Gemeente Utrecht).
2 Heeger, J (1986), Lecture: *Replanning and Redesigning Public Space in Dutch Towns.*
3 For example, the De Strijp development area on the eastern flank of Rijswijk, completed *c* 1998. See section 5.7.
4 Personal communication.
5 *http://www.archined.nl/vinexonsite/manifestatie/english.htm*, pp 8, 9.
6 Conversation with the development control engineer, Rainco van Egmond.
7 Head of Charging and Local Transport, 29 January 1999, formal invitation to traffic authorities and district councils, DETR.

5.10   References

CAIRNS, S, HASS-KLAU, C and GOODWIN, P (1998), *Traffic Impact of Highway Capacity Reductions: Assessment of the Evidence*, London, Landor Publishing.

CHILDREN'S PLAY COUNCIL (1998), *Homezones: Reclaiming Residential Streets*, London, National Children's Bureau.

CROW (1989), *Van Woonerf tot Erf*, Ede, the Netherlands, CROW (Centre for Research and Contract Standardization in Civil and Traffic Engineering).

DETR (Department of the Environment, Transport and the Regions) (1998), *A New Deal for Transport: Better for Everyone*, Norwich, HMSO, p 126.

DIRECTOR GENERAL FOR PASSENGER TRANSPORT (1999), *The Dutch Bicycle Master Plan: Description and Evaluation in an Historical Context*, The Hague, Ministry of Transport, Public Works and Water Management, http://www.archined.nl/vinexonsite/manifestatie/english.htm, pp 8, 9.

GROTENHUIS, D TEN (1978), *Planning for Cyclists and Pedestrians*, Delft, Municipality of Delft.

HILLIER, B (1987), 'Against enclosure' in N. Teymur (ed) *Proceedings of Rehumanising Housing*, London, p 46.

HILLIER, B and HANSON, J (1984), *The Social Logic of Space*, Cambridge, Cambridge University Press.

MATHIJSSEN, M (ed) (1985), *Reclassification and Reconstruction of Urban Roads in the Netherlands*, Leidschendam, the Netherlands, SWOV (Institute for Road Safety Research) and DVV (Road Safety Directorate).

MINISTRY OF TRANSPORT (1963), *Traffic in Towns: A Study of the Long Term Problems of Traffic in Urban Areas* (Buchanan Report), London, HMSO.

MINISTRY OF TRANSPORT, PUBLIC WORKS AND WATER MANAGEMENT (1997), *Sustainable Road Safety Programme*, The Hague, Ministry of Transport, Public Works and Water Management.

MITCHELL, C and LAWSON, S (1999), *The Great British Motorist 2000: Lessons from European Transport and Travel*, Basingstoke, Automobile Association.

NETHERLANDS SECOND CHAMBER OF THE STATES-GENERAL (1990), *20 922, Second Transport Structure Plan*, The Hague, Second Chamber of the States-General.

PHILLIPS, J (trans) (1995), *Cities make Room for Cyclists*, Ministry of Transport, Public Works and Water Management, The Hague.

RAC (2000), *Report on Motoring*, 12th edition, Feltham, RAC Motoring Services, p 58.

SISSONS, JM, SENIOR, V and SMITH, GP (2001), 'A diary study of the risk perceptions of road users', *Health Risk & Society* 3 (3), 261–79.

VAHL, H (1985), *Prikkel in de Stedebouw, Zwart-witboek voor een groene woonwijk*, Rotterdam, Stichting Ruimte.

WHITE MOUNTAIN SURVEY CO INC (1991), *City of Portsmouth: Traffic/Trip Generation Study*, Ossipee, New Hampshire.

6

Developing healthy travel habits in the young: Safe Routes to School in the UK[1]

Jo Cleary, Cleary Hughes Associates, Hucknall, Nottingham

6.1 Introduction

Perceptions of, and attitudes to, young people's travel habits in the United Kingdom have changed significantly over recent years. Gone are the days when adults alone made decisions on behalf of children – deemed to be 'acting in their best interests' – while young people themselves were almost entirely excluded from shaping the policies and the physical environment that affect how they get about.

Several factors were instrumental in delivering this change. First was the study by Mayer Hillman *et al* (1990) of the travel habits of 7–11 and 11–15 year-olds over the period 1970–1990. The authors found that there had been a significant loss of independence, coupled with a growth in car dependency and a disquieting fall-off in equipping children with the skills to become confident and independent travellers later in life. Then, pioneering researchers into children's travel actually started to consult the children themselves[2] on their needs and desires – and found a marked mismatch between what adults perceive to be 'best' for children and what the children would really like. Ironically, parents had drifted into a vicious circle: an understandable desire to keep their children safe led them to ferry them everywhere by car, so creating less safe conditions for children to travel by other means. On the other hand, researchers generally found that children want to explore their capabilities and environment, for example by walking and cycling.

The realisation gradually became more general that trends were counter-productive. It became recognised that children were suffering: they were becoming fatter, less physically fit and less independent; they had less opportunity to develop road sense; they were less aware of their environment; they had fewer opportunities for interaction with friends, adults, even their parents. There was far less spontaneity in their lives: they were becoming couch potatoes – battery-reared rather than free-range.

Parents, too, were doing themselves no favours. Far more of their time was being spent in escorting children to ever-farther-flung destinations or activities. It became apparent that children were less streetwise, evidenced by a rise in accidents among secondary pupils when they started to demand independent travel, but who hadn't learnt basic road sense at an earlier age. Parents themselves were also becoming more car-dependent for their own journeys; if they were taking children to school by car, they tended to continue to work by the same means. Meanwhile, the presence of fewer people on foot on the streets led to a perception (not borne out by statistics) that stranger-danger was increasing – a further incentive to carry children in the apparent security of the car.

In the wake of this came more general problems: more car journeys to carry increasingly car-dependent children to school – at a time of day when roads are busy anyway – merely added to rush hour congestion. It is estimated that in the UK something like a fifth of morning rush hour car journeys are just to take children to school and, since there are vast differences between affluent regions and the less well-off, the proportion is much higher in many areas. (It is worth noting that the average distance children travel to primary schools (4–11 years) in Britain is 2 kilometres and to secondary schools (11–18 years) just over 5 kilometres. Some 64% of trips in the 1.5–3 km band to primary schools are made by car.[3])

Although school travel in the UK accounts for a mere 2% of the distance travelled by the whole population, it has an importance far beyond this. It is widely accepted that many patterns of adult life are shaped during the formative years of childhood: children who habitually walk or cycle are likely to mature into adults who consider these to be normal and accepted forms of travel.

6.2   Redressing the decline

The response to the growing car-dependent trend was led by campaigners – often, to begin with, lone voices and then more frequently campaign groups and organisations – and independent researchers such as Hillman (1990). Typically, their first action was to highlight the problem through research and then ensure that the findings were disseminated as widely as possible by the media. Their second move was to initiate or commission further researches to investigate the problems in detail and to offer possible solutions. On the basis of these they then lobbied local and central government to address the problems.

Two early case studies serve to illustrate this type of response. The first was a three year Safe Routes to School demonstration project, which was initiated by the sustainable transport charity Sustrans and partially funded by the UK government's Environmental Action Fund. Nine schools, of different size, in three different areas of England and with pupils in differing age ranges were selected to take part. Initial surveys of pupils' travel patterns and fitness were carried out to determine the base levels before the project got under way. Pupils were then encouraged through their school work to design safe routes and more secure cycle parking, to contribute to campaigns supporting sustainable transport and to participate in exchanges with schools in Denmark and the Netherlands. All these tasks were to be compatible with the requirements of the English National Curriculum. Fundamental to the project were the aims of increasing the proportion of pupils cycling and walking to school, and consequently reducing car-borne numbers, while at the same time improving pupils' general health, safety and security. While the post-project surveys did not exhibit the ambitious degree of change in travel mode that Sustrans had hoped for, the project has served as a model for extending similar work to schools nationwide. Sustrans now has staff whose work is dedicated entirely to Safe Routes to Schools projects and a dedicated website on the subject.[4]

In the second study independent consultant Adrian Davis, acting as an adviser to Sustrans, highlighted the links between transport and children's health, and the health benefits that would accrue from a reduction in motor traffic. Reduced traffic levels, he pointed out, would lead to fewer casualties from traffic accidents (the leading cause of death in under-15s in the UK), less air and noise pollution, and – by introducing safe travel and play areas away from traffic – lessen the observed health inequalities between children from affluent and less well-off areas. He also pointed out the direct long term health benefits, particularly in lowering the risks of heart disease, stroke, obesity, diabetes and skeletal weakening in later life. Important benefits in well-being and mood and in generating children's feelings of independence were also felt to result (Sustrans, 2001).

A number of local government authorities responded early on to these initiatives by setting up their own projects. Nottinghamshire County Council, in response to a concern about the economic, health and environmental impacts of the growth in motor traffic commissioned sustainable travel consultants Cleary Hughes Associates to examine the nature and extent of car dependence for school travel in Nottinghamshire. As already noted (note 2) the views of 27,000 pupils were canvassed to determine how they travelled to school – and how they would prefer to, given a completely free choice. The results were enlightening: around a third of all pupils would prefer to cycle to school, with figures as high as 54.5% for 11 year-olds; currently, fewer than 2% do so (see Figs. 6.1, 6.2 and 6.3).

A survey of head teachers also showed 90% to believe that more active school travel would be beneficial, although support for cycling in current conditions was less than for walking. Some 94% of head teachers also considered that their own school had traffic problems – congestion and danger – associated with the numbers of pupils arriving by car (Nottinghamshire County Council, 1995).

Nottinghamshire followed up this survey by commissioning the development of a teachers' pack of classroom materials (Nottinghamshire County Council, 1998) relating to sustainable transport and school travel. The pack contains material

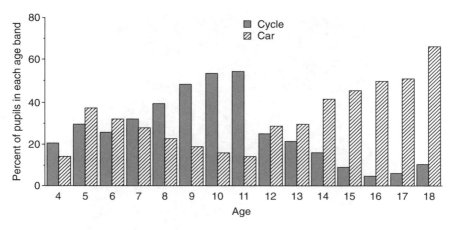

Fig. 6.1 Pupils' preferences for cycling and car travel to school, by age: percentage of pupils in each age band. A demonstration of the need to instil travel patterns at an early age. The graph shows the travel preferences of 27,000 school pupils in Nottinghamshire. Each was asked to indicate how they would prefer to travel to school, given a completely free choice. Up to the age of 11 the enthusiasm for cycling to school increases; by the late teens a preferences for car travel has taken over. The minimum age at which one can drive a car in the UK, subject to passing a driving examination, is 17. Source: Nottinghamshire County Council (1995).

for lessons directly linked to seven subject areas of the English National Curriculum aimed primarily at pupils aged 7–11, with supplementary information extending the age range to 14. The lesson material is complemented by a set of fact sheets on transport and travel.

Nottingham City Council commissioned a series of studies which were to lead to school travel plans (see section 6.4) for a number of schools in less affluent areas of the city, ranging from infant and nursery schools (3–7 years) to a community college (11–18 years, plus adult evening students).

Hertfordshire County Council pioneered the concept of the 'walking bus' in the UK with a pilot scheme in the city of St Albans. Walking buses formalise the *ad hoc* groups of walking children and parents that accumulate members as they travel to the school. A walking bus comprises a group of children accompanied by a number of adults – one a 'bus driver' to lead the way, with others perhaps handling trolleys for carrying pupils' books, sports gear, etc – following a set route and a set timetable each day, and picking up additional 'passengers' at recognised 'bus stops' along the way. The concept has now been extended to other towns in Hertfordshire and has been adopted by other areas of the UK as well.

Buckinghamshire County Council was one of the first to appoint specialised Safe Routes to Schools officers whose work was to develop safe routes projects throughout the county.

Fig. 6.2   Two contrasting views of travelling home from the same school in Hampshire, England. Source: Tim Hughes.

Fig. 6.3   Given a free choice, young children are eager to travel actively to and –
especially – from school. Source: Jo Cleary.

6.3   Central government action

An indication of central government commitment came in 1998 when the government White Paper (a parliamentary report giving information and outlining government proposals) on the future of integrated transport underlined the importance of school travel: 'Our policies will help reduce the need for children to be driven to school by encouraging safer routes for walking and cycling. . . . We will take further initiatives to encourage more children to get to school other than by car' (DETR, 1998). The White Paper then went on to give more detail of these initiatives.

One of the initiatives projected in the White Paper was the establishment of the School Travel Advisory Group (inevitably to be known as STAG), which came into being later in 1998. The group comprises representatives from central, regional and local government and school communities. From central and regional government, departments concerned with transport, health and education are involved, and their involvement has survived changes of government and several revisions of the names and terms of reference of individual departments.

The aims of STAG are twofold: 'to lead the dissemination of best practice, principally by working to raise the profile of school travel issues' and 'to contribute to the development of a coherent approach to travel principally by identifying practical means of influencing behaviour'.

A number of resources have been made available to promote the School Travel Advisory Group's objectives. These are discussed below.

6.3.1   Database of classroom materials

The government department responsible for transport commissioned consultants to identify and evaluate teaching resources dealing with sustainable travel and closely related matters. The resources could be physical (such as books and leaflets, videos or CDs) or web-based. Each was examined for its suitability for particular age groups and particular subjects within the English National Curriculum. The net was spread somewhat wider with the additional inclusion of materials such as on-line encyclopaedias on sustainable travel or the environment, not primarily designed for classroom use but which nevertheless offered useful information for teachers which they could use in compiling their own classroom materials or to supplement other resources. The database is revised twice a year and the information is made available on the government department's website.[5]

6.3.2   Site-specific advisers

A team of government-funded consultant advisers was appointed to supply advice specific to individual schools throughout the UK. Schools may apply for a certain amount of advice (generally five days) tailored to their particular requirements, at no cost to the school, to help them tackle school travel problems, usually through the development of school travel plans (see section 6.4).

6.3.3   Funding

In a further move in 2001 central government announced the launch of over 100 bursaries for local government authorities to allow them to recruit travel co-ordinators, much of whose time would be dedicated to school travel work.

Other organisations, too, have produced resources which further STAG's aims. These include extensive work by Sustrans, which – in addition to the specialised website already cited – publishes leaflets and a newsletter, holds seminars, publicises the issues and raises awareness generally. 'Living Streets' (formerly the Pedestrians' Association) has a Walk to School campaign, while the transport pressure group Transport 2000 has published a guide for school communities (Transport 2000 Trust, 1999). As a result of the burgeoning demand for specialist advice, independent consultancies have started to adapt and hone their skills in this field.

In 2000 a further indication of central government commitment was given in guidance to local government authorities on preparing the newly introduced Local Transport Plans (LTPs), five year integrated strategies used by local authorities in bidding for funding from central government sources (DETR, 2000). The guidance advised that LTPs 'should set out an integrated strategy for reducing car use and improving children's safety on the journey to school, taking account of health and education impacts'. Overall targets were set of 80% of primary pupils (4–11 years) and 90% of secondary pupils (11–18 years) cycling, walking or using the bus by 2010.

6.4   How are the issues being addressed?

The principal weapon in the armoury is the school travel plan (STP), which schools are being encouraged to develop. In some cases preparation of an STP is made a condition of planning permission being given for development, for example extending school buildings.

A good school travel plan examines the initial situation with regard to pupils' and staff travel and travel problems, identifies barriers to travelling other than by car and suggests possible means by which such barriers may be overcome. It also sets attainable targets for shift from car use to travel by sustainable means and makes provision for monitoring the plan's working and evaluating its progress. Ideally, STPs should be developed by – or at least with the assistance of – the school community in its widest sense, involving widespread consultation not only with those who teach or learn at the school but with parents, local residents, highway and health authorities, and transport providers.

In support of the preparation of school travel plans the government department responsible for transport produced in 1999 a school travel resources pack, providing information and materials for use in the development of a plan. The pack includes information and 'how to' leaflets, suggested survey questionnaires, posters and presentation materials, and a list of other resources available.

One of the main roles of the School Travel Advisory Group has been to commission research in fields connected with the production of STPs. Three examples follow.

6.4.1   Evaluation of site-specific advice

Before the national programme of site-specific advice for schools was launched (above) the relative merits of different means of assisting schools used on a trial basis were evaluated by independent researchers. No less than nine different techniques were assessed, although, as the consultants' report concedes, in practice the boundaries between the nine can become blurred. The nine techniques are: telephone help-line; school working group organisation and co-ordination; single site visit; assistance to local government authority school travel co-ordinator; out-of-school training for a school's own travel co-ordinator; use of best practice guides; school travel survey, followed by an 'issues, problems and options' report; a media and community approach; and classroom work.

The key conclusions of the evaluation were that:

- Assistance to schools needs to include at least one visit by those providing the help to see the problems and issues for themselves.

- An STP generally takes at least 12 months to prepare and implement.
- Schools prefer continuing help, rather than a 'one-off' intensive period of assistance, as this helps to maintain the momentum of work on the plan.
- Practical advice generally needs to be supported by resources to implement the recommendations of the STP.
- A good deal of the success of STPs depends on changing attitudes and perceptions.

6.4.2   How best to promote alternative travel modes
This research examines ways in which it may be ensured that the efforts and resources of those involved in raising general awareness about school travel issues, and in particular the need for a move towards sustainable travel patterns, are best deployed. This research continues.

6.4.3   Training needs of travel plan co-ordinators
This research has three objectives: the investigation of the training needs of travel plan advisers, co-ordinators, etc, and others such as planners also involved in the process; investigation of current provision; and identification of gaps in current provision with recommendations for future development. The research covers workplace travel plans as well as STPs. This research also continues.

6.5   Routes to a school travel plan

There are several valid options in drawing up a school travel plan; which is appropriate or possible depends to a considerable degree on the resources available to the school or to the local government authority:

- The local government authority does the work with or for the school. This may show an overall saving in resources if several schools are involved since there are often factors common to different schools, while a local authority staff member used to dealing with them may work more expeditiously than somebody less practised.
- The local government authority appoints a consultant, perhaps if the school has particular problems. Much the same applies as to the preceding point: a consultant familiar with the problem can often suggest a solution relatively rapidly, based on experience elsewhere.
- The school appoints a consultant – perhaps when the provision of an STP is a planning condition.
- The school community prepares the STP. This will obviously make considerable demands on the time of staff or volunteer helpers; nevertheless the school will accumulate a reservoir of expertise as those involved become familiar with the issues and procedures.
- A government-funded site-specific adviser assists the school.

However the STP is developed it is important that the school takes ownership of it. For this to happen the school community must become very much involved in the preparation and implementation, even if outside help is obtained. Ways in which the school community can become involved include:

- Questionnaires: preparing, distributing, collecting and analysing. Analysis and the presentation of findings can often be made a classroom IT or mathematics exercise for pupils.
- Interviews/focus groups involving parents, the school board members (known as 'school governors' in England), local residents and health and safety representatives.
- Audit of school grounds and catchment: a detailed examination of danger and problem points.
- Curriculum work with pupils, such as mapping exercises of school routes and the points shown up by the audit.
- Designing cycle parking facilities.
- Drawing or painting artwork for traffic signs/publicity.
- 'Planning for real'-type exercises: this involves making a large scale three-dimensional model of the school and its surroundings on which issues arising from some of the preceding points can be examined, so making them accessible to all for discussion and comment. The initiators of the technique provide training, packs, guides and videos to encourage full participation.[6]

6.6   Cycling to school

So far this chapter has been concerned with general sustainable travel issues relating to schools, albeit with an emphasis on active modes – walking and cycling. For a number of reasons, not least entrenched attitudes, most schools are likely, initially at least, to place a greater emphasis on walking than cycling (apart from such admirable exceptions as Kesgrave High School below). Nevertheless cycling has a prominent role to play. It is generally agreed that people assess the burden of a journey in terms of the time it will take using their chosen mode rather than the distance. From this point of view, cycling at, say, 15 km/h instead of walking at 5 km/h can either reduce the time taken for a journey by two-thirds or increase the radius of possible journeys by a factor of three – which amounts to a ninefold increase in the appropriate journey catchment area.

6.6.1   Barriers to cycling

The main obstacles to the use of bicycles for school journeys have been, and still are:

- Parental and school concerns regarding the safety on the road of pupils cycling to school. To avoid any possibility of a school being held legally liable in the case of an accident to a pupil quite a number of schools strongly discourage or even ban pupils from cycling to school. Only in late 2001 was official advice issued that schools could not legally prevent pupils cycling to school, although they had no obligation to allow bicycles on to school premises nor to offer secure or indeed any cycle parking facilities. Schools were also stated to have no general responsibility for the safety, with certain exceptions, of pupils travelling to school – whether cycling or by other means.

- Closely connected with these safety fears is a general perception that the infra-structure – particularly general purpose roads – is not cycle-friendly.
- A lack of adequate opportunities for training in basic cycle handling skills, and procedures and techniques to adopt when riding on the road. Nevertheless, quite a number of UK local authorities and other outside bodies do offer cycle training.
- The lack of secure cycle storage at schools. Fear of bicycle theft is one of the main deterrents to adult 'utility' cycling as well.
- The lack of lockers at schools for pupils to store equipment and books. It is now common in UK schools for pupils not to have an established 'base' at the school and to move from one more or less specialised room to another for different subjects. The result is that – unless locker space is available – they have to carry quite considerable loads of books, etc with them, and to take them home for security.
- Lack of a 'critical mass' of children cycling: children like to do what their friends are doing, to be seen to be 'part of the gang', and are only likely to be persuaded to cycle when enough members of their peer group are doing it too.[7]
- Finally, and above all, the absence of a universal cycling culture, such as exists in several other countries in northern Europe, where cycling is accepted as a suitable mode of travel over appropriate distances for all ages, classes and professions.

6.6.2   Encouraging cycling to school

The following are instances of the successful encouragement of cycling to school and show at least a way round some of these obstacles, if not their complete removal. The measures have all been implemented as part of a wider school travel plan.

- The provision of secure cycle parking – some designed by pupils in the Design and Technology subject area of the National Curriculum.

    Pupils at **St Luke's Primary School, Sway**, Hampshire (190 pupils, 5–11 yrs) used a £1000 environment grant to make a 30 space cycle shed them-selves out of wood – entirely appropriate for a rural school in a forest. The shed is regularly full.[8]

    At **Southgate School, Enfield, London** (1600 pupils, 11–19 yrs) individ-ual secure cycle lockers have been installed – initially 10, with 40 more within six months owing to demand. The lockers have proved extremely popular with pupils because bicycles are safe from tampering.[8]

    Pupils' ideas were used in the design of two cycle shelters, capable of housing 50 cycles, at **Hafren Junior School, Newtown, Wales** (280 pupils, 7–11 yrs).[8]

- School cycle parking permit scheme: this is a form of agreement between the school, pupil and parent. In exchange for the allocation by the school of a space in a designated secure cycle parking area, the pupil and parent undertake that bicycles are properly maintained and roadworthy, and that the pupil will ride sensibly and follow the Highway Code, and undertake cycle training where it is available. Statements are incorporated of the school's and parents' responsibilities for security and insurance.
- Providing more cycle-friendly access to school grounds. Would-be cyclists can be put off by having to share an approach road with heavy stop-and-start traffic, such as cars dropping off or picking up other pupils, or by having to walk with their bicycles along a narrow and crowded footpath to reach the cycle storage area.
- Traffic calming measures on key desire lines – sometimes reinforced by signs designed by pupils.

  > The approach roads to a number of schools in the city of **Nottingham** have been traffic calmed with speed cushions, road narrowings and raised junctions. The 'entry gates' to these roads are marked by traffic calming notices incorporating artwork produced by pupils at the schools.

- Provision of lockers for pupils who cycle to store equipment and books. This is to remove one frequently reported disincentive to cycling: 'too much to carry'.

  > When **Judgemeadow School, Leicester** (1200 pupils, 11–16 yrs) was granted a fixed sum by the local authority for cycle facilities including cycle parking, pupils asked for part of the grant to be retained for book and equipment lockers which are allocated to those cycling to school. The popularity of cycling to school is now such that an extension to locker provision is planned.[8]

- Provision of drying and storage area for coats.
- Provision of appropriate cycle training – on-road and on routes pupils use.

  > Pupils at the **Catholic High School, Chester**, developed their own cycle training course in collaboration with Cheshire County Council's Road Safety Unit. The six week practical and theory course has won awards in road safety competitions.[8]

  > The cycle training course devised by **Warwickshire County Council** involves six one hour sessions, of which all except the first are held on the road. The intention is that every pupil in the county should be offered cycle training.[8]

- Dedicated cycle routes/cycle-friendly infrastructure, e.g. toucan crossings, bridges, underpasses.

  > A new cycle and footbridge crossing the busy A701 in **Dumfries**, Scotland, links the formerly severed halves of the catchment area of

> **Locharbriggs Primary School** and provides improved cycle access
> to the school.[8]

- Opening up and formalising of 'desire lines' currently being used informally by cyclists, even if cyclists are required to give way to pedestrian traffic. These desire lines often indicate useful short cuts linking two sections of a formal route or quiet road network.
- Using bicycles for school trips or as transport to sports fixtures.
- Having cycle rides as a school activity. The British Schools Cycling Association[9] acts as an umbrella body for schools which have cycling as a recognised activity; although many BSCA promotions are on the competitive and sporting side of cycling, their activities also include some leisure riding and travel. The British Cycling Federation (under its British Cycling brand name) has an expanding 'Get Set' scheme to give after school cycle skills training for children up to about 12 years old.[10] The Scottish Cycling Development Project is introducing a Cycling Activity Leader qualification, with appropriate training, for those intending to organise cycling activities, as part of educational programmes delivered by the Scottish Cyclists' Union.[11]
- Setting up 'cycling trains' – the cycling analogue of the 'walking bus'.

> The pioneering 'cycling train' in the UK was established in 1999 for pupils at **Woodford Primary School** in the rural village of **Woodford Halse, Northamptonshire**. The 'train' of about 14 children and four adults runs to a pupil-devised timetable and gathers 'passengers' on its route through the village.[8]
>
> A similar 'train' at **Watchfield Primary School** in **Oxfordshire** has 10 regular 'commuters' with up to twice as many on fine summer days. The organisers stress that keeping to a strict timetable is essential, and equipment includes a train whistle to signal departure and guide progress![12]

- Provision of reduced cost bikes and accessories through the school. This is preferably done by means of an arrangement with a local cycle dealer who will be able to offer after sales service and also servicing. A popular way of encouraging pupils and others to have their cycles checked for roadworthiness is to hold free 'Dr Bike' clinics at which bicycles are given a thorough mechanical check after which faults that need rectification are pointed out.
- Talks from police on theft prevention. These may be associated with 'postcoding sessions' at which pupils' cycles are stamped with their home postcode. Some electronic tagging systems for bicycles are now becoming available.
- Raising relevant issues through the school curriculum, for example, cycle or related design in D&T (see the references to cycle parking design by pupils above), or the exploration of safe routes to schools through geography.
- Finally, and probably the most important: tackling negative attitudes.

6.7   Notes

1 It should be noted throughout that while traffic legislation applies uniformly to all the constituent countries of the United Kingdom, there are separate school curricula and education systems for England, Scotland, Northern Ireland and, to a certain extent, Wales.

2 For example, in a study for Nottinghamshire County Council carried out in 1994 consultants Cleary Hughes Associates surveyed the opinions of 27,000 children in the county on school and other travel matters (see also Nottinghamshire County Council, 1995).

3 www.transtat.dtlr.gov.uk/personal.

4 www.saferoutestoschools.org.uk.

5 Currently www.databases.dtlr.gov.uk/schools/ or www.local-transport.dtlr.gov.uk/schooltravel/index.htm.

6 The Neighbourhood Initiatives Foundation; tel +44 (0)1952 590777; fax +44 (0)1952 591771; email nif@cablenet.co.uk.

7 An interesting UK illustration of this is the case of Kesgrave High School (11–18 years) in Ipswich, Suffolk, where the proportion of pupils cycling to school rose from about 45% in 1995 to 61% in 1999: cycling to Kesgrave High School is now the majority – hence 'normal' and accepted – choice.

8 Further information on this work is available from schools@sustrans.org.uk.

9 British Schools Cycling Association (BSCA) may be contacted at susanknight@bsca.fsnet.co.uk or their website is www.bsca.org.uk.

10 The appropriate department of the British Cycling Federation is development@bcf.uk.com; their website is www.britishcycling.org.uk.

11 More information is available from cathyscott@scottishcycling.co.uk.

12 More information on this project is available from travelwise.team@oxfordshire.gov.uk.

6.8   References

DETR (Department of the Environment, Transport and the Regions) (1998), *A New Deal for Transport: Better for Everyone*, London, The Stationery Office.

DETR (Department of the Environment, Transport and the Regions) (2000), *Guidance on Full Local Transport Plans*, London, Department of the Environment, Transport and the Regions.

HILLMAN, M, ADAMS, J and WHITELEGG, J (1990), *One False Move . . . A Study of Children's Independent Mobility*, London, PSI Publishing.

NOTTINGHAMSHIRE COUNTY COUNCIL (1995), *School Travel: Health and the Environment*, West Bridgford, Nottingham, Nottinghamshire County Council.

NOTTINGHAMSHIRE COUNTY COUNCIL (1998), *The Nottinghamshire TravelWise Teachers Pack*, West Bridgford, Nottingham, Nottinghamshire County Council.

SUSTRANS (2001), *The Health Benefits of Safe Routes to Schools*, Bristol, Sustrans.

TRANSPORT 2000 TRUST (1999), *A Safer Journey to School: A Guide for School Communities*, London, Transport 2000 Trust.

7

The UK National Cycle Network: a millennium project

John Grimshaw, Director, Sustrans

7.1 Introduction

What has the UK National Cycle Network achieved and where is it going? The NCN has a relatively short history, short that is for a public work extending for 16,000 kilometres across the whole of the UK. Its first elements were created in the late 1970s when a number of traffic-free paths were built along the alignment of abandoned railways, most notably the High Peak and Tissington Trails in the Peak District National Park and the 27 km Bristol and Bath route (see Fig. 7.1). This last was specifically built by a local cycling group, Cyclebag, to demonstrate that the public would cycle if given the chance – a chance that was being increasingly denied by ever busier roads and in the absence of any sort of government or local authority encouragement or provision for cycling.

The Bristol and Bath path was an immediate success and to this day remains Britain's busiest route with 1.7 million trips per annum and almost the same number by pedestrians. Its obvious value led to the establishing of Sustrans, a registered charity, to create similar routes wherever possible. This work continued through the preparation of numerous studies including nationwide reviews of the opportunities on disused railways in England, Wales and Scotland (Grimshaw, 1982).

Sustrans was aiming to take cycling levels up to those found in neighbouring European countries which had achieved a five to ten times greater level of cycling (see Fig. 7.2). Any hope of matching this achievement required a strategy for winning over those who did *not* cycle rather than focusing on those who did. The latter aim was the more natural viewpoint of cycling pressure groups, including the Cyclists' Touring Club (CTC), which mostly sought to represent their

Fig. 7.1   View of Bristol and Bath path. Source: Kai/Sustrans.

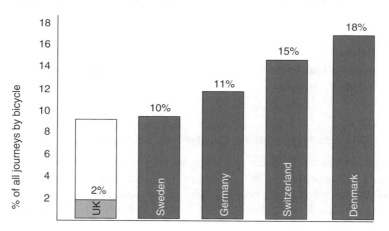

Fig. 7.2   Cycling levels in various countries, 1996. Sources: Department of Transport (1996) and ITS Leeds.

members who in the nature of things – having survived the traffic – were experienced cyclists.

European success had not come from this direction, but rather from a generation of careful planning and provision for both the committed and the occasional cyclist. The example here (Fig. 7.3) from Switzerland shows how the network of cycling routes has grown over a period of 50 years, with consequent high levels

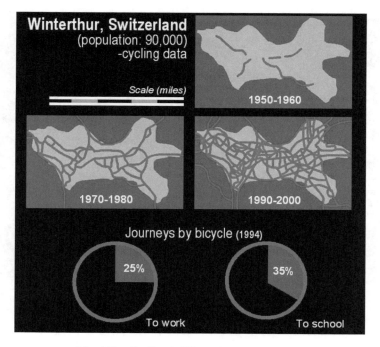

Fig. 7.3   Cycling in Winterthur, Switzerland.

of cycling and low levels of accidents, all integrated facets of a transport policy to encourage cycling.

7.2   The development of a national Network

It was clear that a renaissance of cycling would require many strands of activity, not least the appreciation by government that cycling was a valuable form of transport for convenience, health, independence and the environment, and deserved to be encouraged. This was finally set out in the National Cycling Strategy in 1996 (DOT, 1996).

But 10 years before, without these guidelines, Sustrans had set out the following strands for its own work programme:

- Every urban area needed at least one high quality traffic-free cycle route where novices of every age could learn to cycle again, could enjoy the experience and could convince themselves that the bicycle was still a valuable and appropriate means of transport for modern times.
- Adults returning to cycling were likely to first do so on a leisure trip. So signed routes were developed, thus striving to avoid the disproportionate focus on the journey to work, which was prevalent at that time.

- Children are tomorrow's adults and unless high levels of cycling were achieved at a young age then there would be few who were even able to cycle into adulthood. Sustrans launched its Safe Routes to Schools programme (see Chapter 6) specifically to bring about change in this most crucial of journeys.
- Partnerships are crucial; progress could only be made if other interest groups came to see cycling as a valuable component of their own programmes. The clearest example of this is the whole area of health where regular exercise, day-in-day-out, was seen by the British Medical Association and others as the key to combating our increasingly sedentary lifestyle (BMA, 1992). The only way to incorporate physical activity into most people's daily routine is to exercise on the daily journey to work or school, namely walk or cycle. Another crucial partner was the local authorities, all of which were struggling with the consequences of excessive traffic, but unable to take firm action because of the perceived universality of ownership and apparent benefits of high mobility. But gradually councils came to recognise that the bicycle is a most appropriate means of making many local journeys, a large proportion of which are quite short.

Sustrans built steadily on these principles through to the mid-1990s when the newly launched National Lottery gave a once in a lifetime opportunity to create something greater than the sum of its constituent parts, that would raise the profile of cycling and give it a status that would encourage the public to cycle. This was the chance to make a National Cycle Network with examples of best practice in every town, reluctant authorities won over by the national scale and popularity of the project, and a chance to invest substantial funds into an area that hitherto had to survive on the scraps left over from other transport development and budgets.

Sustrans submitted its bid to the Millennium Commission in March 1995 and on 8 September that year the Commission announced its first major grant award: €70 million (£43.5 million) for the National Cycle Network.

This announcement was widely supported by the public and by local authorities and ushered in a five year period of intense activity across the country which has been an immensely satisfying and very demanding experience for Sustrans' 120 staff. In December 2001 the last millennium claim was processed and it is time to finally move on from the millennium years.

7.3   The growth of the Network

In assessing what has been achieved over this period it is easy to record what has been built on the ground but harder to measure its effects. The physical Network has expanded beyond even Sustrans' most optimistic hopes (see Fig. 7.4). Prior to 1994 Sustrans had built short lengths of route all over the country and had opened just one long route, the 225 km (150 miles) C2C route from Whitehaven to Sunderland. This route, incidentally, went a long way to demonstrate the value of cyclists as visitors in local areas with 15,000 whole length trips being made

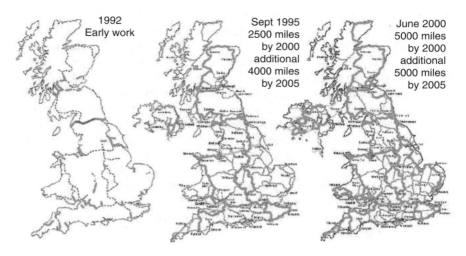

Fig. 7.4   Development of the National Cycle Network. Source: Sustrans.

each year and some €3.2 million (£2 million) spent in the local rural economy (Cope, 2001).

Figure 7.4 shows the original plans for the National Cycle Network – 4000 km planned to be opened in 2000 – and the actual achievement of 8000 km. Although as every cyclist will say there are still details that are poor and sections that are inadequate we have the framework in place, which must now be improved year by year.

The Network is approximately one-third on traffic-free paths. Interestingly this corresponds very closely with the actual breakdown of cycling trips on the very successful Québec network (Vélo Québec, 2001) La Route verte – one of the most successful programmes of recent times to increase cycling from low historic levels (see Fig. 7.5).

The cost of the works has been considerable. As well as the hundreds of miles of on-road improvements and new route construction, we have recorded the construction of over 250 new bridges larger than a five metre span, including some major new millennium bridges – the Lune at Lancaster, the Ouse at York and the Gateshead Millennium Bridge, all separately funded millennium projects. The Millennium Commission grant has amounted to less than 20% of the estimated cost, the balance of which has been made up from innumerable sources (see Fig. 7.6). In some cases where the NCN works were incorporated into other developments the sum apportioned to the cycle route can only be approximated. Altogether Sustrans made 996 millennium grants.

As well as construction, great effort was put into the character of each route so as to make the cycling journey attractive and memorable. Landscaping, clearing views, planting and some 500 milepost sculptures – seats, drinking fountains, gateways, bridges and markers (see Fig. 7.7) – were commissioned along with

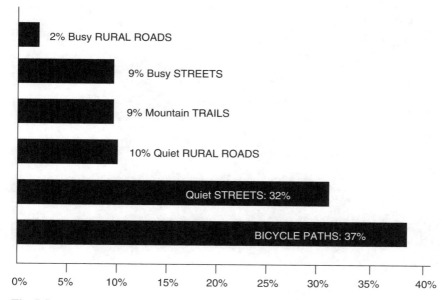

Fig. 7.5   The breakdown of mileage by type of route in the Québec experience. Source: Vélo Québec (2001).

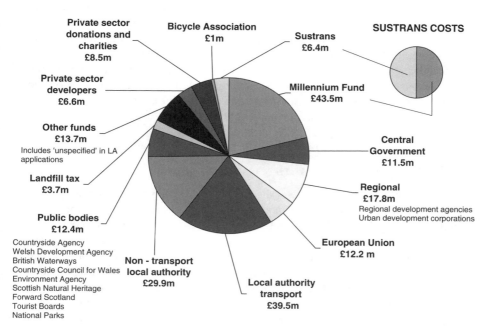

Fig. 7.6   Breakdown of funds. Source: Sustrans.

Fig. 7.7   'Terris Novalis', by Tony Cragg, with cyclists.
Source: Barry Wilson/Sustrans.

1057 cast iron Millennium Mileposts donated by the Royal Bank of Scotland, to mark out the whole and give an aesthetic dimension to the NCN (see Fig. 7.8).

Route numbering, signing and maps were all devised and agreed with the Department of Transport, Local Government and the Regions, and its various antecedents, and technical standards published (Sustrans and Ove Arup, 1997). The whole Network continues to be advertised with each latest update on the National Cycle Network website (www.sustrans.org.uk), where would-be travellers can go on a virtual journey anywhere in the UK.

7.4   Usage of the Network

So much for the physical nature of the Network. What have been its effects? Certainly it has put cycling much more firmly on the map of almost every local authority in the UK, with most sections of the eventual 16,000 km (10,000 mile) Network shown on strategic plans. Usage is the key outcome, because if patronage of the NCN routes does not quadruple by 2012 (in line with the national targets in the National Cycling Strategy) then the Network will have failed in its own right, and will have failed to become the catalyst for change at a much more widespread and local level where the bulk of cycling journeys will take place.

Sustrans set up a Monitoring Unit in 1998 which currently assesses data from over 350 automatic counters, mostly managed by local authorities, and has conducted detailed surveys on various routes for comparative purposes. There

Fig. 7.8 Four Millennium Mileposts in place.

is indeed evidence that usage is increasing on the Network, more so on the traffic-free sections than on-road. We allow ourselves to hope that this performance shows the Network leading a resurgence of cycle use in the UK (see Table 7.1).

Sustrans also holds data on routes 'before' and 'after' construction. For example, we can be quite confident that the current works under construction in Stoke will result in the same sort of levels of use as already achieved in similar circumstances in the Spen Valley in Kirklees (see Fig. 7.9). Over the next few years we expect to see usage increasing to match the effort made and thus be able to assess the scale of endeavour and the extent of the works needed to reach the target of quadrupling of cycling in the UK.

This quadrupling is in terms of trips. It will do nothing to change the phenomenal and still increasing mileage travelled each year, mostly by plane and car. The bicycle is a slow and so essentially short range vehicle. Its renaissance will succeed beyond expectation if the public can be persuaded to value short trips,

Table 7.1 Comparative usage data

	Total	Cyclists	Pedestrians		
Elswick				*Eastbound*	*Westbound*
1998	749	342 (45.7%)	383 (51.1%)	48.0%	52.0%
2001	1071	451 (42.1%)	539 (50.3%)	45.4%	54.6%
Change	+43.0%	+31.9%	+40.7%		
Fremington				*To Instow*	*To Barnstaple*
1998	2499	2154 (86.2%)	345 (13.8%)	53.0%	51.1%
2001	3752	2657 (70.8%)	1086 (28.9%)	47.0%	48.9%
Change	+50.1%	+23.4%	+214.8%		
Rishton				*To Rishton*	*To Gt Harwood*
1998	1285	338 (26.3%)	880 (68.5%)	56.2%	50.7%
2001	1667	427 (25.6%)	1123 (67.4%)	43.8%	49.3%
Change	+29.7%	+26.3%	+27.6%		

Note: The table is derived from manual counts conducted concurrently with the surveys, and shows the number of users, the number of cyclists and walkers and the proportion of the total sample comprised by each ('other' users are not included), and users' direction of travel. The changes in usage volume over the survey period are also shown.

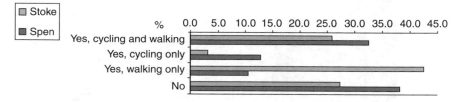

Fig. 7.9 Route usage compared with route open.

to substitute a local bicycle trip to the countryside nearby for a long car trip to a National Park. Success will be measured by the decision of individuals to adjust their lifestyle to enable short trips to be the norm, to live near to work and shops, to travel lightly, carrying only what is needed then and what can be conveniently taken by cycle.

The benefits for local communities of such a change would be considerable: less traffic, slower traffic, local facilities; the benefits to individuals would be equally wide-ranging, including less stress, better fitness and health and more life nearby and local.

If society values these characteristics then we must find ways of positively encouraging individuals to cycle, not only by providing the infrastructure but also through fiscal benefits – reduced entry to attractions, additional holidays to reward the increased output from fitter and less stressed staff, financial benefits for under-taking to do without a car in urban areas, reductions in accommodation costs in holiday areas to reflect less degradation of the environment. In short if this supremely efficient and benign form of transport is to flourish then we need to search out many ways in which the public can be rewarded for making the effort required and eventually the changes to lifestyle which are necessary.

7.5   The future

While these wider matters are evolving, Sustrans has the privilege of continuing to develop the Network, with its regional and local routes. The charity owns 970 km (600 miles) of route and over 1000 structures, many of them wonderful Victorian railway bridges. It manages extensive travelling landscapes through its network of 1500 Community Volunteer Rangers and it has the ongoing reward of seeing the public take great pleasure from a project that it has initiated and developed (Wickers, 2000).

It has often been said that the bicycle is transport at the human scale (Illich, 1974) and this is true also of the small scale works needed for its use. All of us in Sustrans have had the opportunity to design, negotiate, build and open numerous works in a short time. Simply because even the best cycling routes and details are small in scale many can be created in this crowded country of ours. This immediacy is rare in our times, which perhaps explains the enthusiasm and commitment of the whole Sustrans team, whether based at the Head Office in Bristol or scattered around the UK looking after local areas.

We do not know which way transport will develop but Sustrans will continue to strive to put in place practical demonstrations of sustainable transport for the twenty-first century.

7.6   References

BMA (British Medical Association) (1992), *Cycling: Towards Safety and Health* (report of a BMA Working Party edited by Mayer Hillman), Oxford, Oxford University Press and London, BMA.

COPE, AM (2001), *Monitoring Cycle Tourism on the C2C Cycle Route during 2000*, Bristol, Sustrans (unpublished).

DOT (Department of Transport) (1996), *National Cycling Strategy*, London, Department of Transport.

GRIMSHAW, J (1982), *Study of Disused Railways in England and Wales for Potential Cycle Routes*, London, HMSO.

ILLICH, I (1974), *Energy and Equity*, London, Marion Boyars Publications.

SUSTRANS AND OVE ARUP & PARTNERS (1997), *The National Cycle Network Guidelines and Practical Details Issue 2,* Bristol, Sustrans.

VÉLO QUÉBEC (2001), *Bicycling in Québec in 2000*, Montreal, Vélo Québec.

WICKERS, D (2000), *Millennium Miles: The Story of the National Cycle Network*, Whitley, UK, Good Books.

8

Cycling with public transport: combined in partnership, not conflict

Dave Holladay, Transport Management Solutions, Glasgow

8.1 Introduction

It has to be recognised that the bicycle cannot be considered in isolation, and that the quality of, or returned benefit from cycling activity is not necessarily reflected in purely raising the numbers of cyclists. Promotion of cycling as a suitable mode for travel may be a less effective or acceptable choice for some trips, or some travellers, than the alternative low impact options of walking or using public transport, or even not making a physical 'journey' at all. However, the frequent presence of a car, available at the starting point for a trip, tends to predetermine the choice of a car and thus the intrusive and dominant place it takes in city traffic. Unfortunately this means that it is often used where completely inappropriate, both in terms of the user's cost of running the vehicle, the use of resources (road space, parking) and in terms of less tangible effects of safety, noise and degradation of the environments in which we have to live and work.

In managing travel resource consumption, known in the UK as a Travel Plan, and in the USA as Travel Demand Management, it soon becomes clear that no single mode of travel will suit all travellers on all occasions. Both individual and institutional Travel Plans will offer flexibility of choice over a range of modes of transport. Ideally this can be managed on a continuously adjustable basis, to use each to maximum potential.

Working within this framework the bicycle really stands out as a mode offering speed and low effort required to cover moderate distances, of up to eight kilometres when compared to walking, but – especially with compact folding bikes – the opportunity to link to any other mode as easily as a pedestrian can. The facility to link intermodally, retaining the advantage of cycling when this becomes the best value choice over public transport or private car, is the foundation for the material covered in this chapter. The author's own policy of never cycling when the option of a faster or more efficient journey can be made by bus or train, is reinforced by the review and promotion of these practices through the magazine *A to B* (formerly *The Folder*), and its website (www.atob.org.uk). This is supplemented by the CTC's provision of information on bike carriage on bus, train, ferry and other modes. Links are available from www.ctc.org.uk for UK activity, and the site www.bikemap.com provides users with information in the USA.

The attraction of choosing the bicycle, and thereby delivering the environmental benefits and the goals of increased cycle use to which many policies aspire, is influenced by many factors. These can include topography (often handled by choice of contour line routes), extremes of weather (but local cyclists tend to be dressed appropriately for other outdoor activity), social and demographic profiles and local geography. In older towns where commercial and cultural activities were originally based on walking everywhere, the density of paths and streets, and patterns of travel between work, dormitory and service locations will be promulgated by this urban network.

On the other hand, when the development has grown with the private car as principal mode of travel, car use is inevitably selected as the only way that the journeys can be made effectively over the distances involved. Only when cycling and walking work almost seamlessly with public transport can the journeys be achieved in competitive times. This may be reflected in the success of Bike on Bus schemes in the USA, and other cycle links with public transport. This does not *per se* exclude access to city bus stops and train stations in high density developments, but users will tend to self-regulate by choosing to walk the shorter distances when the parking or mode transfer delays for a bike make cycling a slower means of making the trip.

It is vital, with the inertia that leaves us with the private car dominating areas, where it clearly is not the best transport solution, that the planners and politicians work with businesses and users to recognise the importance of developing city and urban spaces to be more easily used by those cycling, walking and using public transport. The incentives – increasingly from the financial gains from reducing the loss of valuable land to car parking and access roads (Newman and Kenworthy, 1989) – have been covered in Chapter 2. There are many intrinsic details which make the changed travel regime work, and for most situations actually provide the same travel itineraries at considerably reduced cost. Subsidising services is an indication that some detail of the system is being artificially forced, and leaves the whole arrangement vulnerable.

In considering appropriate policies it is perhaps most important to recognise that the real measure of travel is to look at the movement of people (or just their image, and influence via electronic transmission), rather than vehicle counts.

Typically for the user of public transport, information and access are the key to enhanced use. The research of Werner Brög with the Socialdata consultancy (see Chapter 18) has shown that use of public transport can increase without the need for extra resources if targeted marketing puts the information in front of the non-user showing them what is available. Similar awareness can generate activity in the cycle–public transport interface. Access is equally crucial in making the seamless connection, and the 1996 Cycle Challenge (DETR, 1999a) was able to note a clear correlation in having well-placed cycle parking *en route* to the train. The visible bicycles, and convenient sequence of arrive–park–board saw cycle trips to Eastleigh station increase by 500% and other locations experience lesser but still significant increases, many exceeding the targets set by the National Cycling Strategy for increasing cycle trips. Further experiences are highlighted later in this chapter and the figures illustrate relevant examples.

8.2   Journey profile – assessing the potential

In terms of mobility the pedestrian mode scores over all others, being able to cover almost any terrain and divert at will. Cycling is in practical terms the next best thing, offering a faster journey over greater range, and the same ability to run to an individual schedule, eliminating the need to build in overlap times for short connecting journeys which often involve a second bus, or taxi. Various presentations by Sustrans (e.g. Velo-City conference 2001) on their Safe Routes to Stations project describe the obstacles and components of these fundamental first, last and occasionally intermediate stages of a journey by public transport.

Typically a range of 2 to 3 kilometres is a 10–15 minute ride (including preparation and parking) which is just on the lower range of being viable when compared with briskly walking the same distance. This compares with the average figures of a 350 metre walk to catch a bus, or 850 m for a train (National Travel Surveys (NTS)) and underlines the way that pedestrian and cycle journeys are measured in terms of time rather than distance by those describing their travel itinerary. This also shows why topography, direct access, and barriers like major road crossings can make the prediction of the catchment range for a rail, tram or bus stop vary from 5 to 10 kilometres.

For short journeys (up to 3 kilometres) the bike will almost always be able to achieve shorter journey times than bus travel, allowing for a typical walk of 400–500 m to the bus stop and usually a short wait for the bus (Grabe/VOV, 1985). This is reinforced by the results for the (usually) annual London Travel Survey where a central London bus journey is considerably longer than the cycle one, with just 20% of the bus trip spent in motion on the vehicle, compared with over 95% of the cyclist's time. However, it is obvious that a frequent or highly reliable service, with users able to minimise waiting and ticket buying delays, will offer equivalent convenience, and as the distances grow beyond 5 kilometres, unless the service is sparse, the bus, along with tram and train show faster overall journey times.

Several US bus operators have profiled their bus routes to establish those which would benefit most from Bike on Bus and Cycle Parking investment. Dade Co (Florida) and San Diego used scoring to highlight routes where the convenience and potential demand were likely to generate highest uptake of new schemes.

To maximise this potential for combining cycling and public transport it is essential that each detail is worked on as part of the total transport chain involved. As with any chain this transport one is only as strong as its weakest link. For example, the weakest link may often tend to be the final stage, between arrival station and workplace, and especially if there is no provision for secure storage, convenient for coming and going at either end. Parking is very important – and the experience of the Cycle Challenge noted previously is followed up later in the chapter. In the UK there is a perceived need also for arrangements to adjust from cycling to working, which may require showers and dedicated changing facilities if the journey is likely to involve great exertion. These are rarely demanded by European utility cyclists, who tend to ride bikes (with integral weather and dirt protection) at a leisurely pace in the clothes they will be working in.

At the start, parking is equally important. Not all cyclists can store their bikes safely and conveniently where they live. This tends to be more of a problem in cities and student accommodation, where a large proportion of the population live in flats (Luers, 1985) with no separate garage space. Work by the author with the City of Edinburgh Council on tenement cycle parking, where bikes can obstruct stairways and passages, where no formal options are offered, found one user who cycled less because her current flat was 200 m from where she could secure her bike.

The journeys to and from the public transport system interchange points form a core element in developing a cycle route network which thrives through daily use, by introducing regular and purposeful journeys. Some detail will be evident from the ability previously noted of cyclists and pedestrians to form their own informal routes along lines of least delay. The adoption of these and designed-in links to areas of journey generation should meet quality standards appropriate to cycling as transport (alignment, sighting distances, etc), which when correctly applied reduce conflict between modes and manage use through design. Directness and capacity are especially relevant, noting that many routes to the 'stop' will be shared with pedestrian activity which delivers up to 50% (returns for rail operators' surveys) of the passengers to trains, and about 80% of passengers to bus stops (NTS extrapolation). Ironically this gets a far smaller share of investment than the access provision for a minority arriving by car. The aim should be to provide safe, direct, convenient and comfortable cycle access right up to and from the next link in the chain, the cycle parking area. One often missed detail here is the desirability of a route which 'divorces' the rider from the bike at a point where conflict with the concentrated pedestrian circulation at a station is neutralised. York and Stirling stations have good examples of separate ride-in access delivering the cyclists to the train without mixing with movements on the main concourse.

Some sites enable cyclists to share subways or ramps, but this does raise an issue of speed. Some provide lifts, a measure which, like ramps, may also improve access for the full range of passengers who find stairs a hindrance or absolute barrier. Appropriate solutions will vary according to local circumstances

Fig. 8.1   One Dutch operator has been offering discounted folding bikes and running an express service with dedicated stowage space.

Fig. 8.2   A wheeling ramp delivers cyclists directly to the undercroft – Winterthur. A large capacity cycle park is in this area. Cyclists enter the station up stairs on to their departure platform.

Fig. 8.3 Ride-in direct access to parking on the roof of trading units – Winterthur. Cyclists then descend steps to trains. Over 3000 bikes park on this site on roofs and in the undercroft.

but where a direct set of steps has sufficient width, a wheeling channel provides a simple, low cost and minimum delay solution, an alternative to using a lift or long ramp route, releasing the former for those who really need it at busy times. Typically a wheeling channel for a footbridge can cost €2500–€3300 (£1500–£2000) (e.g. the footbridge at South Gosforth, Newcastle upon Tyne, 1999).

Signing is important, especially for locations likely to be used by visitors to the area, but just as a good cycle route should be laid out in such a way that the natural flow takes you onwards, the obvious flow route, so arrival at a station should direct cyclists to the places they need to be (parking, ticket purchase, joining a service). Promotion may be needed through external media, but with security it is best delivered by making the cycle parking and route to the station a prominent feature for all passengers to see.

Cycle parking at interchange points is dealt with in detail later, but note that bikes left regularly in the same place for long and predictable periods are a prime target for theft and vandal attack.

8.3   Cycle access to and from railway stations

Dutch transport policy is now looking to increase the role of the bicycle at the expense of some bus trips for distances of up to 5 kilometres (Welleman, 1991). In that densely populated country, 60% of all trips by bus, tram and metro are of this distance or less. Official thinking now regards these short trips as determining the capacity required during the rush hour and also that they cause a major part of public transport operators' losses. Moreover, as part of the Bicycle Master Plan 1991–95 (Ministry of Transport, Public Works and Water Management, 1999) there is a specific aim to improve cycling to and from stations. About one-third of rail passengers typically arrive by bike, and sometimes 50%, but the aim is now to achieve an increase in the number of cycle-borne passengers by 15% by 1995.

Many Dutch people also use a bike when they complete their rail journey, to go on to their destination, according to a survey by Bracher *et al* (1991, p 88). This kind of travel habit, bike–ride–bike or 'sandwich' travel, is also very common in Denmark, with people often reducing the risk of theft of the bicycle left at the destination station by leaving an old bike there and keeping their best bike at home, for home-based trips. It occurs at certain London rail termini – notably Waterloo and Paddington, where several hundred bikes are left overnight by commuters – although this is small scale when measured against the several thousand spaces in valet car parking buildings – plus an equivalent outdoor parking site for casual use.

For the cyclist such combinations offer the advantage of being able to complete their journeys with a faster door-to-door journey time than if they relied on public transport alone, or public transport and walking. As the same Dutch survey showed, average distances of all kinds of trips to stations are likely to be significantly longer (3.7 km) than those at the end of the rail trip (2.9 km). In efficiency terms recent Dutch studies (Bouwman, 2000) indicate that the combination of rail and bike provides the most energy efficient overall journey package for journeys exceeding 10 kilometres, with the bicycle delivering the required results below this distance.

8.4   Providing short and long term cycle parking at stations

There are nominally three simple levels of provision for cycle parking, which most working in the field recognise, and in a report for Southampton City Council (Holladay, 2001), these were subcategorised further. One need is rarely addressed – parking a bike whilst tickets are purchased, prior to boarding a train with a bike. For this function a wheel grip, or resting rail, within sight of the booking counter makes the process fast and convenient, as the bike does not need to be locked or secured against unwanted attention.

For longer periods the user will be balancing the convenience of use with the level of security offered. Casual users will accept open access to parking stands which provide a solid anchor to which a bike can be fully secured, and where the bike is properly supported. There will be a strong preference for stands which are protected from weather, and (on many stations) pigeon droppings. If the location is poor, with low perceived surveillance, lack of weather protection, etc then cyclists will use other secure items in more convenient locations – as recorded by an MVA study (Hampshire Stations Cycle Challenge) and the Southampton report.

Fig. 8.4   This situation in a busy station is where no-lock cycle parking would help the cyclist and protect the décor from damage.

Even the right designs of parking stand can be installed badly. Spacing between racks is made too tight (i.e. less than 700 mm) which makes it very hard to get bikes in and out without getting entangled with each other or getting in to place a lock properly. Some are set too close to walls, leaving the bikes vulnerable to being knocked over.

Although weather protection is also important this can be given too much weight at the expense of convenience of use. Often the designs provided have insufficient headroom to reach in to park and lock the bike. Fortunately the development of standards such as the Dutch Fietsparkeur are beginning to deliver installations which have a guarantee of being dimensioned to fit and be usable by a majority of cyclists.

Open stands which are well lit and visible, with obvious CCTV surveillance, provide a strong deterrent to theft, and also promote themselves. Management is vital though, as useful parking capacity is taken out by abandoned bikes creating an air of neglect and poor security. The placing of cycle parking at the principal entry point to a station provides high levels of convenience for cyclists, proclaims the cycle facilities available, and can also help to reduce the menace to pedestrians of cycles left in a disorderly manner. It may not satisfy the full demand for parking, and a large dedicated area (as in for example Amsterdam or Cambridge) is often needed. Clear marking of rows helps here, much as in floor identity on a multi-level car park. Moreover, as Brunsing comments (Brunsing, 1990), cycle parking facilities need to be designed in such a way that they blend into surroundings but also remain in full view of the public.

A further way to provide weather protection is to use space under canopies or in the station complex but this does have potential for problems – debris (notably pigeon droppings) and water from overhead, and the conflicts induced by leading cyclists through circulation areas or service roads.

In the Netherlands, the standard of cycle parking at stations has for many years been fairly high and in western Germany major improvements have been made in some areas in recent years, as for example in the cities in North Rhine–Westphalia covered by the Rhein–Ruhr Transit Authority, VRR (Gyukits *et al*, 1986). At many stations, including both main and suburban ones, and including new park-and-ride sites on the edge of the urban area, extensive, attractive and fairly secure covered cycle parking has been provided. The German experience – good bike-and-ride facilities at some stations in Munich generating large increases in the number of cycle-borne passengers in a place with a weaker cycling tradition than many cities elsewhere in Germany – has parallels with the growing feedback from UK projects.

8.5   Parking for regular users – all day/all night

Whilst it is impossible to guarantee the complete elimination of cycle theft from any areas used for cycle parking, several management regimes with associated hardware have shown levels of security which allow the site operator or hardware supplier to carry the liability for loss or offer very low cost theft cover. This issue is vitally important for the valuable core business generated from regular commuters, who leave bikes for long periods on a regular basis in the same location. Research by Michael Groll which formed part of his Velo-City paper (2nd prize 2000 Velo-Mondial) (Groll, 2000) showed that regular cyclists would pay for the peace of mind that a locker would give them. His findings in the West Midlands (Centro) area are matched by the demand for lockers in place at stations (Wolverhampton, Sutton Coldfield, etc) often being over-subscribed, and with a waiting list. Falkirk High station provides a clear example of demand suppression as the original installation of four lockers has had to be increased twice by a further four units each time within a year of being installed. Lockers do permit the use of a bike at either end of a journey, and Caltrain in California has offered users two lockers for the price of one in an attempt to take pressure off for on-train spaces. The possible ultimate user is a Berkeley-based professor with five bikes located at stations on the BART system in the San Francisco area, to suit his regular travel needs.

Lockers require a higher level of management than open parking stands, but where the user is committed to regular travel from that station the investment can be justified, and some operators (Nexus) link the offer of a bike locker to possession of a season ticket. Security is an issue, and the experience from major rail operators such as Nederlandse Spoorwege (Dutch Railways) is that every locker should be allocated to a known and traceable user. The use of a locker rental agreement has enabled some locations in the UK to remain operational

Fig. 8.5   Fiets BV – franchised operators inside a major station in the Netherlands.

Fig. 8.6   Fiets BV – casual parking outside a major station in the Netherlands.

under conditions of security alert. Transec, the rail security regulator, has worked closely with passenger transport executives (PTEs) and some locker manufacturers to develop designs which are easy to inspect without weakening the structure or compromising the security of the contents by using a mesh (which displays the contents, and is easily cut out), and to ensure users are managed through a standard form of rental agreement, to capture appropriate details, and set work-

Fig. 8.7   A typical managed cycle locker, which carries its own advertising.

able conditions. Locker design is crucial in minimising site preparation costs, and resisting attack. Some manufacturers (e.g. Cyclesafe) can point to years of theft-free operation. Earlier research (Froitzheim, 1990; Schafer-Breede, 1987) showed the popularity of lockers, provided that the charges were thought reasonable by potential users, and noted the management problems of the early systems. They too highlighted the problems of coin operation, of all types, which can attract vandals, and also the desirability of not offering lockers to casual or short term users – the minimum period for most agreements is one month. Most opt for three or six month renewals.

Lockers are not the only solution though. Southampton University's cycle compounds have had no reported theft in over six years of operation with the University carrying the theft liability for bikes correctly parked in the swipe-card access cages and rooms. These offer a higher density parking over lockers, and lower costs, with a full audit trail of access by authorised users.

Of course the greatest convenience for joining the train is to have a valet parking system, or Cycle Station. Nederlandse Spoorwege has franchises for operation of some 80 large units designed in to their major stations, and some smaller operations contracted with local community operators at secondary stations. There are two large sites at Amsterdam Central (interview with Frans van Buuren – NS Stations). No units operate in the UK, but US transit systems have steadily opened low cost centres (FHWA, 1998). These will not be viable if they exist solely on commuter parking, as the activity will be very limited for long periods. The NS contracts set minimum hours of operation, and national rates for parking and bike hire, but open the opportunity for the franchisee to offer bike sales, repairs and extended hours to maximise the return for local circumstances. It is likely that any UK development will be supported or cross-subsidised by activities such as a cycle shop, station café or news-stand, or offered as an extension of left luggage provision.

8.6 'Bicycle stations' and cycle hire facilities

The most ambitious types of cycling provision at stations offer a wider menu than just cycle parking. In the Netherlands, the cycle parking at even quite small urban stations has for years been combined with cycle hire and repair facilities. By 1989 there were 92 of these 'bicycle stations', which in large urban areas often offered cycle hire, advice to cyclists about choice of routes for daily and touring purposes, and also the sale of maps, guides and accessories. The concept of bicycle stations has grown, but often with the lack of a robust commercial outlook – hence many fail when initial funding or enthusiasm run out, or the subsidy ceases. In Germany, the first such bicycle station was developed in the early 1980s on the square in front of Bremen main station, and several other cities have now made similar provision, usually the result of initiatives from cycling groups rather than from Die Bahn or other public transport operators.

Bicycle stations, especially if linked to extensive and guarded cycle parking areas, are excellent for promoting this enhanced security for parked cycles, and providing this in a compact space. The obvious presence encourages the cyclists to stop leaving bikes at random causing nuisance to other users. Lack of a focus, or management scheme creates the 'Don't Park Bikes Here' problem, and typically a popular major station can have up to 1000 bikes parked on any convenient fence or lamppost. An extreme example in Germany of this 'bicycle pollution' is Munster where the problem of the proliferation of parked bikes became so great, with well over 3000 bikes (Froitzheim, 1991) being left in and around the station area, that the city council has built a special multi-storey cycle parking facility, with financial support from the *Land* (state government) (see Chapter 13).

This is the first multi-storey cycle parking facility in Germany, although such facilities have been provided for some years in Japan, where both the lack of space and the increase in cycle usage in the 1980s became even more serious (Replogle, 1983; Schafer-Breede, 1987). Kuranani and Bel (1997) described the situation at some Japanese commuter stations where cyclists abandon bikes on the street or in unofficial parking spaces – which are then impounded. Over 2.3 million bikes a year are impounded and disposed of by the rail operator or local authority. At Tokyo's Kichijouji station between 7500 and 8800 bicycles daily try to park in an area with just 5800 official places. Some locations have reduced the incidence of illegally parked bikes by promoting a universal bike hire scheme or compact and light folding bikes, and evidence is that this is having a measurable effect. The bicycle may be succeeding here for two reasons – driving from homes too distant to walk from is not possible, and the average speed of cycling now exceeds the decreasing average speed of local bus services.

Fig. 8.8 The Long Beach bike station: a steel prefabricated unit, using corrugated sheet panels to form the café/hire and sales office with a secure bike parking area in full view (the tubular frame is strung with HT steel wires – almost invisible at a distance).

Cycle hire facilities are very common at stations in several continental countries, and not only at main railway stations. Widespread cycle hire helps casual users, as well as commuters, and the nature of the service will vary to suit the relevant market – utility bikes for town use, or leisure hire in tourist areas. The NS operations – now known as Fiets BV – set a standard hire tariff for hiring from their franchised operators. The Swiss tradition has even greater coverage (Tschopp, 1988) where Swiss Railways, SBB/CFF, introduced cycle hire in the 1950s as a service for travelling salesmen. In recent years demand at Swiss stations has grown steadily, with the growth of leisure.

By 1991 cycle hire facilities were offered by no less than 600 Swiss stations, that is, most stations in the country (Froitzheim, 1990), but the dominant operation is now branded as Rent-a-bike with 250 outlets in Switzerland, and 120 in Austria. The Japanese have gone further to handle the major demand for cycle access from the station to final destination, and have fully automated dispensing 'towers' (Hanaoka, 1997), which deliver a fits-all utility bike, when a user swipes their stored charge card to debit for a bike hire. The consideration of free bikes is one which should not be taken lightly. There is a history of failure in achieving a scheme through which the bikes can be maintained and users made responsible for their return.

The US bike stations mentioned previously offer electrically assisted bikes and cafés as well as community initiatives to maximise the income streams and activity around the centre throughout the day. In the UK the only bike hire facilities linked to any public transport to date have been leisure-based, the largest being at Brockenhurst in Hampshire, where two operators compete on the same site, and there is a slow development of further locations.

8.7   Cycle parking at suburban stations and bus/tram stops

Informal cycle parking at any location is a good indication of demand – as are NO BIKES signs. Away from central stations, the popularity of cycling to the station will drop significantly where the time taken to take out the bike, and park it on arrival, reduces or negates the time advantage over walking. This will be a range of about 500–700 m as German research shows (Bracher *et al*, 1991), although it is more likely to be measured by time, to take account of the route topography. The bike really begins to show its benefits in outer suburban areas, and a range of up to 2 kilometres is generally considered acceptable. In the UK this can be seen from travel patterns for the stations on the main London–Southampton rail line, where there are very high levels of cycle use at stations like Woking. Here the catchment is large, users benefit from riding further to a better-served station, and the cycle access and parking enable a more direct route to the train. The option is aided by a greater preponderance of relatively spacious one family houses, with garages and often garden sheds, making it easy to store the bikes. On this basis, encouragement of bike-and-ride can increase tenfold the catchment areas of local stations (Bracher *et al*, 1991).

By contrast, in residential areas closer to the centre, public transport frequencies, especially of buses, are often higher, and thus the bus can fill the cycle range journey requirements without the hassles of parking, or keeping bikes in flats or other homes with difficult access and limited space.

Minimal transfer distances and times between cycle parking and the waiting area can encourage greater public transport use. Cyclists know they can rely on finding a secure and convenient place to leave their bikes, without having to resort to railings and poles. Better cycle parking at stations can also encourage greater train use by cyclists (Gyukits *et al*, 1986).

The Woking effect shows how demand for cycle parking at particular stations can also be influenced by the fare zone structure operating in certain city-wide or conurbation transport systems. In a further example, from Zurich, it has been observed that there is a heavier demand for cycle parking at the last station within a particular fare zone, as people use their bikes to make a rather longer trip to catch their train, rather than go to a nearer station and pay a higher fare!

8.8   Park-and-ride schemes – and the corruption of their use

Bike-and-ride provision may also be made in connection with bus-based park-and-ride, as well as at railway stations and tram stops. A good example of this in England is York, with a well-sited cycle parking area both near the main entrance to its major park-and-ride site to the south-west of the city, and close to the bus stops. It also has cycle path access. Experience in Oxford has shown two patterns of cycle use, those who come on their bikes and then continue by bus and also those who come by car with bikes in the boot, who then park their cars and complete their journey by bike!

Fig. 8.9 Cycle parking at BLT (Basle). The tram stop is visible from platform (platform in foreground). The covered stands are paid for through advertising revenue. Note the dominant utility bikes.

Not least of the many potential advantages from the promotion of bike-and-ride is a reduction in space requirements compared with the major park-and-ride developments now being pursued in many European cities. Park-and-ride developments may help to reduce pressure for commuters to drive into city centres, but still may be of limited value in reducing inessential car use by people who have no need to use their car in the course of their work. German experience shows that 60% of park-and-ride customers live less than 4 kilometres away, little more than the longest distance, 3.5 km, that most cyclists are prepared to ride to a bike-and-ride site (Froitzheim, 1991). Although park-and-ride is promoted as a very important way of luring drivers from their cars, experience shows that it may even attract people who previously made the whole of their journey by public transport. This has some interesting parallels with the observed user profiles polled from permit holders using bikes on bus/tram services in Portland, Oregon (interview regarding internal review on progress, 1994).

In so far as park-and-ride developments do succeed in enticing drivers away from their cars for part of their total journeys this may be at the expense of traffic and living conditions generally in the area around the park-and-ride site. Large park-and-ride sites entail the allocation of large tarmac areas for drivers, often on valuable land that might be better used in other ways. In Germany some sites now include special multi-storey car parks, and even smaller sites may themselves generate substantial extra traffic in the vicinity in peak hours, often in areas that have hitherto been relatively unaffected by the intrusion of motor traffic. Any time benefits that drivers gain from using public transport for the main part of their trip may be eroded by congestion when they return and want to drive away from the park-and-ride site. This spreading of congestion, unless linked to bus priority signals and other special measures, may also add to the risk that park-and-ride promotion extracts patronage from other public transport services, and retards their promotion and improvement. Froitzheim (1991) has commented that 'Park and ride is a luxury solution for relatively few drivers.'

Furthermore, improvements in terms of car access to park-and-ride sites may be at the expense of creating safer access for cyclists to the same bus or train stops. At many public transport stops of course there will simply not be room for official encouragement of park-and-ride, while bike-and-ride provision, needing only about one-tenth as much space per parked vehicle, may be much easier to accommodate. Indeed, encouragement of bike-and-ride may help to reduce the problems of unofficial park-and-ride that, with serious increases in traffic congestion, have often increasingly plagued suburban railway stations, often to the intense annoyance of local residents. Where bike-and-ride is promoted alongside official park-and-ride it is important to give preferential siting to the cycle parking provision rather than to locate cycle stands in peripheral parts of the car park where they will be inconvenient, insecure and thus poorly used by cyclists.

Preferential siting of cycle parking also gives a message to drivers of what is officially regarded as most important, especially if reinforced by good publicity and marketing. By contrast, badly sited cycle parking and indeed poorly designed cycle parking facilities in general, with unsafe, inconvenient and ill maintained cycle access, give a message to cyclists that their needs are not being taken seriously and that they are just second-class transport users. Bike-and-ride must be well publicised and signed, and its use will also be encouraged by a perception among drivers that they stand a poor chance of getting a car parking place.

8.9   Carriage of bikes on trains

The provision of cycle parking at stations may be a relatively problem-free form of bike-and-ride promotion compared with the improvement of arrangements for carrying bikes on public transport. In many countries, and perhaps especially in Britain, these arrangements have got much worse in recent years, with the relatively low investment in railways (and other public transport) and with the fact that the new rolling stock which has been introduced has often made the carriage of bikes impossible or extremely difficult, even outside peak hours.

Different services, local and Intercity, have their own regulations and restrictions, and the growing complexity of these has made the carriage of bikes on trains very difficult. These have driven away potential passengers and helped to generate a large increase in sales of cycle carriers for cars! In most continental countries the picture is not so bleak and indeed major improvements have been introduced. In the Netherlands, for example, the Nederlandse Spoorwege imposed severe restrictions and charges in order to discourage passengers from taking their bikes with them but has in recent years revised this policy, especially on

Fig. 8.10   Cab car of FS (Italy) push–pull set for local services. DB (German Railways) also opt to use the compartment behind the cab to take hanging bikes.

long distance trains. Denmark also has done much to lift restrictions, especially outside peak hours (Koop, 1990). The changes over the past decade have been significant, as revealed in the ECF report on bicycle carriage on mainland Europe's rail network (European Cyclists' Federation, 1994). Notably Die Bahn (DB) (German Railways) has seen a healthy return on accommodating cyclists, for a fee, on all trains, and carries details in English on their website. Similarly SBB/CFS/CFF (Swiss Railways) have experienced massive growth in international bookings from passengers with cycles. SNCF, the French carrier, whilst receiving a poor review from the ECF report, does have internal activity for delivering the service.

The enlightened attitude of DB has led to co-operation with the ADFC (German Cyclists' Club) in producing a report following their research into the most suitable designs of train for cycle carriage. This concluded that, apart from special accommodation on popular tourist routes, or for large parties, the ideal option is to stack bikes on supporting systems along the sides of a coach, where the seating can be folded away (or tips up) when not required. This concept of flexible space is often the only practical one which can be provided within the constraints put on UK rail operators by the need to maximise peak hour seating capacity on their investment-starved and thus limited fleet size.

In terms of urban cycling, the main theme of this book, the reduced capacity and inflexibility of modern train designs is regrettable, especially when the alternative option for use of bicycles on rail-based journeys is also discouraged by a lack of secure and convenient cycle parking at origin and destination stations. In

Fig. 8.11   Caltrain: low cost conversion for bike carrying: 24 or 48 bikes/train; 2000+ bikes/day; 68 trains.

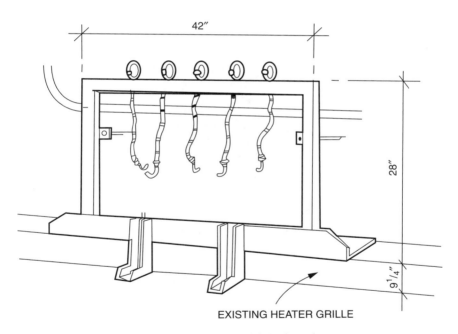

Fig. 8.12   Diagram of Caltrain rack.

continental Europe the ability to travel with a bike on local trains is dispelling the fears of some railway personnel about the delays for train timetabling from loading and unloading bikes. In the UK there has been a similar softening of attitudes on local services, with most frequent urban routes having no bike charges, or pre-booking of space, but generally imposing peak hour bans to maximise pas-

Fig. 8.13   The Calyx system can cost from €3200 (£2000) to install with tip-up seats. Scotrail class 158 DMU.

Fig. 8.14   Low cost UK system class 158 DMU. Arriva Northern trains. Up to three tip-up seats.

senger capacity. One feature which has been welcomed, however, is the provision of information, in the form of a National Rail leaflet collating details of all operators' charges, and rules, and sponsored by the Brompton Bicycle Company.

The US operator Caltrain has been cited in the FHWA report referred to above (FHWA, 1998) and has delivered dramatic results at relatively low cost, with the help of 50% funding by California's Mitigation of Air Quality scheme (CMAQ).

Over a three year period the number of bikes being carried daily rose by almost 1400% from barely 100 to nearly 2000, when the leading driving coach on their push–pull operated sets was modified to take six sets of four bikes in place of a four seat bay (see website www.caltrain.com).

Enthusiasm for combining bike and rail travel has seen one-third of the enquiries for season tickets on the new Altamont commuter route to San Jose asking about bike carriage or parking. Elsewhere, operators of intermediate Metro style systems such as WMATA (Washington DC) and BART (Oakland, California), have now quietly discontinued their initial permit system, which ensured that any cyclist travelling on the trains had an identity card, and a knowledge of the rules for using the trains. The progressive removal of a control system, as the operator gains confidence in user behaviour and system capacity, is a valuable tool for taking many schemes forward, in gentle and palatable stages.

In Germany, Berlin has shown a lead with incentives which carry financial benefits (Bracher *et al*, 1991), offering free bicycle travel for regular ticket holders, an important incentive for cyclists to buy season tickets, even if they are less regular public transport users. There has also been a general easing of restrictions on both its S-Bahn and U-Bahn (overground and underground urban railways), with several positive initiatives to accommodate more bikes, especially on the S-Bahn, which is particularly important for leisure traffic.

8.10   Carriage of bikes on trams and light rail

Some of the most impressive examples of arrangements for carrying bikes on public transport services are to be found in Basle in Switzerland. For some years tram drivers have been quite willing even in peak hours to carry folded bikes, and the carriage of bikes outside peak hours is now well established. In the case of two tram lines, 10 and 12, bikes can be carried at any time, along with prams and wheelchairs, in a special low floored compartment in the centre of the tram. Appropriate logos painted on the platform indicate where cyclists and wheelchair and pram users should wait to use these. The nature of operation of some routes allows the use of a special 'bike-and-ride' tram trailer, for carrying up to 25 bikes, that can be rented by groups at weekends on tram line 10 which runs out into the nearby attractive countryside of Alsace. The use of old tramcar chassis for special vehicles is an option only available on the older networks, where such fully depreciated but serviceable vehicles are to hand. In Stuttgart, a particularly hilly city, this is very much encouraged on the rack railway to the southern suburb of Degerloch, where a special platform designed to carry bikes, loaded by cyclists themselves, is carried in front of the driver on the line from Marienplatz in the city centre. A similar facility exists on the Postlingerbahn in Linz in Austria (Rauh, 1990).

New systems and those where modern vehicles require the traction and coupling systems to be compatible are less likely to offer this facility. In Germany the carriage of bikes on trams has faced more prohibitions but is now becoming

Fig. 8.15   Up to eight bikes can be suspended vertically in the centre section of this BLT (Basle) tram.

more common, and not just outside peak hours (Bracher *et al*, 1991). New vehicles on demonstration, as shown by Siemens in Barcelona in 1997, even include a bike carrying space designed in.

US and Canadian tram systems also carry bikes and in many cases the initial operation has been closely controlled, with later relaxation. Portland (Oregon) has provided a video for instructing MAX light rail users on correct behaviour. No incidents have been recorded.

In 1992 the ATA (Advanced Transport in Avon) project commissioned Sustrans to report on the cycle/tram interface, and data was collated from Europe and the US, although the aborted project meant that this work was not widely circulated. Subsequently both the MVA and the CTC produced reports in 1998 which tended to overlap, although neither covered the issue of the external interaction in great depth. Recent papers have included work by Richard Smith of Sustrans in Newcastle (Smith, 2001), tracking the on-vehicle issues.

8.11   External interface between bikes and trams

The major issue affecting cyclists is not the parking and access to tram stops, or the on-vehicle facilities, but the interaction between the cyclist and the tram tracks where there is on-street running. On-street running of trams has traditionally been feared by some cyclists who worry about getting their wheels caught in the tracks

(Davies, 1989) but this is not the way in which a cyclist is brought down. Essentially the railhead and flange guide offer narrow and smooth surfaces of a width comparable with the maximum acceptable for tar banding, another slippery surface when wet. Cyclists will instinctively seek a path that crosses an irregularity such as a dropped kerb, or any street ironwork, including tram rails, with no side thrust on the contact patch between tyre and road. In designing a new system the on-street sections need a rigorous safety audit to eliminate such hazards.

The danger is thus greatest where cyclists cross rails at an acute angle, and reports from Sheffield and Croydon, where extensive street running takes place, highlight a very limited number of locations where practically all falls take place. Sheffield casualties have been analysed (Cameron *et al*, 2001). By comparison the Metrolink, Birmingham and Blackpool systems, with very limited on-street sections, tend to have casualties where the tram lines cross the main traffic flows. The potential for risk is thus much reduced, as casualty rates show.

The flangeway also can grab a tyre or retain leaves and other debris which will 'lubricate' the adjacent surfaces. The proper tramway rail profile is, however, preferred to the assembly of check rails and infill, but Corus – the UK supplier of most rail profiles – advises that some categories include special preparation of the rail to limit flange squeal (i.e. greater slipperiness) and a bonded polymer fillet which compresses to form the bond between road surface and track. The necessary gaps at point blades and frogs are further details to keep cyclists clear from.

Observing the standards for installation it is noted that HSE (Health and Safety Executive Railway Inspectorate) set a tolerance for rails to be flush with the road surface, or up to 6 mm above it. In Freiburg where there are high levels of cycling, the rails tend to sit slightly below the road surface, allowing the rail to be bridged by the tyre contact patch, which is then more likely to be bearing on the higher friction road surfaces on either side.

In Freiburg there is also regular flangeway cleaning which removes debris that will create slipperiness for both the cyclist and the tram, as it is displaced or washed out onto the railhead. A further issue is the monitoring of railhead wear, as the flange guide will gradually end up sitting above the railhead with a narrow edge which will act like a machinery skid on any items rested on it. Evidence to date is that there is no closely managed regime to monitor and maintain the track profile. Even if one existed, there is a risk of divided responsibility for maintaining the rails (operator) and maintaining the road surface (roads authority). The interface between the two seems to escape any attention or responsibility for its condition.

A number of proprietary street track units do offer a soft infill for the flangeway, but this does not hold up for regular use with speeds above a slow crawl. These do not last in use on a transit system – as evidenced in several US cities. Fundamentally the design of areas where cyclists meet with trams and tram tracks should either provide a clear direct crossing of the track at a safe angle, or provide sufficient road space for the cyclist to take a suitable line through the hazard.

Fig. 8.16 Tram track in Freiburg being cleaned. Note that the rails are set lower than the street surface.

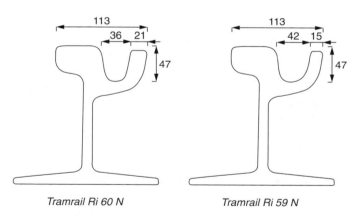

Fig. 8.17 Common tram rail sections used in the UK. The Ri60 is more usual. Reports suggest that it can trap tyres in the 32–37 mm range. If this affects the front tyre the bike stops, throwing the rider off.

Sheffield's most notorious location puts the cyclist in a position on the carriageway where they cannot easily claim space to swing out and across the rails, and are thus forced to cross at an unsuitable angle.

Toronto's system has a large number of routes and resulting 'turn-backs' in the Central Area as lines criss-cross the grid system of streets. Strong enforcement of parking bans at junctions and the functional concrete slabs along the outer edge of the rails provide a space for cyclists to pull to the right, and avoid hitting the turning rails at too shallow an angle. A zone of differing colour and texture, which defines the swept path for parked vehicles to observe, and reminds the cyclist they are about to cross a tram rail, if they stray further, is very important.

It is important when considering any traffic management measures to give advantage to trams and, as discussed later, buses, to consider the potential for applying the same priority to cyclists, as for example with traffic lights that let buses or trams start ahead of motor traffic at junctions. Links between 'cell' systems in city centres, designed to restrict through movements by private motor traffic, as in Bremen or Gothenburg, can be maintained for cyclists as well as to make the journeys more direct and convenient. Generally the speed of cycling with no stop dwell times matches closely that of city centre bus and tram, a point not lost on the cyclist who keeps leap-frogging a bus at each stop.

Finally there is the management of passenger boarding at stops which Toronto – and other systems running along the centre of the street – address by making overtaking moves on the inside of a stopped tram illegal.

8.12   Carriage of bikes on buses and taxis

Comprehensive consideration of the potential for combining cycling and public transport also needs to include other means of public transport such as buses, ferries and taxis. This is especially important in cities with less dense rail networks. There have been several experiments in Europe, although not always whole-hearted and successful, in the carriage of bikes on buses, either in the boot, on prongs or in trailers, for both leisure and commuting purposes (Brunsing, 1990; Karl, 1986). Cities like Bremen and Brunswick in Germany have introduced successful experiments in converting buses to carry bikes for weekend leisure trips. A similar experiment, supported by the former Countryside Commission, ran from the English

Fig. 8.18 SYPTE Dennis Dart with bike rack.

Fig. 8.19 Express commuter coach services.

Fig. 8.20 Cardiff Bus vehicle tows 24 bikes on trailer (Sundays – pre-bookable). The loading time and safety dictate the service only accepts cycles at specified stopping points.

Fig. 8.21 Rochester, New York was one of the first east coast systems to adopt the Sportworks rack. This is an all-stainless version.

Fig. 8.22   This bike is inside an Optare Solo, supported by and secured to the wheelchair backrest.

Fig. 8.23   1976: early rear racks on the San Diego Bay Bridge crossing service.

cities of Liverpool and Manchester in 1990 for recreational bike and bus access to the Peak District National Park. This proved to be unviable.

However, existing conditions of carriage for rural buses in Scotland in the 1980s still permitted carriage of bikes in underfloor lockers, and bike-carrying

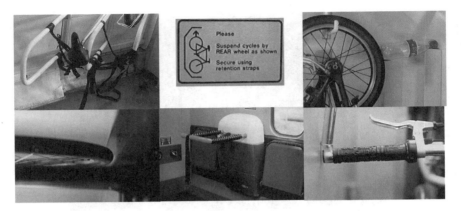

Fig. 8.24 Various niggling problems with some current designs.

special vehicles had been used on the Forth and Dartford river crossings in and since the 1960s. The real developments came in the mid-1970s, when US transit agencies acted to resolve problems in policing estuarial bridges which were not open to cyclists, but the gains from breaking the law were immense, often saving over an hour of cycling.

The earlier facility in the UK for old-style single deck buses survives in the acceptance of cycles on many express coach routes. The latter actually works well in the UK, given the typical distances involved, and Oxford Tube/Oxford Citylink coaches both offer this facility, providing a useful customer choice – for reaching the intermediate stops (e.g. Lewknor) and not having a problem of having no provision for parking a car, or facing a very long walk.

The way in which existing vehicles can be used with minimal cost to trial a service has seen widespread informal use of space in low floor accessible vehicles, especially in rural areas.

The US systems developed over many years and are reported in the paper *Integration of Bikes and Transit* (TRB, 1994) and the FHWA review *Improving Conditions for Bicycling and Walking* (FHWA, 1998). Key milestones were the 1990 trials of the Phoenix (Arizona) system, reported for the Surface Transportation Project, which in turn encouraged Sun Line (Palm Springs) to trial home-made units, which led to their current levels of use (8000 bike trips/month with 24 bus minimum vehicle requirement to cover services). In 1993 Sportworks won a competitive contract to supply a fast loading rack for 100% fitting to the 1200 bus Seattle fleet. Current estimates are that with 100% equipped fleets in many cities, over 25% of the US urban bus fleet is fitted with front mounted cycle racks, and a conservative estimate is that 0.5 million 'bike on bus' trips are made every month. There is no reported safety concern nor any significant incidents for the US system, but in Europe, vehicle legislation and a different operating environment (more pedestrian activity around buses and streets) has prevented a direct transfer of the technology.

In Denmark it is compulsory for taxis to be able to carry a bicycle, and in Germany it is common in some very hilly places for cyclists to ride their bikes downhill and, when going uphill, to take their bikes by taxi! The UK specifications for accessible cabs and the London Hackney Cab minimum requirement of being able to carry two bales of hay (for the horse) which has left us with such a spacious design, all act to offer cabs in most cities which can accommodate up to two bikes and riders with cycling luggage, inside the vehicle.

Even though the main arrangements for enabling cyclists to carry their bikes on buses (and some long distance trains) have tended to be for leisure rather than daily urban cycling there is an important relationship between the two. People who take up leisure cycling may feel safer than launching out on their own in a busy urban area, especially if they are part of a group. This experience can then give them the confidence and skill to try riding a bike for some of their daily trips as well.

But the bike bus in the mould of the US system, with capacity for just 5–10% of the full passenger load, is focusing UK operators on the potential of the daily user, and, with some taking advantage of the bike to return downhill, or to avoid a long wait on an infrequent service and thus being premium fare single journeys, they can be a welcome market for filling off-peak services in the same way that buggy users are. For example when low floor vehicles were introduced, 400% was added to the traffic volumes on Tyneside. One caveat very relevant to urban operation is that loading and unloading bikes for short city centre journeys is a time consuming detail, which many US operations ban for this very reason, with an added benefit that for these journeys the bus' overall length (i.e. with a trailer) is reduced for pulling in at bus stops.

For the same reasons, the use of trailers is generally inappropriate for urban bus routes. These are difficult to manage (illicit riders) and take time to load. They also require extra space and driver skills, especially when used in the restricted space of city streets. They are, however, popular for longer runs to leisure destinations (for example the Cardiff–Brecon bike bus in South Wales) but often require a second person to manage loading at stops.

Interior carriage is possible too, and a few services are permitting this officially (Chingford–Picketts Lock) with extensive informal use of low floor easy access vehicles, and of course the projects based on wheelchair carrying buses, used at weekends for leisure destinations. These have rarely been commercially viable, as they do not run as a commercial/supported bus service in the first instance, and a bike-carrying one as an incidental additional feature. Unfortunately the design and operation of UK buses can create conflicting demands – for example with just one entry/exit route often with a 'throat' limited to 80 cm by the front wheels, or with grab rail spacing only wide enough for one person to comfortably pass through. This is an inhibiting factor on the internal carriage of bikes in the bus saloon, where there is great (but unproven) fear of injury and damage claims. The 1990 Peak Park and 1996–1999 Devon bike bus services used half of a two door model Leyland National vehicle, from which seats had been removed, so that the bikes could use a separate entry from general foot access. The operation of a 45 seat vehicle with around half the seats removed on just one round trip/day was not a recipe for making a profit.

8.13   Sharing road space with buses

It is now generally expected that a dedicated lane for buses will include cycling, and even in locations where bikes are officially not permitted, such bans are almost impossible to enforce, and the users adapt their behaviour to respect each other's different performance and needs for road space. The US practice for a curb lane is to make this 12 feet (3.7 m) wide, rather than the 10 feet (3.05 m) into which a 2.55 m bus or truck just fits. Research in the UK on road narrowings (Field, 2001; DETR, 1999b) makes it clear that the better facility will improve both perceived and actual safety details.

Shared use of bus lanes by bikes and buses is widespread, and the variety of acceptable practices allows for some flexibility in how the local demand is handled. The guided bus facilities now operating in Leeds and Ipswich also tempt cyclists to ride the slabs, and bypass the same traffic queues. These may have to be treated like a tram system, and generally the implementation of current schemes has included parallel cycle route development (Ipswich and Leeds – early proposals for Edinburgh have now been shelved).

8.14   The paperwork – audits and conditions of carriage

Clear conditions of carriage are desirable to establish confidence for the user in knowing what they can expect from a service, and a positive position for staff to apply in the field. Just as good information on timetables and actual running times greatly increases patronage, clear and simple information on how and when you can travel, or where you can park or hire a bike, will generate use. Many people are unaware of how to use a bicycle as part of a travel itinerary other than by cycling the whole journey, and officers often express surprise that cyclists are using the bike lockers overnight, when councils have installed these.

Permit systems for on-vehicle carriage offer a useful polling list of users, as well as increasing the operator confidence that users know how to behave, or use the bike securing equipment. It is also easy, at the beginning of any innovative trials, to be able to contact users and advise on the adjustments which may be needed in the light of operating experience. Permits and other controls set at maximum initially can be relaxed as many have been, but open access is considerably more difficult to restrict as with the restrictions of UK bike carriage on trains. This followed a period of free, almost unrestricted access.

Rental agreements for lockers or secure parking offer similar opportunities, and peace of mind for the operator. The free carriage of cycles on buses and trains has a key problem when it comes to justifying future systems, and perhaps underwriting the existing service. To get data on this Devon County Council required a €0.16 (10 pence) ticket to be issued for bikes, which then recorded the journey to aid analysis of both total trips and the popular stages of the route.

8.15   Promotion

Where there is vigorous promotion of public transport as a choice with obvious linkages between modes the connections will begin to be made, but there will be some subtle pressures which influence the users' decision to cycle, and the distances ridden. Nigel Coates conducted an interesting analysis of behaviour changes in Oxford, tracking the switch to the bicycle when bus fares increased and restricted access by car increased the journey costs in varying terms of time and money according to the mode from which the cyclist transferred. This change was further driven by additional provision of more secure cycle parking in the city centre (Coates, 1997). The Nerima Ward in Tokyo exhibits the same variations – no car parking capacity close enough to the station and too far to walk, with the option of the bus having an average speed lower than cycling, and costing more.

The actual infrastructure in terms of kilometres of cycle route is of much less importance. Oxford has, in the last 20 years, done more than most British cities to restrain the growth of motor traffic in its city centre. Transport policies have included the provision of bus lanes and the development of peripheral park-and-ride sites to reduce traffic on narrow central streets. While motor traffic overall in the city has increased steadily, the levels in the city centre have remained constant, but the amount of cycling has doubled (Mathew, 1990).

An even more impressive example is Freiburg, just west of the Black Forest. Concern about the dangers of *Waldsterben* (death of trees) from motor traffic emissions produced particularly strong local pressure in the mid-1980s to reduce motor traffic growth. Transport policies were then changed, with greatly increased investment in both cycleways and new and better tram lines. Within a few years the shares of the total modal split accounted for by both cycling and public transport grew significantly, with cycling increasing to 27% of all trips (Gobel, 1988). The share of daily trips undertaken by private car has also been reduced by the introduction in the early 1990s of increasing varieties of 'environment tickets' giving large discounts for regular public transport use in the city, and now extended to the whole region. These have been based on similar concepts pioneered in Basle and other Swiss cities and, like those, have been heavily marketed. As in other areas of cycling policy, this underlines the need for policy makers not to take a too blinkered approach but to cultivate a wide awareness of the opportunities for promoting cycling in the course of implementing measures with other primary purposes. These may in turn provide the cycling access or other facilities without demanding a special cycling budget.

8.16   Conclusion

Given the increasing evidence of the very adverse social and environmental effects of the domination of Western cities by the motor car it is essential that cycling and public transport are seen as complementary rather than rival modes of transport. The aim of this policy should be to build on the complementary strengths of both 'environmentally friendly' modes of transport to offer an attractive combined package that will offer a serious alternative, in terms of convenience and door-to-door travel times, to car travel. This means, above all, good quality and well-designed cycle parking, with easy connections to train, tram and bus services. It also means paying attention to the scope for other possible improvements in each link in the chain of a trip. Furthermore, it means encouraging the provision of cycle hire at transport interchanges and bicycle stations. It also means increasing rather than decreasing the opportunities for passengers to take their bikes with them on both short and long distance trains and also buses and trams, and introducing imaginative and well-publicised initiatives to encourage this. Finally, in terms of shared use of road space, cyclists, buses and trams can benefit, especially in a context of general restraint of private motor traffic.

8.17   References

ALRUTZ, D, FÉCHTEL, HW and KRAUSE, J (1989), *Dokumentation zur Sicherung des Fahrradverkehrs*, produced for the Bundesministerium für Verkehr by the Bundesanstalt für Strassenwesen, Bereich Unfallforschung, Bonn.

APEL, D and LEHMBROCK, M (1990), *Stadtvertragliche Verkehrsplanung: Chancen zur Steuerung des Autoverkehrs durch Parkraumkonzepte und Bewirtschaftung*, Berlin, Deutsches Institut für Urbanistik.

BOUWMAN, M (2000), *An Environmental Assessment of the Bicycle and other Transport Systems*, paper presented to the Velo-Mondial Conference, Amsterdam, June, Utrecht, Fietsersbond.

BRACHER, T (1991), *The Bicycle and Public Transport Mode*, paper presented to the Velo-City '91 Conference, Milan, November.

BRACHER, T, LUDA, H and THIEMANN, H-J (1991), *Zusammenfassende Auswertung von Forschungsergebnissen zum Radverkehr in der Stadt*, Forschung Stadtverkehr, Band A7, Bergisch Gladbach/Berlin/Bonn, Bundesministerium für Verkehr (Federal Ministry of Traffic).

BRUNSING, J (1990), 'Public transport and cycling: experience of modal integration in West Germany', in RS Tolley (ed), *The Greening of Urban Transport: Planning for Walking and Cycling in Western Cities*, London, Belhaven Press (first edition).

CAMERON, IC, HARRIS, NJ and KEHOE, NJ (2001), 'Tram-related injuries in Sheffield', *Injury* 32 (4), May, 275–7.

COATES, N (1997), *Parking Policy and Bicycle Promotion in Oxford*, Velo-City Conference, Barcelona, September.

DAVIES, D (1989), 'Light rapid transit: implications for cyclists', *Cycle Touring and Campaigning* (CTC), June/July.

DETR (Department of the Environment, Transport and the Regions) (1999a), *Traffic Advisory Leaflet 11/99: Improved cycle parking at South West Trains' Stations in Hampshire*, London, DETR.

DETR (Department of the Environment, Transport and the Regions) (1999b), *Traffic Advisory Leaflet 15/99: Cyclists at Road Works*, London, DETR.

DSB (Danish State Railways) (1990), *Cykelparkering og cykelcentre: et idekatalog (Cycle parking and cycle centres: a catalogue of ideas)*, Styregruppen vedr. cykelparkering, Copenhagen, DSB.

DSB (S-Togsdivision) (1991), *Handlingsplan for forbedring af cykelparkering ved S-stationer (Plan for promotion of cycle parking at S-train stations)*, Copenhagen, DSB Styregruppen vedr. cykelparkering, Copenhagen.

ECF (European Cyclists' Federation) (1994), *Bikes on Trains*, Copenhagen, ECF c/o Danish Cycling Federation.

FHWA (Federal Highway Administration) (1998), *Improving Conditions for Bicycling and Walking: Best Practices Report*, Washington DC, FHWA.

FIELD, R (2001), *Are You Being Squeezed at Road Narrowings?*, Godalming, Surrey, CTC.

FROITZHEIM, T (1990), *Fahrradstationen an Bahnhofen: Modelle, Chancen, Risiken*, Düsseldorf, ADFC-Nordrhein-Westfalen.

FROITZHEIM, T (1991), 'Kuckkucksei und hassliches Entlein: Park-and-Ride contra Bike-and-Ride', *Verkehrszeichen* 4, 15–20.

GOBEL, N (1988), *Freiburg: Kommunalpolitische und verwaltungstechnische Durchsetzung der Verkehrsumteilung*, in Fahrrad-Stadt-Verkehr, Darmstadt, Conference Report.

GRABE, W and VERBAND OFFENTLICHER VERKEHRSMITTEL (VOV) (1985), 'Das Fahrrad als Erganzungsverkehrsmittel des OPNV', *VOV-Schriften* 1.68.2, Reihe Technik, Düsseldorf.

GROLL, M (2000), *Falco Lecture Second Prize Winning Paper*, paper presented to the Velo-Mondial Conference, Amsterdam, June, Utrecht, Fietsersbond.

GYUKITS, H et al (1986), *Planung und Betrieb von Fahrradboxen im VRR*, Gelsenkirchen, Germany (Verkehrsverbund Rhein–Ruhr).

HANAOKA, S (1997), *Present Bicycle Traffic Situation in Japanese Cities*, paper presented to the Velo-City Conference, Barcelona, September.

HANEL, K (1986), *Sachexpertise Infrastruktur – Servicestationen – Fahrradverleih – Fahrradhandel, Modellvorhaben Fahrradfreundliche Stadt*, Berlin, Werkstattbericht 16, Umweltbundesamt.

HANTON, A and MCCOMBIE, S (1989), *Provision for Cycle Parking at Railway Stations in the London Area*, London, London Cycling Campaign.

HEYNEN, P (1992), *Travelling Cleaner*, London, Transport 2000.

HOLLADAY, D (2001), *Bikes with Buses and Other Intermodal Opportunities*, paper presented to the Velo-City 01 Conference, Edinburgh/Glasgow.

KARL, J (1986), 'Mit dem Fahrrad in Bus und Bahn', *Verkehrszeichen* 3, 46–50.

KOOP, E (1990), 'On the recent engagement of bicycles and trains in Denmark', in N Jensen (ed), *Velo-City '89 Proceedings*, Copenhagen, National Agency for Physical Planning.

KURANANI, C and BEL, DD (1997), *Bicycle Parking in Tokyo: Issues, Policy and Innovation*, Washington DC, TRB (Transportation Research Board) Meeting.

LUERS, A (1985), *Reiseantrittwiderstande, speziell der Einfluss wohnungsnaher Abstellmöglichkeiten auf den Verkehrsanteil des Fahrrades* (Resistance factors at the start of journeys, with particular reference to the availability of cycle parking facilities near residences) in *Perspektiven des Fahrradverkehrs: Internationaler Planungsseminar auf Schloss Laxenburg bei Wien*.

MATHEW, D (1990), 'New way ahead for Oxford: a balanced transport policy', *The Surveyor*, 175 (5126), 3 October.

MINISTRY OF TRANSPORT, PUBLIC WORKS AND WATER MANAGEMENT (1999), *The Dutch Bicycle Master Plan: Description and Evaluation in an Historical Context*, The Hague, Directorate-General for Passenger Transport, Ministry of Transport, Public Works and Water Management.

NATIONAL CYCLING FORUM (2001), *Model Conditions of Carriage: Accommodating the Bicycle on Bus and Coach*, London, Department of the Environment, Transport and the Regions.

NEWMAN, PWG and KENWORTHY, JR (1989), *Cities and Automobile Dependence: A Source Book*, Aldershot, Gower.

PATSCHKE, W (1987), 'Entwicklungspotential der Systemverknupfung Fahrrad und Schiene', in K Schafer-Breede (ed), *Kombinierten Personenverkehr: Bike and Ride, Beiträge und Materialen zu einer Tagung in Essen 1986*, Bremen, Allgemeiner Deutscher Fahrradclub (ADFC).

PLOEGER, J (1988), 'Access to the city', in T de Wit (ed), *Report of Proceedings of the Velo-City '87 Conference, Groningen*, Ede, Netherlands, CROW.

RAUH, W (1990), *Das Fahrrad im Verkehr: Wegweiser zu einer fahrradgerechten Organisation des Strassenverkehrs*, Vienna, Arbeitsgemeinschaft umweltfreundlicher Stadtverkehr (ARGUS) and Verkehrsclub Osterreich (VCO).

REPLOGLE, M (1983), *Bicycles and Transportation: New Links to Suburban Transit Markets*, Washington DC, Bicycle Federation of America.

SCHAFER-BREEDE, K (1987), *Kombinierter Personenverkehr: Bike and Ride, Beiträge und Materialen zu einer Tagung in Essen 1986*, Bremen, Allgemeiner Deutscher Fahrradclub (ADFC).

SMITH, R (2001), 'Bikes on Light rail: Mind the Gap in Provision', paper presented to the Velo-City 01 Conference, Edinburgh/Glasgow.

TRB (Transportation Research Board) (1994), *TCRP Synthesis 4: Integration of Bikes and Transit*, Washington DC, Transportation Research Board.

TSCHOPP, J (1988), 'Bike and ride and the introduction of the green reduction card: Basle, a success story in stimulating use of public transport and the bike', in T de Wit (ed), *Report of Proceedings of the Velo-City '87 Conference, Groningen*, Ede, Netherlands, CROW.

TSCHOPP, J (1991), *Massnahmen für den Veloverkehr, Herzogenbuchsee*, Basel, Switzerland, Verkehrs-Club der Schweiz (VCS).

UMWELTBUNDESAMT (1983), *Fahrrad und Oeffentlicher Verkehr, Modellvorhaben fahrradfreundliche Stadt*, Werkstattberichte No 4, Berlin, Umweltbundesamt (Federal Office of the Environment).

WELLEMAN, AG (1991), *The Netherlands National Cycling Policy and Facilities for Cyclists at Signalled Junctions*, paper given to the meeting of the Local Authorities Cycle Planning Group, York, 17 May.

9

Planning for more cycling:
the York experience bucks the trend

James Harrison, City of York Council

9.1   Introduction

Historically, York has long had a high level of cycling. The city is compact, meaning that most trips are well within cycling range. It is also flat and enjoys a relatively dry climate. The conditions are therefore right for cycling. The city's primary employment, based on the traditional factory settings of the rail and chocolate industries, has also tended to favour cycle use. Like the rest of the UK, however, cycle usage greatly declined throughout the 1960s and 1970s.

Huge growth in car use over the same period resulted in the inevitable congestion problems common to most urban areas. This was aggravated by the ancient street and land use pattern of the city. In the 1970s plans were brought forward for a major inner ring road project which would have required the demolition of a large number of properties including many listed buildings. Fortunately this was eventually turned down following a public inquiry.

In the mid-1980s a new political administration, faced with increasing public dissatisfaction with congestion, undertook a comprehensive review of the city's approach to transport planning. This came at a time when academic and professional opinion was shifting away from the prevailing attitude of building additional capacity to solve congestion. An opportunity therefore existed for a radical rethink of policy which led to the adoption of a new Transport Strategy in the late 1980s.

9.2   The Transport Strategy

The new Transport Strategy, for the first time, explicitly sought to promote cycling, together with other socially and environmentally sustainable transport modes. At the heart of the strategy lies a hierarchy of road users which was (and still is) used to guide both design and funding priorities. The hierarchy (now slightly modified) is set out below:

1   pedestrians;
2   people with mobility problems;
3   cyclists;
4   public transport users (includes rail, bus, coach and water);
5   powered two wheelers;
6   commercial/business users (includes deliveries and HGV);
7   car-borne shoppers and visitors;
8   car-borne commuters.

Note: 'pedestrians' includes especially those with mobility difficulties.

Another key part of the overall Transport Strategy was a Cycling Strategy. This formally set out the council's policies to promote cycling. The two main objectives of this were to improve conditions for existing cyclists, especially their safety, and to encourage a transfer of journeys from private cars to cycles. To achieve these objectives a policy framework was developed. The fundamental policy was to develop a city-wide network of safe cycle routes. This proposed network was adopted in 1988 and is complemented by the provision of secure cycle parking facilities. Since the adoption of the network, its implementation has been overseen by a dedicated cycling officer whose remit also involves further development of cycle policy and advice to other departments throughout the council as well as other organisations throughout the city.

In 1996, following the reorganisation of local government, York became a unitary authority with new boundaries including a rural hinterland around the built-up area. This new area is 10 times the area of the old city and its population of 175,000 compares with the old city's 100,000. This new responsibility meant that the adopted network needed to be reviewed to include the new area. The full proposed network is some 200 km of which 100 km has been constructed.

9.3   Funding of cycle infrastructure

The majority of funding for new cycle facilities now comes from central government via the Local Transport Plan settlement. Thanks to the government's recent change in emphasis in terms of transport, the money available for cycling projects is much enhanced meaning that progress will take place more rapidly over the next few years than it has in the recent past. Roughly £600,000 (€970,000) is programmed in the financial year 2001/2002 for cycle specific schemes whilst many other schemes included in the capital programme will also have significant benefits for cyclists. This figure compares with an average of nearer £200,000 (€323,000) over the first few years of the Strategy.

Another important source of funding for cycle infrastructure is new development. Whenever developments of any significant size are proposed, developers are required to implement access improvements to the site. Emphasis in York is for pedestrians and cyclists to be considered first in this respect. This allows more cycle schemes to be built than would otherwise be possible but also means that some sections of the network get built in isolation due to the location of a particular development. Forward planning of the proposed network means that advantage can be taken of these developments and that these improvements eventually tie in with the city's strategic network.

9.4   The cycle network

The network of proposed routes was developed to link all major generators and attractors of cycle trips (see Fig. 9.1). These include residential areas, schools and colleges, workplaces, shops and leisure facilities. The ultimate aims of the network are to allow all cycle trips to be made on safe routes and for these trips to be as convenient by cycle as they would be by car. This means that routes need to be direct and continuous as well as safe. To achieve this, a combination of on-road and off-road routes have been identified. Additionally, locations posing particular difficulty for cyclists, such as large junctions, are either avoided or modified.

Over the years it has become apparent that cyclists do not form a single homogeneous group. There are those who are prepared to take a small detour to avoid busy routes or to get away from traffic altogether and those who will take the shortest and quickest route regardless of traffic levels and speeds. On top of this, many people who enjoy using quiet, traffic-free routes during daylight hours are loath to do so after dark. This is due to the feeling of insecurity engendered by remoteness. A dual network is therefore emerging to cater for different types of cyclists and different trip purposes. Unsurprisingly, the most popular routes are those which are traffic-free, well lit, in full public view and more direct than the alternative road.

Since the proposed network was developed, 100 km have been constructed. This comprises completely traffic-free routes, off-road tracks adjacent to the highway and on-road cycle lanes together with special facilities at junctions. This is made up of roughly 60 km off-road and 40 km on-road.

As with most transport networks, the cycle network is largely made up of radial routes to the city centre. This is not entirely the case, however, as cycle trips by their nature tend to be fairly local and, therefore, are often entirely suburban. Nevertheless the city centre remains the focus for very many trips and so the major routes follow this pattern. The single most important feature of the central area is the large pedestrian priority zone which encompasses over 30 streets in the city centre. This is known as the footstreets. In this zone, motorised traffic is strictly controlled throughout the day. During the core shopping hours it is excluded completely with the exception of certain permit holders who are allowed on specified routes. In effect the zone is pedestrianised at these times. Cyclists are allowed special privileges; outside the core hours they are allowed

KEY

① Existing Cycle
   Route Network

② Proposed Cycle
   Route Network
   (inc. Millennium
   Routes)

③ City of York
   Council Boundary

   Former York City
   Council Area

   Built-up Areas

④ River Ouse /
   River Foss

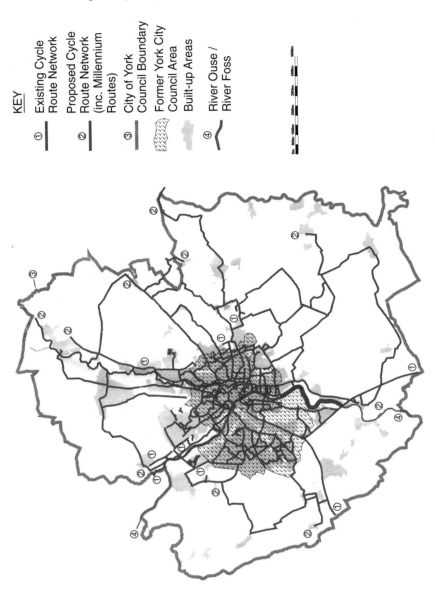

Fig. 9.1   York cycle route network, showing the extensions in the expanded area.

to cycle unhindered within the footstreets. Within the core hours, cycling is not permitted but cycle parking facilities are provided both within the footstreets and, in larger numbers, at the edge of the zone. There are over 1000 secure cycle parking spaces in the central area.

In addition a virtually traffic-free route has been provided for cyclists skirting around the edge of the footstreets. This serves as the equivalent of an inner ring route allowing cyclists to traverse the city centre on the north–south axis without doing battle with the busy traffic that inevitably surrounds the pedestrian priority area. This route is made up of several residential streets which are closed to through traffic except cycles. As well as providing the north–south route it also gives good access to a large cycle parking area on the edge of the footstreets.

9.5   City centre bridges

Also in the central area, but outside the footstreets, is the River Ouse. This has three bridges all of which are available to all traffic and which are very busy. Cyclists have not been specifically catered for on these bridges due to the lack of available width. This means that in crossing the river, cyclists have to mix with general traffic. A recent scheme which recognises this and conforms to the council's policy of promoting cycling is on Skeldergate bridge – one of the three central bridges. Until recently this bridge had a two lane approach to traffic signals at one end. One of the general traffic lanes was removed to allow the introduction of a dedicated cycle lane on the approach to the signals for cyclists turning right and a lane which allows left turning cyclists to bypass the signals altogether. This had inevitable consequences for the traffic capacity of the signals leading to much greater queuing on the approach to the bridge. However, several months later this has now been accepted by the motoring public and has huge benefits for cyclists. This is typical of the approach now being taken to help cyclists. At many locations the only way to provide meaningful help for cyclists is by reallocating road space away from motorised traffic. A piecemeal approach has been taken to avoid an outcry from motorists which would be likely to result if many such schemes were introduced at the same time.

This approach is more radical than would have been politically acceptable a few years ago. At that time, space was often made for cyclists by either allowing them to share footways or by narrowing footways to allow for a cycle lane to be introduced. Recent changes in both local and national emphasis now allow these difficult decisions to be made. To create space for cycling, the presumption is now that reallocation is away from motorised vehicles and not from pedestrians.

9.6   The 'magic roundabout'

Another recent innovation in design is the Heworth Green roundabout (Fig. 9.2). This has replaced a complex priority junction on a main radial about 1 kilo-

Fig. 9.2   The 'magic roundabout' at Heworth Green.

metre from the city centre. Emphasis in the design was again to create a junction which would allow safe passage for the 800 cyclists who use the junction each day. Roundabouts on fast, busy roads are not renowned for their cycle-friendliness so traffic signals were first considered. Modelling of a signalled junction revealed that they would result in an unacceptable amount of queuing for motor vehicles. A roundabout design was then developed which incorporated annular cycle lanes. When this has been tried elsewhere in the UK it has been conventional to set the cycle lanes tight against the kerb on the outside of the roundabout, a position which puts cyclists away from the field of view of following drivers and in a vulnerable position at each entry and exit point. In this design the cycle lanes have been moved nearer to the centre of the roundabout. This puts cycles more directly in the sight line of drivers.

Additionally, on the approach to each exit, the lanes split into two so that it is clearer whether cyclists are turning off or continuing round the roundabout. As well as the innovative cycle lanes, the roundabout also features a geometry which encourages low vehicle speeds – the so called 'continental design'. This effect is further enhanced by the cycle lanes which make the roundabout look smaller. Average entry speeds have been brought down to 28 km/h compared with 50 km/h before. The roundabout has been in place for eight months and no accidents have been recorded in this time. In the five years prior to its introduction, 18 accidents had occurred. Queuing has increased significantly on the inbound approach to the roundabout but has been kept within reasonable limits. Outbound queuing has been reduced.

Fig. 9.3   The Millennium Bridge. Source: Richard Bryant/Arcaid.

9.7   The Millennium Bridge

The highest profile addition to the cycle network in recent years is the Millennium Bridge (Fig. 9.3). This crosses the river about 1 kilometre south of the city centre. This is the only river crossing south of the city centre before the outer ring road which is a high speed dual carriageway about 3 kilometres from the centre. Until the bridge was constructed therefore movement between the south–east and south-west parts of the city was difficult. The journey was both circuitous and involved busy roads including one of the busy central bridges mentioned previously. The new bridge for pedestrians and cyclists was built with funding from the Millennium Commission, the council, the Joseph Rowntree Foundation and local businesses. The bridge links important cycle routes on either side of the river and the network is planned to be further augmented by the construction of a major new route linking to the bridge from the western suburbs. The bridge was opened in spring 2001 and is already attracting 2000 users per day. Just over half of these are cyclists. Because of the difficulty in making these cross-river journeys before the bridge was built, it is expected that it will lead to a significant transfer of trips to cycling. Comprehensive monitoring is taking place to assess this effect.

Because of its high profile, the Millennium Bridge has helped to enhance the image of cycling in general.

9.8   Marketing

Marketing, publicity and promotion are vital elements of a strategy to encourage more people to cycle. Simply providing the infrastructure on its own is unlikely to have as great an effect. In York a number of initiatives have been undertaken. These include direct publicity for the network and promotional campaigns as well as associated activities to raise the profile of cycling.

Publicity for the network is mainly done in three ways. The first is to publicise individual schemes as they get built through press releases. The second is through a district-wide cycle route map. This not only fulfils the conventional role of a map but also makes people aware of the extent of the facilities. The map is made available widely and not just in the usual cycle outlets. In this way it is hoped to spread the message to non-cyclists. Information about the cycle network also appears on the council's website. This is proving to be a useful means of disseminating information.

The third is a city-wide cycling marketing campaign, targeting car drivers. The campaign uses media space that is most likely to reach the target audience, such as the reverse side of car park tickets and bus rears. Although there are many personal and societal benefits from increasing levels of cycling, listing them all in a communication campaign would lead to a confusing and complex message that would be unlikely to register with the target audience. Consequently, the campaign focuses on one motivation for cycling at a time. In 2000, the campaign focused on the health benefits, particularly reducing the coronary heart disease risk factor, which is Britain's largest single cause of premature death. Follow up interview surveys revealed that this campaign was successful in getting the message across to the car driving target audience.

A significant aspect of this campaign is that it has been run in partnership with the Selby and York Primary Care Trust. This is important because it moves away from the negative feeling that the council is trying to bully people into not using their cars towards a much more positive message.

9.9   Reducing social exclusion

An unusual and inspiring project which also involves partnerships with a number of agencies is the Recyclist Project. This project affords young unemployed people an opportunity to gain work experience in a supervised environment. Trained staff help them to refurbish old cycles which have not been reclaimed after being stolen. An unusual feature is that at the end of the five weekly sessions each person is allowed to keep the bike they have worked on. This promotes pride in the work they have carried out but also gives them more freedom to travel to jobs or education opportunities in the future. This project clearly has benefits across a range of fields including education and social inclusion as well as transport.

9.10   Maintenance

For cycle networks to be useful they need to be well maintained. This has proved to be a problem over the years as no system existed which was well suited to this task. Over the last few years maintenance has been carried out as part of the general highway maintenance work. This has its limitations as the staff are not familiar with the levels of intervention needed to provide for safe and comfortable cycling. This is added to by the fact that a system, and staff, dedicated to highway maintenance are not physically well equipped for cycle route inspection. Often urgent attention is needed but on a very small scale. For example, if broken glass is reported on a cycle path, it needs to be dealt with the same day. To employ standard highway maintenance staff and machinery is very difficult.

Over the last few months a new system has been set up which has had a profound beneficial effect on network maintenance and illustrates how the scale of the operation needs to match the scale of the task. Two part time rangers have been employed who are dedicated to cycle path maintenance. Between them they patrol the entire off-carriageway network using bikes equipped with trailers and hand tools. This enables them to carry out all day-to-day maintenance including cleansing, removal of broken glass and control of vegetation. They are also able to report more major problems to the highway maintenance team. The rangers have a high profile due to their trailers and high visibility clothing which means that users are aware that the maintenance issue is being taken seriously. It is now intended to equip the rangers with mobile telephones so that they can respond even more quickly to complaints and also contact the appropriate authorities in the event of any misuse of the network.

The value of this new arrangement cannot be overstated. The network is now consistently in good condition and free from broken glass whereas in the past many users were put off using certain paths due to these problems.

9.11   Monitoring

Monitoring of the network is important to both the council and central government. We need to provide evidence of the value of the network to ensure future funding but also to allow us to prioritise how future expenditure is allocated. The principal tool is surveying the level of usage.

Historically, monitoring has been carried out through manual counts. This has now been augmented by automatic traffic counters using induction loop technology. The benefits of these are twofold. First, they allow us to collect much more data than would be possible through manual counts. Secondly, they allow us to monitor trends and daily variations which are important due to the effects of weather; once per year we count all traffic at a cordon surrounding the city and also crossing all the river bridges. As this is a one day count only, and particularly as it takes place in October, it means that the cycle counts are very vulner-

Table 9.1   Recent trends in cycle casualties in York

	KSI*	Slight	All Severities
1996 to 1998			
All cyclists	15	109	124
Adult cyclists	13	89	102
Child cyclists	2	20	22
1991 to 1993			
All cyclists	38	109	147
Adult cyclists	33	87	120
Child cyclists	5	22	27

* Killed and seriously injured.

able to fluctuations caused by the weather. With the permanent counters in place, these data can be adjusted to take this into account. We are therefore able to see patterns emerging over time.

Every 10 years the national census provides reliable data on journeys to work. The last of these for which data is available was carried out in 1991. In 2000 we carried out household interview surveys to inform the Local Transport Plan process. This gave us an opportunity to update the journey to work data.

9.12   Safety

Whatever success the strategy achieves in terms of cycle usage, it is important that this is not attained at the expense of cycle safety. Table 9.1 shows recent trends in cycle casualties which reveal that safety has, in fact, been improved. It is particularly pleasing to note that the figures for KSI (Killed and seriously injured) have gone down by more than 50%.

9.13   Usage and targets

One of the main objectives of the Cycling Strategy is to increase usage. Until the installation of the automatic traffic counters, our most reliable source of data was the census information. This revealed that from 1981 to 1991 we had maintained a largely stable level of cycling to work. This, in fact, is thought to mask a drop in the early eighties followed by an increase later. For the built-up area (which made up the whole of York's administrative area until 1996), the journey to work figure was 22.1% in 1981 and 20.3% in 1991.

In 1996, following local government reorganisation, the boundary change means that we have inherited a much larger geographical area meaning that commuters with longer journeys are now included in the data. It also means that people living near to the new boundary are more likely to work in nearby

Table 9.2   Targets for cycle usage in York (%)

	2000	2006
Journey to school	5.2	7.2
Journey to work	18.6	20.6

towns and cities rather than in York itself. The 1991 figures were therefore recalculated to reflect the post-1996 boundary. This revealed that 15% of all journeys to work were by bike. The household interview survey carried out in 2000 showed that this figure had risen substantially to 18.6%. This is a very encouraging outcome.

Experience elsewhere has suggested that long term investment in cycle infrastructure and promotion is needed to induce a significant modal shift and this is borne out by experience in York. A significant point in this respect is that the first sections of the network to be constructed were inevitably isolated whereas over the last few years a genuine sense of a coherent network is beginning to emerge. This means that many journeys are now catered for on a continuous basis promoting confidence in even the least experienced cyclists.

An interesting monitoring exercise has just begun. This relates to the Millennium Bridge. Because this creates distinct, new journey opportunities, there is the potential for significant modal shift. To examine this properly we fitted automatic counters on all the paths in that vicinity over a year before the bridge was completed. These have now been complemented by counters on the bridge approaches themselves. In due course, we will carry out interview surveys on the bridge to assess not only origin and destination information, but also to ascertain how such journeys would have been made before the bridge was built.

We have five targets relating to cycle usage. The central ones are for the proportion of journeys to work and school. The current and target figures are shown in Table 9.2.

9.14   Future aspirations

The next few years will see big improvements in the cycle network. This is because of more generous funding from central government with a stronger emphasis on green transport issues. For York this means that we will spend roughly three times as much per year on cycle infrastructure over the next few years as we have in the past. We are concentrating on two specific areas. The first is to identify remaining gaps in the existing network and to fill these. This recognises the need for continuity, which was discussed above. In this context very good value can be achieved by the implementation of quite small schemes that overcome a particular difficulty in an otherwise continuous route. The second is to link the larger outlying settlements to the city centre. In both cases priority will be given to those schemes which have the greatest potential for modal shift away from cars.

9.15   Summary

Experience in York has shown that it is possible to promote cycle use whilst also improving cyclists' safety. Key lessons which have been learnt are that isolated cycle facilities will not affect people's modal choice on their own. However, sustained investment at a realistic level can bring about significant changes in people's travel habits. The cycle network needs to be based on strategic planning to ensure that coherent, continuous routes are created. Security has also been shown to be important – both in terms of personal security and safe parking facilities.

Finally, 'soft' measures to promote cycling can complement the physical infrastructure. Partnerships with other organisations have proved very worthwhile to enhance the overall image of cycling. Relating cycling to health, in particular, has allowed us to move towards a very positive message which people can relate to their own lifestyle aspirations.

10

Planning for cyclists in Edinburgh

Richard Williams, City of Edinburgh District Council

10.1 Introduction

In 1870 one of the world's first cycling clubs was established in Edinburgh and
five other clubs were formed in the Edinburgh area in the following 10 years.
Move on 100 years and people are still cycling. However, there was sufficient
worry about traffic conditions in the area that SPOKES, a cycling campaign
organisation, was formed in 1977. In its early years SPOKES successfully lobbied
for a number of traffic management measures. In 1986, European Road Safety
Year, its members converted a section of old rail route into a cycle path because
they despaired of the rate of progress of official action.

SPOKES lobbied for the council to establish a cycle team, devoted exclusively
to improving measures for cycling. During the 15 years since that team was established much effort has been expended by council staff and by campaigners to
improve the lot of the ordinary cyclist. This chapter details much of the results
of that effort. However, cyclists still represent a small proportion of road users.
In the early 1980s a 30 year blueprint for action was produced, of which a significant amount remains to be reviewed and progressed if still appropriate. Will
fuel prices, concern about the environment and still more cycle facilities encourage some people to restart cycling and others to cycle more often?

10.2 Conversion of old railway lines

During the nineteenth century Edinburgh was served by what became two railway
companies which competed for custom on both selected longer distance routes
and passenger and freight routes serving a variety of local suburbs. Even before

the Beeching era of the 1960s some services had been withdrawn. During the 1970s and 1980s the local authorities for the Edinburgh area negotiated with the British Rail Property Board and bought up many of the routes for recreational use by pedestrians, cyclists and sometimes horse riders too, or for other transportation purposes with works gradually undertaken so they could be used in the meantime by pedestrians and cyclists.

The first route of any substantial length was the route to the south-western suburbs of Currie and Balerno. The converted section starts at Slateford where the former railway had commenced with a junction on the Edinburgh–Carstairs–London line. In more historic times there had been a spur from Balerno back to the main line at Dalmahoy but that section had been partially infilled and built over or passed back to the adjacent landowners.

A lot of the original conversion work was carried out by means of a government employment scheme and their skills were limited to a whin-dust surfaced track with timber edging. Sections have been resurfaced over time as the actions of users and water led to deterioration. An opportunity to get the surface further upgraded to roads standard was not accepted by the then local member of the district council.

Some 1.5 km from the city end the route goes through a relatively tightly curved tunnel below part of Colinton village. Here lighting has been provided. Route users can see the remains of a couple of stations along the line of the route together with some of the sidings to former paper mills sited because of their proximity to the water power and resources of the Water of Leith. This route now forms part of the National Cycle Network Route 75 Edinburgh–Glasgow.

The then Lothian Regional Council bought up sections of the 'Innocent Rail Line' which first ran from St Leonards, 1.2 km south of the city centre southeastwards towards collieries in the Dalkeith area. The rail route first opened in 1832 with a rope hauled section through a tunnel under the fringes of the hill named Arthur's Seat. This tunnel is straight and when converted for use by pedestrians and cyclists lighting was installed. Housing developments have occurred in the last 15 years at the urban end of the route but a line of passage has been retained for pedestrians and cyclists, in part segregated from other traffic.

South-east of the tunnel the route is unlit where it passes beside Bawsinch Nature Reserve but lighting is provided beyond that where it passes between a travelling people's site and a golf course and then along the edge of housing schemes.

After 3 kilometres the rail line is no longer disused as it has become absorbed into what is currently a freight only rail loop through the southern suburbs and the cycle route follows a burn to Bingham. Most of this cycle route has a 3 metre wide asphalt surface. There are two at-grade road crossings in the section southeast of the tunnel, one of which has puffin traffic signals. This route now forms part of National Cycle Network Route 1/The North Sea Cycle Route. Those interested in industrial archaeology should note that just west of the Duddingston Road West crossing there is a small bridge over the Braid Burn which dates back to the beginning of the railway.

Whilst the southern rail loop through Edinburgh's southern suburbs is still in use the two Victorian rail companies formed a loop of routes in the northern suburbs which were, over a period of years in the 1980s and 1990s, mostly converted into a network of shared pedestrian and cyclist routes which, because the road crossings remain grade separated, presents a route length of approximately 8 kilometres. There are a further 9.5 km of spur routes which can be linked by short on-road sections.

The oldest route in north Edinburgh ran north from Waverley station in a tunnel at right angles to the present rail lines down to Newhaven where a ferry took rail passengers across the Firth of Forth to Fife. The tunnel, which lies below Scotland Street, is not open to general public use but a short adjacent open air section is available at King George V Memorial Park. The next short tunnel section is also not open because of a roof fall.

Beyond Broughton Road the line of passage is open, initially adjacent to a Tesco car park, then over the Water of Leith. After 1.4 km there was a grade separated crossing of rail lines. This has been removed and there is now a five way junction of cycle routes. Continuing northwards under Trinity Road tunnel one comes to the former Newhaven station. Initially passengers passed to the east of the station to the pier at Newhaven. Subsequently the rail line was extended to a new ferry terminal at Granton Harbour. Opening of the Forth Rail Bridge in 1890 resulted in the loss of long distance rail passengers on this route but freight use of this section continued until the mid-1980s. The northernmost section of this route follows the foreshore of the Firth of Forth and has good views up, down and across the estuary. There is one at-grade road crossing and a section of on-road route too. Alternatively from 'five ways' junction a loop will take the cyclist westwards along a grade separated route, largely in cutting, through residential areas, past a sports centre at Ainslie Park. A new bridge has been provided over Ferry Road near Crewe Toll as the road below was widened to accommodate traffic expected to be generated by development of the former rail-served gas works. Cycle facilities northwards to the Forth are also being provided along what used to be another rail line to Granton Harbour. Details of the precise nature of the facilities are expected to change as development proceeds.

Continuing south from the Ferry Road Bridge the cycle route passes through residential areas before crossing the Water of Leith at Roseburn, a bridge over the A8 to Glasgow and then a ramp takes the cyclist down to below the Edinburgh–Glasgow rail line (the former rail bridge there was dismantled several decades ago). Continuation of the former rail route is not currently feasible except for one short section parallel to Dalry Road as the former terminal station behind the Caledonian station has been redeveloped by offices and a road link.

An eastward turn at 'five ways' junction would lead the cyclist along an old rail line to Leith Docks. Part of the 3 metre wide route is lit, however the most easterly 0.4 km leading to Lindsay Road has only a blaes surface at present.

There are other, shorter lengths of former rail line which have been converted to shared use pedestrian and cycle facilities including:

- Balgreen to Corstorphine in the western suburbs;
- Craigleith to Barnton in the north-west suburbs;
- Warriston to Leith to the north of Edinburgh (see section 10.4);
- Seafield to Leith to the north-east of Edinburgh;
- Brunstane to Newcraighall in the eastern suburbs;
- Two short sections in Gorgie in the south-west suburbs;
- Newbridge to Dalmeny in the countryside west of Edinburgh;
- Dalmeny to South Queensferry, in the countryside north-west of Edinburgh.

The longer lengths of converted routes with roads passing over or under them are especially appreciated by parents with young children as they are traffic-free and so relatively safe for walking or learning to cycle.

10.3   A 30 year plan

In autumn 1981 the Highways Committee of Lothian Regional Council invited comments on a network of possible cycle routes. One consequence was a decision to commission John Grimshaw and Associates to prepare a long term plan for the development of cycle facilities in the region. This commission was at a time when the same company was also reporting on the conversion to cycle use of a considerable number of disused rail lines in various parts of Britain. The consultancy brought in their knowledge and experience of conversions already undertaken in the Bristol area as well as on-road measures to assist cyclists implemented both in Britain and elsewhere in Europe.

When completed the report had several themes:

- Four longer distance cycle routes linking Edinburgh to neighbouring towns, and in some cases to places outside the Lothians.
- Safe Routes to School proposals based on travel surveys to most of the region's secondary schools.
- Traffic management measures on a corridor running south from Edinburgh's city centre through the residential and educational areas used by students. This was designed to reduce the large number of road accidents involving cyclists.
- Traffic calming on side roads to permit a network of routes suitable for cyclists throughout the Edinburgh area.

This network and the other proposals could be phased in over a 30 year period if spending was sustained at a particular level. There were several volumes of appendices accompanying the report. These showed in more detail how the proposals could be implemented and provided information on costs and benefits.

During the 1980s political control of Lothian Regional Council underwent a number of changes, however support for cycle facilities was across all parties and included the personal commitment of the Highways Committee convenors and the opposition highways spokespeople. When a Cycle Project Team was established in 1987 work could truly begin on implementing the plan. Now in 2002, whilst the world has moved on could the work be said to be half completed?

The sections below on Safe Routes to School (10.7), traffic calming (10.8) and the longer distance cycle routes referred to in the section on new projects for a new millennium (10.11) describe the additional facilities provided in areas of work with which the council had been little engaged prior to commissioning of the Grimshaw report.

The report is a truly useful blueprint and one still worth referring to for considering the basic principles of what additional facilities should be provided and where.

10.4   Following the water

Edinburgh lies on the southern shore of the Firth of Forth. Through it run a number of watercourses which tend to run in a south-west to north-east direction. The River Almond runs through the rural western part of Edinburgh; the Water of Leith starts in the Pentland Hills south-west of Edinburgh and enters the Forth at the port of Leith, 3 km north-north-east of Edinburgh's city centre. The Burdiehouse Burn–Brunstane Burn runs through the southern suburbs and enters the Forth at Edinburgh's boundary with East Lothian. There are also several smaller, natural watercourses.

The Union Canal was built about 180 years ago and runs generally west from the city centre to Wester Hailes and then through the countryside to Ratho on its way to Falkirk. Further details on how cyclists have been enabled to use its towpath are detailed in section 10.11.

In this part of Scotland, following a watercourse means the gradients are not severe and there are pleasant nearby scenes and occasionally longer distance views. Some watercourses have long had a footpath along them. Routes connecting the mills in the Water of Leith are an early example of this. Where the path is of adequate width measures can be taken to allow for shared use by cyclists. In places ramps have been built alongside weirs or the path has been cantilevered out below bridges if the bridge piers leave little width of bank for a path. Over the years the local planning authority has had a strategy of providing a path along the Water of Leith, so when adjacent ground has been developed it has often been a planning condition that a section of path should be constructed along the river frontage.

As these routes were often seen as being of recreational advantage they have usually been constructed to a lower construction standard than those routes adopted by the local roads authority, i.e. with a blaes surface and little or no lighting. Both the planning and recreation parts of the local council have been involved in progressing works of this nature. Where the path adjacent to a burn runs through a park the park managers have an ambivalent attitude to permitting cyclists.

The Burdiehouse Burn–Brunstane Burn has considerable sections where cycling is permitted. Sometimes money has come from other sources such as the Roads Authority or the Green Belt Trust or developers to provide or upgrade the facility. A short distance to the north the Braid Burn–Figgate Burn has far fewer sections where cycling is permitted. Maybe because the path network in the Hermitage of Braid is promoted for walkers, most of the paths not alongside the burn are quite steep and narrow and there is a blanket ban against cycling.

This situation may be compared with the River Almond paths on the north-western edge of the built-up area of Edinburgh. Cycling is permitted here, although two sections near Cramond are so steep there are sets of steps so bikes have to be carried over these. Continuing upstream the river flows along the boundary of Cammo Park and the Dalmeny estate. Levels of use are less than in the Hermitage of Braid and in wet conditions the route is not very satisfactory for cyclists or for 'Sunday walkers'. Still further upstream the river forms the northern boundary of Edinburgh Airport. There is an informal route here; for-malising it has so far not been possible for security reasons. In dry weather this does not deter an intrepid cyclist, but in wet times the crossing of the Gogar Burn, where it flows into the Almond, is a hazard which only provision of a footbridge could overcome. The potential of this route which could be connected to the Newbridge–Kirkliston–Dalmeny railway path merits further efforts to achieve its establishment.

An extensive length of cycle route in Edinburgh follows the shore of the River Forth, although this is not continuous between Queensferry and Musselburgh. The Dalmeny estate at the western end does not permit cycling in its grounds so a shore-side ride has to start at Cramond from where there is a wide promenade to Granton. At Granton there is a lot of shore-side development until Granton Square is reached. From here until Newhaven a railway used to run along the coast and much of this has been converted into a route for pedestrians and cyclists. From Newhaven via Leith to Seafield there is more coastal development before another promenade through Portobello. Cyclist use of the busiest section of the promenade in summer is not appreciated by some of the local citizens. From the eastern part of Portobello there is occasional development on the seaward side of the main road so cyclists have to follow the road until they are over the bound-ary into East Lothian when they can regain the coast at Fisherrow harbour.

10.5   Following the bus

Out of all the people in the UK those in the Edinburgh area make about the most number of bus journeys per year. On most main roads there is better than a 10 minute frequency during the working day and in the evenings and on Sundays those routes with both town and country buses have a combined frequency of at least three to four buses per hour.

The first bus lanes were introduced in 1976 and were with the flow of traffic and 3 metres wide. There was no objection to bicycles also using these lanes. At peak times the lanes are quite full of buses so faster cyclists prefer to use the general traffic lane(s). Slower cyclists will find that they can overtake buses at bus stops, whilst passengers alight or board, but then they are overtaken just before the next bus stop. Use of the bus lane does segregate cyclists laterally from general traffic with the greatest risk of a traffic accident being at side road junctions when traffic crosses the lanes to enter or leave the main road.

In one early traffic management scheme a short section of a wide side road was made one way except for buses and cyclists and marked out as a contraflow cycle lane. The exit from the lane is under traffic signal control and is a left turn into a bus lane. The entry to the lane is a right turn from a two way street, but with some protection by a large splitter island.

After the initial set of bus lanes was installed on one corridor a year elapsed before the principle was followed through on three other routes. Twenty years elapsed before additional bus lanes were established under the marketing name of 'Greenways'. They generally had a green pigmented surface layer. Under the Greenways principles the needs of pedestrians and cyclists are also incorporated into the route action plans, so if in a particular section of route there was insufficient width for a bus lane to be marked out there might be enough width for a cycle lane. This would have a red pigmented surface layer. Installation of Greenways also saw the introduction of additional crossing facilities including some combined pedestrian/cyclist signalled crossings and traffic calming on some of the residential side roads.

The Greenways principle has been implemented in a number of phases, partly for financial and logistical reasons. Currently some of the corridors where bus lanes were first installed are being revisited and, in the light of current traffic conditions, the lengths of bus lane are being extended. In so far as many cyclists travelling to work will be using main roads for much of their journey, provision of bus lanes is a net gain for them too.

For some years there have been proposals for a guided bus link from the city centre westwards to the Airport and the Wester Hailes housing scheme. The successful introduction of bus lanes in these corridors has reduced the time advantage such a link would have and the original expectation that a private finance initiative could implement and run the facility has fallen through. In these original proposals there would have been a cycle route adjacent to the bus guideway. It remains to be seen what will eventually be implemented.

In some places you can find cycle racks near key bus interchanges, but Edinburgh only has these at most of its rail stations. In some parts of the country cycles are permitted on buses, perhaps carried on a special rack outside the vehicle. In Edinburgh this has only existed on certain of the buses on the cross-country route to Dumfries.

10.6   Other traffic management measures including cycle parking

Cycle parking is a lively topic for cyclists. Where can cyclists leave their bikes? Is the bicycle physically stable at such a site? Can it be securely locked there? Is there any weather protection? If the parking is on-street are traffic regulation orders needed to legalise the situation? Can, or should, anyone other than a public authority provide cycle parking facilities?

In the beginning of Lothian's cycle project there were few bicycle parking facilities within the public realm except informal leaning or padlocking of cycles to street furniture such as lighting columns, traffic signs, guardrails, etc. There were some cycle parking facilities at a few venues visited by cyclists but they would usually be of the 'wheel twisting' variety such as a slot in a concrete block or a piece of angle iron with a wire bracket to accept the tyre of one wheel at its half diameter. This latter style did have the advantage that a chain could be passed through the cycle wheel and the parking frame to decrease the likelihood of theft.

In 1982 the regulations governing the parking of vehicles in Edinburgh were amended to legalise the placing of cycle parking facilities on the footway at specified places. The number of cycle places designated in that traffic regulation order was four. This may be compared with the 1000+ cycles taken into central Edinburgh for travel to work and the many more taken there for education, shopping and other purposes.

The order was significantly extended in 1993 by designation of another 44 parking areas, each taking at least a pair of bikes and in some places as many as a dozen. This is still insufficient to meet the demand in overall terms or in locations.

SPOKES, the Lothian cycle campaigning organisation, and the Lothian Cycle Project Report recommended that cycle parking places should be of a style called the 'Sheffield rack'. This is a simple hoop whose uprights are separated by a distance similar to that between the front and rear axles of a bicycle, and whose horizontal bar is similar in height to the handle bars and saddle. A bike can then be leaned against it and a chain or U-padlock passed around each wheel, part of the cycle frame and the rack.

Over a period of time organisations other than the local authorities have provided cycle parking. The Tourist Board put up four racks near its entrance in Princes Street. Tiso's, an outdoor activity shop, placed cycle racks outside its premises in Rose Street before the council placed street furniture (including more cycle racks) along the complete length of Rose Street. The advertising agency, the Moore Group, is committed to providing some cycle parking amongst its items of street furniture which are being funded through a council licence to place a certain amount of advertising opportunities around the city. The council's guidelines for developers were amended in 1988 to require the provision of cycle parking in larger developments for retail, office and leisure purposes.

Other on-road facilities provided for cyclists include signalled crossings, cycle lanes, advanced stop lines and exemption for cyclists from a number of traffic prohibitions.

Over the years the statutory regulations concerning cycle crossings have been eased, so a tour of cycle crossing sites in Edinburgh could be a look back into the past on traffic management concepts. At the south end of the Meadows the cycle crossing of Melville Drive had traffic islands installed to provide a 'landing' for pedestrians and cyclists and also relocate the point where cyclists wishing to turn east or west would cross pedestrian routes away from the area under signal control. At the western end of the Meadows the then Scottish Office gave approval for a modified pelican crossing with the amber and green aspects showing a pedestrian and a cyclist. The red phase is a plain red aspect and not a 'red man'. A few years later, at the south-west edge of the Meadows there is a cycle and pedestrian route across Bruntsfield Links where the crossings of Marchmont Road and Whitehouse Loan were installed as toucan crossings, where pedestrians and cyclists cross together and any cyclist wishing to turn across the path of pedestrians is expected to use his/her judgement.

In Marchmont Road itself cycle lanes have been installed to help cycling students *en route* between places of study and residence. A red pigment surface has been laid and in this instance because of the amount of demand for kerb-side car parking the cycle lane is placed outside the parking areas. Most cycle lanes have been laid out on busier roads. In Edinburgh most are advisory, rather than mandatory, that is there is not a traffic regulation order which says only cyclists may use that area of the carriageway. In practice there was usually an existing traffic order restricting kerb-side operations during peak periods and sometimes during other parts of the working day too. Thus no additional police resources should be required to keep the cycle lane clear. The volume of cyclists, say 100 per hour, or accident rates, were some of the reasons why some roads were selected for installation of a cycle lane.

The cycle lanes in Buccleuch Street, at the eastern edge of the Meadows, contain advanced stop lines at the junction with Melville Drive. This gives cyclists a head start over general traffic when the lights turn in their favour and also makes them more conspicuous for following motorists (compared with the previous situation where cyclists would have been squeezed in between the motorists). The site layout is that originally developed in Bristol which involved fewer signal heads than are specified by government regulations. Once it was found to work here there has been a programme of progressively increasing the number of advanced stop lines across the city.

Cyclists are allowed into some of the semi-pedestrian areas such as Rose Street and The Royal Mile (between George IV Bridge and North–South Bridge). They are also exempted from a number of prohibited turns imposed partly for traffic capacity reasons. However, greater exemption for cycles from one way street orders has received little support from senior council engineers and police. There are a few examples where a short length of street is theoretically two way for all vehicles but only cycles are permitted to enter at one end.

10.7   Safe Routes to School

Over the past 20 years there has been a shift in Britain in the mode of travel used by pupils to get to and from school. Fewer are using the more sustainable modes and more are being ferried around by their parents. There are a number of reasons which could be used to justify this change. When the Lothian Cycle Project

Report was being put together 20 years ago one aspect it looked at was the road safety record of routes used by pupils to travel to school and how they could be improved, especially for those walking or cycling. The hope was that a network of safer routes would encourage more pupils to walk or cycle.

Most secondary schools in the Lothians were surveyed and pupils asked about their mode of travel and the routes they used. Police road accident records include the school attended by any casualties of school age and whether or not they were travelling to or from school at the time of the accident. The location of accidents helped highlight where improvements should be targeted, or alternative routes identified. Pupils were also asked to suggest improvements.

The proposals for each school are gradually being implemented and are involving school staff and pupils. Initially they were implemented as part of the capital budget for cycling; later a separate budget line was established. More recently there have been government initiatives to encourage improving routes to school and Edinburgh has taken up these opportunities.

Within Edinburgh there have also been a few schemes to help students travel to places of further and higher education. Edinburgh University was highlighted in the Lothian Cycle Project Report, but both Napier and Heriot-Watt Universities have cycling students and staff who are wanting improved facilities. Queen Margaret University College and Telford and Stevenson Colleges also have students, some on day release, who could be tempted to use cars if cycling is not made easier. In so far as they are currently located in the suburbs the traffic situation around them may not be too serious, but there are busy roads nearby, some with 40 mph (65 km/h) speed limits, so cyclist diversion to alternative routes has some attractions.

10.8   Traffic calming

A major feature of the Lothian Cycle Project Report was proposals for traffic calming in many of Edinburgh's residential areas. This was based on the situation that whilst most medium or longer distance cycle journeys might well use some of the off-road routes, or cycle priority measures on main roads, most journeys would start and finish on roads without special cycle facilities. Traffic calming in these areas would therefore reduce the severity of consequential injuries if accidents still occurred with the same frequency. If traffic did indeed slow down there might even be fewer accidents as there would be more time to react to potential accidents before the point of impact.

Prior to the mid-1980s there had been a number of traffic management schemes in residential areas but it had been realised that an area-wide approach was required as reducing traffic in one street could lead to increases in adjacent streets. The Lothian Cycle Project Report, and its appendices, included a variety of examples as to how speeding through traffic could be inconvenienced, slowed or diverted. However, initially there was not the support from senior council engineers, or residents, to pursue these schemes.

Within Edinburgh the next programme of traffic calming schemes was drawn up, in consultation with residents, for those areas with poor accident records. Another group of schemes was drawn up where there were relatively high volumes of 'rat running' through traffic. The expectation was that speeds would be reduced and some users would be diverted back onto the main roads surrounding the area. The third group of traffic calming schemes was prepared as part of the bus corridor Greenways scheme. Here, to discourage general traffic from leaving the main road, speed tables, ramps/speed cushions, chicanes, etc have been deployed prior to the completion of the scheme to aid bus flow along the main road.

By the mid-1990s many groups of residents had requested some form of traffic calming scheme for their area so these requests had to be prioritised before a cost-effective scheme could be prepared. The SPOKES cycle map for Edinburgh shows a network of minor road alternatives to major through roads which takes cognisance of where there are traffic calming devices.

10.9   SPOKES

SPOKES is the cycle campaign organisation for Edinburgh and the Lothians. It has now been campaigning to improve the situation for cyclists for more than 25 years. Its first campaign was to formally allow cyclists use of the wide footpath which runs across the Meadows park and links various Edinburgh University buildings to a popular area of flats. It then moved on to campaign for the engineers and planners of the then Lothian Regional Council and City of Edinburgh District Council to think about the needs of cyclists when preparing their schemes.

In the early 1980s the local Labour Party was persuaded to include a promise to establish a Cycle Team for the region along the lines of a team that had been built up in the Greater London Council. However, a year after the election there was still no one employed by the council exclusively on cycle-related activities, so the politicians were called to account and the Director of Highways was prevailed upon to set up a three person Cycle Unit.

Dialogue between SPOKES members and staff of the Unit enabled the SPOKES criticisms to become more constructive and informed. The enthusiasm of SPOKES members was also enrolled to assist in reporting road defects through production of a form/postcard. Council staff also assisted in providing information for a cycling map of the city whose profits helped SPOKES run their affairs. A number of conferences have been run jointly by SPOKES and the council at Scottish level. They co-operated too when the UK Local Authorities Cycle Planning Liaison Group meeting was hosted in Edinburgh, and most recently when the international Velo-City 2001 cycling conference was held in Edinburgh for four days before transferring to Glasgow.

SPOKES members regularly scrutinise planning applications and proposed traffic regulation orders to see how amendments might be made to secure benefits for cycling. As a lobby organisation it is one of just two very effective bodies operating city-wide. There are now regular liaison meetings held with council staff and chaired by a councillor, and the council assists in a number of publicity seeking events such as hosting a breakfast for cyclists who are travelling to work during National Bike Week.

10.10   A network approach

There is a philosophical question on how best to frame a strategy to encourage cycling. Should one start by improving the facilities, and if so where first?

In Edinburgh much of the early investment was put into opening up off-road facilities, partly because the opportunities were there to convert disused rail lines and provide routes along a number of the watercourses. Initially the routes were fragmented, but time has enabled some to be connected together and a few cyclists were willing to divert to these cycle paths from parallel road routes.

Cycle lanes (and bus lanes) on selected main roads enabled another set of routes to be drawn up and key problem points identified where action was desirable to increase route continuity. Where cycle lanes and cycle paths crossed was it easy to transfer from one to the other?

Meanwhile, progression of Safe Routes to School schemes has resulted in local improvement schemes and eventually a sufficiently detailed skeleton of all types of route was built up that it was possible to see where there might be significant gaps in a city-wide network. The Scottish Executive's current advice in *Cycling by Design* is for authorities considering cycle facilities to prepare for a network from day one.

10.11   New projects for a new millennium

For cyclists in the Edinburgh area four major projects have been progressed to celebrate the occasion of the millennium:

- Linking Edinburgh to the UK National Cycle Network.
- Opening in June 2001 of the North Sea Cycle Route, linking Edinburgh with 70 counties/local authorities in seven countries.
- Establishment of the Millennium Link, linking Edinburgh and Glasgow by means of the Union and Forth & Clyde canals.
- Formal completion and signing of the Water of Leith project which helps people enjoy the nature and other surroundings of this river.

The infrastructure provided by these projects extends the opportunities for local cyclists to explore Edinburgh and its surrounding countryside as well as enabling cyclists from further afield to visit the city.

The National Cycle Network routes through Edinburgh are Route 1, which enters from Midlothian in the south and leaves for Fife in the north-west, and

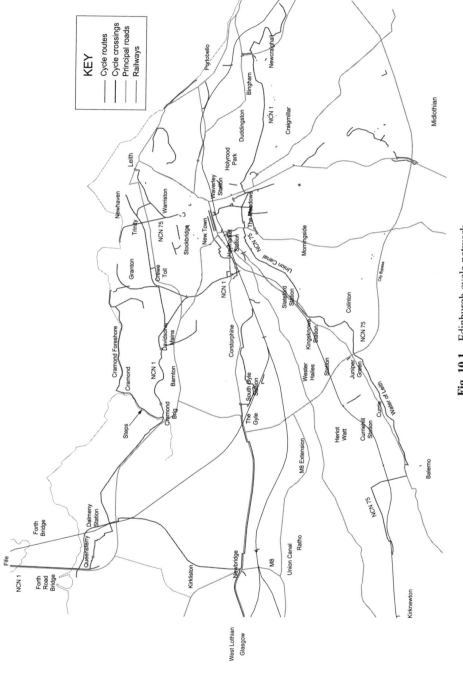

Fig. 10.1   Edinburgh cycle network.

Route 75, which starts in Leith and enters West Lothian on its way to Glasgow (see Fig. 10.1). Whilst there was relatively little new construction for these routes there have been some additional on-road facilities and signing to connect routes through the city centre. The Royal Bank of Scotland has also funded a number of 'Milestone' sculptures at various positions throughout the city (see Chapter 7).

In the Edinburgh area National Cycle Route 1 coincides with the North Sea Cycle Route. For cyclists who do not have several months to do the whole route, Edinburgh is a useful starting point for people arriving by train from their home base. In mid-2002 a daily ferry began crossing the North Sea from Zebrugge to Rosyth, just a short distance away on the northern shore of the Firth of Forth. Maybe cyclists will find this a useful transport link.

For many years British Waterways Board has insisted that cyclists riding along canal towpaths carry a permit. Quite a few cyclists use the Union Canal towpath to travel into the city centre although it is narrow and has poor sight-lines under bridges. A few kilometres from the start, at Slateford, there is an aqueduct carrying the canal over the Water of Leith. The towpath is narrow and made from setts which slope towards the water. Cyclists are advised to walk with their bicycles at this point, or they or any passers-by might find themselves in the canal! There is another narrow aqueduct towpath at the western boundary of Edinburgh where the canal passes over the River Almond. In between, at Wester Hailes, the canal had been piped through the housing scheme until the Millennium Link project paid for its reinstatement and the provision of half a dozen new bridges over the canal. For the time being the towpath surface is of fair quality in the urban section within Edinburgh, although there are puddles on damp days. In the rural section beyond the city bypass the ease of passage along the towpath may depend on when the verge was last cut or any hedges trimmed. The village of Ratho, some 20 km from the city centre, is a pleasant place to pause, with a waterside inn and places to sit.

The Water of Leith millennium project improved the signing along this popular watercourse. In places, both between Colinton and Balerno, and between Canonmills and Leith, people had already been following disused rail lines along the river valley but there are alternative routes possible in most places. A Visitor Centre has been established at Slateford.

10.12   Cycle policies

Over the years council policies towards cycling have been contained in the Regional Structure Plan and the annual Regional Council Transport Policy and Programme submissions. The latter has now been superseded by a Local Transport Strategy.

The 1985 Structure Plan stated that the regional council would introduce measures to improve the situation for pedestrians and cyclists, including the provision of longer distance routes in collaboration with the district councils. The policy of safeguarding disused rail lines within Edinburgh was also stated; outside

Edinburgh studies into safeguarding would be undertaken only if development proposals came forward.

By 1994 the next Structure Plan added in provision of cycle parking. Traffic management measures were also included as one of the ways to encourage cycling. About this time the region developed its 'Moving Forward' transport policy to meet development pressures in the area and the potential effect upon the environment of the area.

After the introduction of single tier local government in Scotland, Edinburgh still worked with its neighbours in formulating transport policy for the area. The Local Transport Strategy, published in autumn 2000, put 6% as the target for use of cycles by all trips in 2010 with a greater proportion for cycle to work. The council also set a target for reducing cycle accident rates by 20% by 2005, from the 1994–98 average.

10.13   Benchmarking

'Best value' is one of the portmanteau phrases in use in public service in the UK at present to cover quality of service, efficiency and value for money. Whilst any activity and place could demonstrate its uniqueness, the process of benchmarking encourages the making of comparisons with a view to establishing best practice.

The CTC (Cyclists' Touring Club) proposed a benchmarking exercise among the UK's local authorities and three series of exercises have been or are being held. Edinburgh took part in the first series together with Glasgow and eight English authorities. Representatives spent two days in each area looking at successes and problems and meeting some of the local practitioners. Topics of especial interest in Edinburgh were the new Local Transport Strategy, cycling's place in that strategy, cycling's relationship with public transport, road user safety audits and how Edinburgh maintains the variety of infrastructure detailed earlier in this chapter.

Edinburgh scored highly for policy and cyclist infrastructure and accident reduction strategy. Less high were the scores for publicity/promotion, education and partnership working.

10.14   Cycle use

Travel statistics from the 2001 National Census have still to be made available. However, figures from the 1991 census showed an increase for cycling to work in central Edinburgh over the 1971 and 1981 figures; see Table 10.1. Table 10.1 shows the increasing proportion of cyclists travelling to work plus the targets stated in the former regional council's 'Moving Forward' policy document.

The city council also has a series of regular counts to monitor the effects of its transport policies. During the 1990s quite a number of people who used to

Table 10.1   Travel to work in Edinburgh (%)

	Census results			'Moving Forward' targets	
	1971	1981	1991	2000	2010
Car, motorcycle	29.1	38.2	45.8	47	36
Public transport	46.5	41.3	32.1	34	41
Walk	19.9	16.9	14.5	16	18
Bicycle	0.7	1.4	1.8	3	5
Other	3.8	2.2	5.8	–	–

Table 10.2   Travel to central Edinburgh, morning peak survey results

	1986		1991		1996		2000	
	No	%	No	%	No	%	No	%
Car	27 097	36.6	26 214	38.1	22 873	41.1	24 240	39.8
Bus + rail	30 932	41.7	29 212	42.5	22 397	42.0	25 662	42.1
Walk	assume	15.7	assume	14.5	6 176	11.1	7 154	11.7
Bicycle	684	0.92	583	0.85	695	1.2	845	1.4
Other	3 745	5.1	2 792	4.05	3 557	4.6	3 011	4.9
Total	74 090		68 772		55 698		60 912	

work in the city centre moved out to offices in Leith or on the western edge of the city. However, the number of cyclists had picked up by the middle of the decade and has continued to increase; see Table 10.2.

The National Cycle Strategy targets of further increases in cycling will require greater changes in cycle use than those seen hitherto. From which modes will the new cyclists come? At present in Edinburgh there is a 'war' between the two major bus companies so fares are cheaper than in many surrounding towns and frequencies are relatively high. Are there still sufficient numbers of car users taking short journeys for whom cycling will become more attractive as car use or parking becomes more difficult?

11

Nottingham

Hugh McClintock, University of Nottingham

11.1   Introduction

Nottingham has had a reputation as one of the foremost British cities for cycling since the early 1980s and, before that, as the home of Raleigh, the bicycle manufacturer. This reputation was increased by the involvement of Nottinghamshire County Council, with Nottingham City Council, in a series of special cycling projects, particularly the largest of the government-assisted major cycling projects in the 1980s, the Greater Nottingham Cycling Project (McClintock *et al*, 1992).

In contrast to other cities with above average commitment to cycling such as York (see Chapter 9) and Cambridge, Nottingham has much lower levels of regular cycle use, little more than the average for Great Britain (Nottingham City and Nottinghamshire County Council, 2001). This is particularly true for Nottingham City which has a population of 287,000, less than half of the 644,000 (1998) (Nottingham City and Nottinghamshire County Council, 2000) in the Greater Nottingham conurbation which dominates the county of Nottinghamshire, particularly its southern half (Ricci, 2000) (see Fig. 11.1). The suburbs of Nottingham, outside the city boundary, have a total population of 219,000.

As the Local Transport Plan (LTP) Progress Report in 2001 showed (Nottingham City and Nottinghamshire County Council, 2001, Table A3), only 1.9% of trips in the city are by bike (compared with 1.4% for Great Britain) and only 2.5% in Greater Nottingham. This is due to higher levels in some suburban parts of the conurbation such as Beeston, in Broxtowe borough, which lies a short distance west of the main Nottingham University campus, in a relatively flat part of the area. The figure for the county as a whole is slightly higher, 3.0%,

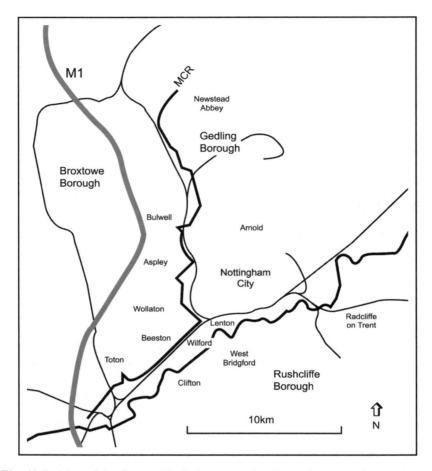

Fig. 11.1   Map of the Greater Nottingham area showing main place names and local authority boundaries.

reflecting the higher levels in some places such as the flat and compact market town of Newark in the east near the Lincolnshire border.

These figures are for all journeys, which probably underrates levels of use of bikes for work journeys. For example, as Table A4 in the same report shows, comparing modes for work and non-work journeys, there was almost twice as much cycling for the latter as for the former, in both the city and the county, 3.4% as against 1.9% (for the city) and 5.8% as against 3.8% (for the county). The same was true for the LTP area (Nottingham City, all of Broxtowe, Rushcliffe and Gedling boroughs and Hucknall in Ashfield District), i.e. 4.8% of trips to work as against 2.5% of all journeys. For other modes the proportions for work and non-work journeys are much more similar, with the exception of walking where the proportion of trips to work (8.7%) in the LTP area is well under half that for all journeys: 20.3%. It should also be noted that the proportion of trips to work locally (24.3% for the city and 26.5% for the county) is significantly higher than the average for Great Britain (19.0%), meaning that cycling is that much more important than might be assumed by focusing on all trips or non-work journeys.

Cycling to school, even secondary schools, is generally low: 1% of secondary school pupils' trips in Nottingham in 1999/2000 (Nottingham City Council and Nottinghamshire County Council, 2001, p Cviii). For Rushcliffe borough the figure was double, i.e. 2% in line with the national average, and only in Broxtowe was it much higher, 9%, in line with the generally much higher levels of cycling in a few wards in that part of the conurbation, already mentioned.

11.2   Development of the Greater Nottingham cycle network

The series of efforts to provide cycle facilities in Greater Nottingham since the early 1980s have been extended since the mid-1990s by the increased involvement of local district councils, particularly Rushcliffe borough council. This, the most southerly district of the county, is a mainly rural area but includes the important and recently much expanded suburb of West Bridgford lying just south of the Trent. Until the late 1990s most of the cycling initiatives were taken by the county council, with a lesser but still important role for the city council. Since then the districts have taken an increasing interest, in response to political changes and the changed national policy context, although it is fair to say that some still lag more than others. Gedling, for example, in the hillier north-east of the conurbation has been slower to show commitment than both Rushcliffe and Broxtowe. The involvement of the district councils has helped to achieve further stretches of river and canalside paths, in particular.

Important in the evolution of the Greater Nottingham cycle network has also been, belatedly, the Highways Agency, in respect of trunk road schemes. Getting commitment from the Highways Agency and their predecessor, the regional office of the former Department of Transport, was often very difficult in the 1980s and early 1990s (McClintock et al, 1992). This did change by the late 1990s, resulting in progress being made after much delay on some very worthwhile schemes such as the A52 trunk road cycle path between West Bridgford and Radcliffe on Trent. However, getting commitment from the Highways Agency still often remains difficult, with several long proposed local trunk road cycle schemes not yet implemented, despite the changed national policy context. The fact that the Agency staff since 1995 have been based in Birmingham and are now no longer locally based may have contributed to this.

Despite these continuing problems it is true to say, as the Local Transport Plan for Greater Nottingham claimed (Nottingham City and Nottinghamshire County Council, 2001, p 40), that the Greater Nottingham cycle route network is one of the most extensive in Britain with over 100 kilometres now in place. Many of the ideas for cycle routes came originally from proposals in the 1980s, in particular from Pedals, the local cycling campaign (Pedals, 1981). The extent of the network was reinforced with the completion in 2000 of the Inverness to Dover Millennium Cycle Route, part of the Sustrans National Cycle Network. Some 24 km of this route passes through Greater Nottingham, between Hucknall in the north and Toton on the Derbyshire boundary in the south-west, just east of Long Eaton.

Giving an exact total length figure for the cycle network is difficult not just because a number of cycle paths and shared paths are short (though still often useful links) but also because several routes include substantial lengths of quieter back street roads. Although these are usually signed when they form part of a designated cycle route it can be the case that other quiet roads become used as feeder routes though not officially signed as cycle routes. Safer cycle crossing facilities of main roads can encourage this pattern.

The provision of on-carriageway cycle lanes in the Nottingham area has been fairly rare, in comparison with off-carriageway cycle paths and shared paths, but a substantial number of bus lanes have been implemented since the mid-1990s, totalling by 2001 13.9 km in the city and 4.4 km in the county. Almost all of these are open to cyclists.

Despite the completion of the Millennium Cycle Route (MCR), which in Greater Nottingham included some upgrading of existing cycle facilities as well as new ones, and also the extension by the late 1990s of the existing cycle paths on the outer ring road along its entire length to Daybrook on the A60 in the north, it remains true, as the LTP commented, that 'the provision of cycling facilities remains distorted with most of the cycle route development concentrated in the flatter south and west with comparatively little development in the north and east.'

Plans for a series of feeder routes to the MCR will soon help to do more to overcome this bias, which probably reflects a number of factors including lower levels of cycling on the north side of the city, weaker political pressure, generally hillier terrain and the more densely built-up nature of the area with less scope for good quality cycle path provision. A greater emphasis on well-designed traffic calming, as well as greater funding for toucan shared crossings and on-road schemes like advanced stop lines and contraflow cycle lanes has now begun to erode this relative neglect. The advent of the MCR in the late 1990s certainly gave a boost to this change. Indeed it was in part of the inner areas north of the city centre that some of the early local traffic calming schemes were introduced, for example Forest Fields and Hyson Green. Some of these early schemes were designed with insufficient attention to the needs of cyclists but this had improved by the early 1990s with some defects modified.

Only recently have firm proposals been drawn up for promoting cycling on Nottingham's ring of inner boulevards in areas like Radford and Lenton. Cycle lanes on several of these should soon help further to reduce the gap in cycling provision with the more southerly parts of the conurbation, provided that their value is not eroded by domination of moving and parked cars and inadequate arrangements at junctions where the real challenges lie.

In terms of physical provision there have also been extensive efforts to provide cycle parking, including, since 1993, cycle lockers in three of the city centre multi-storey car parks. Cycle parking also perhaps is still relatively neglected in the northern part of the conurbation. As the LTP also commented (p 40)

> Cycle parking is generally concentrated in the City and district centres and other large travel generators. Secure cycle lockers have been provided at a number of City Centre car parks. As part of travel plans a number of large employers have considerably improved cycle parking and changing facilities for staff.

11.3   The Greater Nottingham cycle-friendly employers project

The Cycle Challenge project promoted by central government in the late 1980s focused on ways other than cycle routes for promoting cycling. This was a very important further boost to the encouragement of cycling locally, reinforcing the value of the earlier major cycle route network project (Cleary Hughes Associates, 1999; Cleary and McClintock, 2000a, 2000b; DTLR, 2001). This project received €360,000 (£225,000) from the former Department of Transport's (DOT) Cycle Challenge budget in 1995, the second largest of the 62 awards. The involvement of groups representing the public, private and voluntary sectors – together with a commitment to match the Cycle Challenge grant with local funds – was crucial to the success of the Nottingham bid and has also been crucial to its broadly successful outcome. The original government grant was matched by equivalent sums from the employers concerned and some have made further allocations, including some to off-worksite cycle facilities nearby.

The objective was to increase the extent to which people cycle for commuting journeys, and for official work trips such as site visits. The project officially commenced in April 1996, when funding from the DOT was released, to run for approximately two years. In the event, both implementation and evaluation took considerably longer than envisaged. Eight major employers in the Greater Nottingham area were involved as the project partners: Nottingham City Council, Nottinghamshire County Council, the University of Nottingham (see Fig. 11.2), Nottingham Trent University, Queens Medical Centre (QMC) – a large general and teaching hospital, Clarendon College – a centre for further education (and since merged with another local FE college to form New College Nottingham), the Boots Company – the city's largest employer with its head offices in Nottingham, and Experian (formerly CCN) – a credit referencing agency. Between them, at the time of the project, they employed over 32,000 people, and if students are included the numbers using the sites (and thus potentially targeted by the scheme) rise to around 77,000.

While the resources from Cycle Challenge were invested exclusively in workplace measures to facilitate cycling, Nottingham's eight Cycle Challenge partners

Fig. 11.2   Cycle parking at Nottingham University installed under the Cycle Challenge project.

were all committed to producing a broader commuter plan aimed at promoting a variety of alternatives to private car travel for work journeys. The eight employers introduced a variety of incentives to facilitate cycling amongst their employees; these included combinations of the following:

- Workplace showering and changing facilities: both male and female if appropriate.
- Secure cycle parking at the workplace: cycle lockers and communal cages, as well as cycle stands and shelters offering weather protection.
- Cycle mileage allowances for short journeys on official business.
- Interest-free loans of up to €800 (£500) repayable over two years, for the purchase of bikes and equipment.
- Purchase of company 'pool' bikes for communal use for short official business trips and site visits.

- Publicity and information material endorsing the personal and environmental health benefits of cycling.
- Promotional events such as bikers' breakfasts, bikers' barbecues and bike-to-work days.
- The establishment of Bicycle User Groups (BUGs) to advise on the measures required to promote cycle commuting.

Each Cycle Challenge partner implemented a package of cycling incentives tailored to its individual needs and constraints. Most of the project partners consulted their employees on how the Cycle Challenge money should be invested. This was achieved through a combination of staff travel surveys, newsletters, email and discussion groups. Those partners that set up Bicycle User Groups (BUGs) – comprising existing and potential cycle commuters – found them very useful in identifying the most effective focus for investment.

In addition to the Cycle Challenge partners there were three supporting organisations involved in the Nottingham cycle-friendly employers project. The first was Pedals, the local cycle campaign group. The second was the Nottingham Green Partnership, an alliance of public, private and voluntary sector bodies in the city which directs and financially supports green initiatives, including several directly concerned with sustainable transport. The third was Cleary Hughes Associates, a Nottingham-based firm of cycle planning consultants, charged with undertaking the monitoring and evaluation of the project.

The experience of involvement in the Nottingham Cycle Challenge project has for most partners been very positive, with an enthusiastic response from employees and favourable publicity. This led most partners to continue to support and invest in pro-cycling measures even after the deadline for completion of Cycle Challenge funded infrastructure.

From the experience of the project, it is evident that there are a number of conditions which, if satisfied, contribute to the smooth implementation and likely success of initiatives to facilitate cycling for work journeys (Cleary Hughes Associates, 1999). These include the following:

- An enthusiastic facilitator within the organisation, who is committed to the project, to managing it and following it through, as well as getting it off the ground in the first place.
- A forum for discussion and the exchange of ideas between those responsible for implementing measures to facilitate cycling and those who will be making use of them.
- The Bicycle User Groups (BUGs) from different organisations should be encouraged to network to facilitate the exchange of valuable information, ideas and experience.
- Ensure that the needs of non-, but potential, cyclists are taken into account in deciding spending priorities.
- Acknowledge that it is not essential for people to make a complete modal shift to cycling for work journeys.

- A demonstration of commitment from senior management for the project's aims and objectives.
- BUG meetings and newsletters, which offer a useful opportunity for the exchange of information and ideas between senior management and their cycling commuters, as well as for a show of support from the former for the latter. They need to be maintained and supported after the initial flush of enthusiasm.
- There is a need to identify and overcome onerous disincentives to cycle use for work journeys which may arise, e.g. detailed conditions on the recording of bike mileage to justify cycle mileage allowances.
- Employers need to consider permitting greater timekeeping flexibility to those who cycle to work.
- Promote the accessibility, versatility and benefits of cycling, rather than focusing exclusively on what some might perceive as complications.
- Sufficient time needs to be allowed for the implementation of pro-cycling measures, including consultation on what measures are required.
- A simple system of evaluation and monitoring of the project and its various components, e.g. to ensure that facilities like lockers go to those most likely to make good use of them and to assess whether further provision, or other changes, are needed.

In terms of the main aims of the project, to encourage cycle commuting, the results were impressive, if not altogether clear cut in terms of being able to say with confidence that a precise number of new cycle commuters was generated directly as a result of the project. The evaluation found that only 15% of cyclists in the 'after' survey considered that conditions for cycle commuting had deteriorated in this period while 85% disagreed. And 42% of cycling respondents indicated that their level of cycle commuting had increased during the life of the project. The survey also found that there was less of a seasonal drop-off in cycling in winter where good quality showers and changing facilities were provided.

The figure for those indicating an increase in cycle commuting included those who started commuting by bike during this period, of whom 49% said that the level had remained the same and 9% that it had decreased. Of those who indicated that their cycle commuting activity had increased three key reasons were given equal weight, and all mentioned by nearly one-third of relevant respondents. In addition to the provision of workplace cycle facilities these were a house or a job move and heightened concerns/awareness about personal health. The evaluation went on to conclude from this that:

> What can be said with some certainty is that the provision of workplace facilities under Cycle Challenge has attracted some new cyclists, and satisfied a number of long-standing demands and needs of existing cycle commuters. The project has also probably helped to sustain cycle use, as even dedicated cyclists can be deterred eventually if there are limited facilities to support them; for example, if a lack of secure cycle parking leads to cycle theft.

A number of indications emerged from the project which suggested not only that it had a positive influence on the extent to which cycling is used for work journeys, but which also suggested that more needs to be done to achieve a really significant shift in favour of commuting by bicycle.

The employers reported that most of the workplace pro-cycling measures which were introduced have been very well used, particularly secure cycle parking, showering, changing facilities and cycle loans. Cycle mileage allowances were less taken up, due partly to the system for claiming them. The popularity of most measures has encouraged sustained commitment for cycling, emphasising the 'pump priming' effect of the project, as well as provision of similar facilities at other worksites and, in some cases, a broader range of measures.

The indications of increased bicycle usage led to pressure for highway improvements in and around the partners' worksites, as well as on commuter routes. They also had the effect of encouraging other employers in the vicinity to consider introducing workplace cycle facilities. There is no doubt that the project was broadly very successful. It yielded a number of important detailed lessons not only for further encouragement of cycling but more generally in terms of 'green commuting' and the implementation of more sustainable transport measures.

The cycle-friendly employers project arose out of the Green Commuter Plans work which Nottingham City Council had pioneered in the UK in the mid-1990s (McClintock and Shacklock, 1996). A similar initiative, known as 'Steps' (Sustainable Travel Equals Perfect Sense) was adopted by Nottinghamshire County Council a little later. Interest in this work has since been sustained by the Nottingham Commuter Planners Club, a forum for the local authorities, local employers, transport operators and other groups and interested parties which meets regularly to exchange ideas and examples of good practice. This work, in which cycling features strongly, has been extended to include smaller scale employers, via the TRANSact project, and has also resulted in Nottingham becoming involved in European projects to encourage green commuting such as the MOSAIC project in the late 1990s. At the same time such initiatives have been helped by the changes in transport policy at national level (DETR, 1998) and the much stronger encouragement not only for Travel Plans and close working with employers but also travel awareness campaigns and Safe Routes to School projects (see Chapter 6).

11.4   Wider changes in the national and local transport planning context

Changes in national policy also helped directly to see that all local authorities in the area, to varying degrees, increased the profile of cycling. This gave new opportunities for the local cycling campaign Pedals to gain acceptance for some of its longstanding ideas. Nottinghamshire County Council published its Local

Cycling Strategy in 1997 (Nottinghamshire County Council, 1997), in direct response to the National Cycling Strategy (DOT, 1996). The city council followed with its Cycling and Walking Strategy (Nottingham City Council, 1999), soon after it had taken over from the county council as the local highway authority for Nottingham. This change resulted from the adoption of unitary status for the city council in April 1998 as part of the national reorganisation of local government. It also set up a Cycling and Walking Forum. Since 1998 therefore the city council's views on transport have been much more important, while the county council, though still responsible for most highways in the suburbs of the conurbation, has become relatively less concerned with urban traffic issues.

Despite this change the county council has remained keen to maintain its pro-cycling reputation. Before the reorganisation it had introduced district cycling working parties to involve a wider range of interested parties in generating ideas to improve local cycling conditions. These meet three or four times a year and have so far probably made more of an impact in some districts than others, for example in Rushcliffe borough, helped by the adoption of their own Cycling Strategy (Rushcliffe Borough Council, 1995).

At the same time the district councils have, following changed government planning advice, been much more willing to include detailed cycle route proposals in their local plans and to consider how these could complement those by other agencies including the county council, the Highways Agency and Sustrans. Some have also adopted Green Travel Plans for their own staff in which, again, cycling measures feature prominently.

The Local Transport Plan for Greater Nottingham, published in 2000 after extensive consultation (Nottingham City Council and Nottinghamshire County Council, 2000) strongly supported the national targets to double the modal share of cycling by the end of 2002 and double them again by the end of 2012. Recent monitoring figures (pp Dii–Diii) have shown a broadly encouraging trend of increased cycle use, particularly in areas served by the cycle network. The LTP aims to build on this while also maintaining 'a downward trend in the number of cycle related casualties'. It made clear that

> increases in cycling will be achieved within the context of improving the safety and convenience of cycling across the whole highway network. Road safety considerations are paramount and complementary to local safety schemes, particularly those that reduce vehicle speeds.

It also recognised that 'significant increases in cycle trips can be achieved by changing attitudes towards cycle use . . . and that this requires a dual approach between employers and highway authorities, such a system as is established in greater Nottingham.'

The positive health benefits of cycling were also strongly acknowledged, as well as the role of Travel Plans, the quality and maintenance of cycle networks (including upgrading of existing routes after assessment by conducting cycle reviews). The importance of cycling information, cycle parking and monitoring of existing and future cycle use were also recognised, as well as integration of cycle usage, building on improvements for cyclists at Nottingham station and some Robin Hood Line stations, and the bus/cycle lanes. The LTP proposed key cycling targets:

- increase in cycle modal share;
- reduction in cycling casualty rate;
- reduction in cycle theft;
- high quality network expansion;
- implementation of cycle audits.

For cycling to work there is a targeted increase of 6.5% by 2011, with a special target for businesses adopting commuter plans of increasing cycling to work to 20%.

11.5   Other cycling policy initiatives

In addition to the major encouragements to cycling in the Nottingham area in the 1980s and 1990s, represented by the Greater Nottingham Cycle Route Network Project and the cycle-friendly employers project, a whole series of other initiatives have played their part in encouraging cycling. Directly and indirectly central government, the local authorities, cycle campaigners and a wide variety of other partners in the public, private and voluntary sectors, including the (former) Nottingham Health Authority, have all helped.

Within the county council, which pioneered the efforts to revive cycling in the early 1980s, there have been important initiatives in terms of promoting leisure cycling via the authority's involvement in the annual Great Nottinghamshire Bike Ride and the Guided Rides programme, offering a variety of short and longer rides, especially in the spring and summer. As well as cycling forming an integral part of the local councils' Green Travel Plan initiatives it has also been an integral part of the wider travel awareness campaigns such as the Travelwise centre and website. This includes a steadily expanding section of cycling information, including a downloadable version of the local cycling map and, from late 2001, a facility for on-line reporting of defects and making comments on the quality of existing cycling provision.

Recent plans for promoting adult cyclist tuition and bike maintenance classes reflect the strong interest in promoting cycling from several health practitioners in the area. This shows the general increased interest in the health aspects of transport and cycling in particular. Cycling promotion locally has also benefited from the increased links between transport and air quality improvement, as in the Clear Zones project in the city centre, and between transport and social exclusion. This has been demonstrated in the Workwise project, providing arrangements to lend bikes, and give other cycling assistance, to people in several deprived areas to help them find work. Cycling provides one of a range of options.

Free carriage of bikes on trains was provided from the start of operations on the Robin Hood Line (RHL), a heavy railway line that reopened in 1993 and by the late 1990s was extended to Mansfield and Retford in the north of the county. In the late 1990s cycle lockers were provided at RHL stations. The first cycle lockers had been installed some time earlier at Nottingham station and, with funding from the Nottingham Green Partnership, in three of the city centre multi-storey car parks.

Promotion of cyclists' safety on the highway has been gradually more assisted not only in terms of more sensitively designed traffic calming and stronger enforcement of speed limits through digital speed cameras but also through various publicity measures including, in the late 1990s, a special mural on the back of a local bus to encourage drivers to watch out more carefully for cyclists (see Fig. 11.3).

Fig. 11.3   One unusual example of local cycling promotion: a 'Drivers: watch out for cyclists' back of bus advertisement. Source: Nottingham City Council.

Cyclists are now more widely considered in various traffic counts and surveys and the city council, in particular, is greatly extending its monitoring of various aspects of cycling and cyclists' opinions. This new importance has been helped by the support of several cycling champions in different parts of the local authorities (councillors and officers) and other agencies, as well as by the sustained political pressure from local cycle campaigners, and of course the changes in the national transport policy context.

11.6   Cycling controversies

The promotion of cycling in the Nottingham area has, however, not been without its controversies, despite the broad political consensus in favour of encouraging the mode which, in the case of the city and county councils, now goes back more than 20 years. In the early 1980s there was strong opposition from some residents in the Silverdale estate in Wilford to the plans for the Clifton cycle route. It was at first approved by the county council only as a one year experiment.

In 1986, the decision to allow general two way cycle access on the newly pedestrianised High Road in Beeston proved controversial, especially when a pedestrian was run into by a student cyclist hurrying to early morning lectures at the nearby Nottingham University campus. The consequent uproar resulted in the imposition of a ban on peak hour cycling.

In 1991, a decision to allow continued cycling on a cross-city centre route between the Lace Market and the Castle (Castle Gate/Low Pavement), post pedestrianisation, was successfully opposed by a disabled group at a public inquiry in 1992. The ban was gradually relaxed a few years later.

Cycle access to, from and within the city centre remained for many years relatively neglected. A long standing Pedals proposal for a contraflow cycle lane on Goosegate as part of an improved route on the east side of the centre was implemented only in 1998, 18 years after it had first been proposed (see Fig. 11.4). Despite the inclusion of an important signalled cycle crossing of the inner ring road, on Canal Street, in the Clifton cycle route in 1982, little more was done for many years. It has remained a difficult barrier for cyclists.

Since the drawing up of detailed plans for the first line of the light rapid transit system, the NET (Nottingham Express Transit) there has been a new round of comprehensive thinking about radical changes to traffic arrangements in the city centre, taking account also of other objectives such as improving air quality, extending pedestrianisation and allowing for a number of major redevelopments within and adjoining the central core. These include major developments near the station such as the Capital One office project and the redevelopment of the Broad Marsh shopping centre, on the south side of the city centre, to be carried out in the next few years and involving a doubling of total floorspace.

Improving air quality was the main objective behind the plans for the Clear Zone introduced in September 2001 and involving more restrictions on the use of motor vehicles in the city centre. Cyclists have in general greatly benefited

Fig. 11.4   Goosegate contraflow cycle lane, Hockley, on the east side of the city centre. A rare example of improved cycle access to and from the city centre which has remained relatively neglected.

from this although considerable difficulties remain in providing for safe cycle movements on and across much of the inner ring road. These are still to be addressed, partly as an integral part of wider traffic reduction plans. The original Clear Zone proposals involved widespread relaxation of prohibitions of cycling on some central area streets. Two of these in particular, designed to help cyclists avoid part of the NET (tram) Line One route, proved particularly controversial, on Long Row East and High Street. These objections were upheld by the Inspector at the public inquiry and so the two proposals were then dropped in the final scheme proposals.

Some of the sections of off-highway canal- and riverside path have also been controversial, with concerns about increased risks for pedestrians, abuse by motorcyclists, increased risk of theft and vandalism and also general concerns about the effect on the appearance of the countryside of more durable surfaces to accommodate cycling. Even minor proposals such as cycle parking have some-times been controversial, on the grounds of the 'intrusive' appearance of cycle stands. The opposition to cycling engendered by fears of irresponsible behaviour by riders is rarely far from the surface, especially in the case of proposals for shared area solutions with pedestrians. On some of the upgraded rural paths there has also been opposition from horse riders who feel that they have been deprived of 'their' bridleways by upgrading to accommodate cyclists. Some path 'upgrad-ing' schemes have also caused criticisms of the neglect of detailed nature con-servation considerations.

Cyclists themselves have often been critical. The LTP (Nottingham City Council and Nottinghamshire County Council, 2000, p 40), reporting on the con-sultation process, said that several key problems for cyclists stood out:

- the need for better dedicated cycle routes, cycle parking facilities and better public transport integration facilities;
- concerns about conflicts with traffic and pedestrians;
- maintenance issues;
- enforcement of local speed limits.

The LTP progress report (Nottingham City Council and Nottinghamshire County Council, 2001, p Di), commenting on the results of the Personal Travel Survey, said that:

> Whilst 81% of cyclists are satisfied, cycling has the second highest levels of dissatisfaction (only rail is worse), with 11% of City cyclists and 8.6% of County cyclists dissatisfied with their journeys. If such a minority of cyclists find their journeys unsatisfactory because of traffic levels, etc., then the same factors are likely to be inhibiting other people from cycling at all. This highlights the need for the authorities to continue to improve cycling facilities.

The general safety trends in accident statistics are mildly encouraging for cyclists. As the Local Transport Plan monitoring report pointed out (Nottingham City Council and Nottinghamshire County Council, 2001, p Fi), there has been a decline in the period 1979–2000 in the casualties for cyclists (as well as pedes-trians and motorcyclists), in contrast to a dramatic rise for car occupants. However, as the same report makes clear:

> The pattern is much more a reflection on the level of use of a particular mode rather than changes in the levels of safety for these groups. Cyclists remain vulnerable with a casualty rate per kilometre travelled 13 times higher than the car user.

11.7   Monitoring of cycle use

In the late 1980s and early 1990s extensive monitoring of numbers of cyclists was carried out as part of the government sponsored cycle route network project in the south and south-west of Greater Nottingham, and a 15 point screen was established. Unfortunately, no further monitoring was done until 1998, with the results shown in Table 11.1. As the 2001 LTP monitoring report commented (Nottingham City Council and Nottinghamshire County Council, 2001):

> the results have been broadly encouraging, even taking into account a small decline (−1%) in 1999–2000 which seems due to the exceptionally wet weather of Autumn 2000. Over the longer period of 1990–2000 there was an increase of 17.2% in cyclists using the network; this compares favourably with the 22.6% decrease nationally. This clearly demonstrates the success of the network in delivering real increases in cycle use.

11.8   Conclusions: some future challenges

It is now almost a quarter of a century since the revival of cycling started to receive official encouragement by the local authorities in the Nottingham area – an area, it should be emphasised, with levels of cycle use only slightly above the national average and therefore very different from places like York and Cambridge. There is no doubt that many of these varied efforts, now sustained over many years, have been mostly beneficial and in general appreciated. However, several important challenges remain. These include the following.

11.8.1   Reversing longer average trip distances

Arguably the most fundamental challenge in encouraging cycling is to reverse, through land use planning and other measures, the continuing trend for many years towards longer trip distances which are harder to cover on a bike. As the LTP commented in 1991 only 47% of people working in Nottingham City also lived there, compared with 53% in 1981. There are plans for extensive new housing developments on the periphery of the built-up area which, although now scaled down and likely to be built at much higher average density, may well tend to increase travelling distances further. Making it easier to combine the use of bikes and public transport, as well as more local service provision, will be among the changes needed to ensure that cycling is not made more difficult by these possible developments.

11.8.2   Improving the quality of substandard cycle schemes

Despite renewed impetus since the late 1990s much of the cycling provision in the Nottingham area dates from the period of the major cycle route network project in the late 1980s and could be said to show its age, with substandard facilities such as narrow shared paths and poorly designed crossing arrangements. There was little introduction until the late 1990s of newer types of cycle facility such as advanced stop lines, contraflow cycle lanes and well-marked on-road

Table 11.1   Greater Nottingham cycle route network usage trends 1990–2001

Location	2-way flow, 16 hr					% change 2001/1990
	1990	1998	1999	2000	2001	
Clifton bridge	441	466	458	438	502	13.8
Wilford toll bridge	666	740	719	709	561	−15.8
Clifton screenline	1107	1208	1177	1147	1063	−4.0
Welbeck suspension bridge	415	688	862	735	621	49.6
Trent bridge	459	573	805	704	653	42.3
Lady Bay bridge	218	299	292	264	305	39.9
West Bridgford screenline	1092	1560	1959	1703	1579	44.6
Lenton Road (The Park)	94	78	89	78	71	−24.5
Castle Boulevard	614	653	563	581	627	2.1
Canal (adjacent Castle Blvd)	57	279	168	269	189	231.6
City centre screenline	765	1010	820	928	887	15.9
Nottm University (Beeston Lane)	1089	926	874	862	682	−37.4
Univ. Blvd	1133	1513	1448	1358	1495	32.0
Canal (Beeston)	68	315	186	240	308	352.9
Beeston screenline	2290	2754	2508	2460	2485	8.5
Middleton Blvd	331	280	531	429	530	60.1
Charnock Avenue	246	348	417	355	401	63.0
Wollaton Road/ Faraday Rd junction	476	376	358	467	449	−5.7
Wollaton screenline	1053	1004	1306	1251	1380	31.1
Total	6307	7356	7770	7489	7394	17.2
Daleside Road/ Vale Road		388	297	247	382	
Mansfield Road/ Forest Road junction*	572	552	762	637	876	53.1
Carlton Road/ Porchester Road junction*		190	258	243	294	
Alfreton Road*		347	267	340	303	
National (Great Britain)						−22.6

Source: Nottingham City Council and Nottinghamshire County Council, 2000, pp Dii–Diii.
Note: 16 hour flow period = 06.00–22.00 hours.
* Control point off the main cycle route network.

cycle lanes. Only a few substandard schemes have been upgraded and these efforts need to be extended as part of a stronger message to cyclists that their needs are being taken seriously. Improvements to signing and markings are also needed.

11.8.3   Improving maintenance for cyclists, including improved arrangements for reporting problems

As the LTP surveys found, this continues to be a major problem, particularly on facilities that suffer regularly from problems such as broken glass, intruding vegetation, and vandalised or missing signs.

11.8.4   More attention to making the existing road system safer for cyclists

There is a willingness in principle to recognise this need but progress so far has been slow. Much wider use is required of cycle audit and review procedures and also further measures to promote observation of speed limits and other responsible road behaviour by drivers and indeed all road users.

11.8.5   Taking more account of cyclists' needs in bus lane schemes

In general cyclists have benefited considerably from the expanded introduction of bus lanes, especially on radial roads with no special cycling provision. However, cyclists in some cases have suffered from details such as narrow width or indirect implications such as a reduced width of other lanes on the general carriageway.

11.8.6   Improving enforcement (cycle lanes, bus lanes, speed limit enforcement)

Nottingham's involvement in 2000 as one of the eight national pilot schemes for the introduction of digital speed cameras has had a very positive impact on safety but this needs to become more widespread. Improved enforcement is also needed on some bus lanes and cycle lanes regularly obstructed by drivers. There is also a need to tackle frequent obstructions by parked cars on some sections of cycle path.

11.8.7   Extending and improving bike parking in existing and new developments and redevelopments

Widespread cycle parking facilities have been introduced, especially by the major employers involved in the Greater Nottingham cycle-friendly employers project, but things remain patchy elsewhere, especially outside the city centre. It has not yet been incorporated as an automatic requirement for planning permission although this may now change, through changes to development control standards and procedures.

11.8.8 Maintaining and developing Green Travel Plan work with employers

As the workplace parking levy is implemented in Nottingham, from 2004, this will be all the more important as part of a concerted drive to improve the quality of alternative mode provision. Some of the many efforts made in the late 1990s by employers involved in the Greater Nottingham cycle-friendly employers project have not been sustained, for example ensuring that all new building developments are equipped from the outset with good quality cycle parking facilities. Relatively little has so far been done to make small scale employers more cycle-friendly.

11.8.9 Harnessing the detailed local knowledge of BUGs

There is a big need to extend the growing consultation in local bicycle planning to include BUGs who can help for example to identify the need for further improvements of direct benefit to workforce cyclists in the vicinity of particular worksites, helping to improve the quality of bike access and the new multi-mode Transport Assessments. Well-focused suggestions for improvement around particular locations could complement ideas from other sources including local campaigners, area and district Cycling Forums and the city council's Cycling and Walking Forum.

11.8.10 Promoting safer routes to stations and interchanges

Some useful attempts have been made to promote this integration, for example in providing bike parking at stations, but these need to be extended, especially in the plans for major and minor transport interchanges, with particular reference to security as well as the convenience of users. This will be particularly important in the plans for the major redevelopment of Nottingham station.

11.8.11 Maximising the potential for promoting cycling through Safe Routes to School and homezone schemes

Nottingham is the location for one of the national pilot homezone schemes, in Clifton, and two other homezone schemes have been introduced, in the Ladybay area of West Bridgford and Beeston Rylands. Increasingly these opportunities and those for Safe Routes to School projects, can be used to promote a wider range of cycle-friendly road layouts, in addition to cycle audit and review on the main road system.

11.8.12 Working with health authorities, general practitioners and others to promote the extensive health benefits of cycling (and walking)

The Local Transport Plan recognises the need for this close collaboration in which there appears to be increasing interest among health and transport professionals as well as in partnership with several other bodies such as environmental groups. The Internet and the local Travelwise centre, in particular, offer good opportunities to do this, in addition to local radio and television, on a regular basis including daily traffic reports.

11.8.13 Extending cycle routes including continuous safe links to nearby settlements and countryside

There is scope for such extensions, in co-ordination with Sustrans and other county-wide and regional plans. These should include more cross-river pedestrian/cycle bridges, and extended bike access arrangements on trains, e.g. the Robin Hood Line in the north of the county and other proposed local heavy rail service improvements. Getting safely by bike between the urban area and nearby countryside often remains difficult. So far the Nottingham area has seen none of the very impressive new cycle bridges recently developed in some other parts of the UK such as Tyneside and York (see Chapter 9).

11.8.14 Maximising potential for the NET (Line One and extensions) to be bike-friendly

Nottingham has a good opportunity to learn from the earlier tram systems in the UK and abroad, both to achieve integration between bikes and trams and to minimise possible safety problems for cyclists from the NET-related infrastructure. This will be important not only in the finalisation of implementation plans for Line One, due to open in late 2003, but also in the other routes now being planned. The principle of allowing some access arrangements for bikes on NET Line One has been rejected and it remains to be seen whether or not this decision will be reconsidered in future, as indeed on other new British tram systems.

11.9 References

CLEARY HUGHES ASSOCIATES (1999), *Nottingham Cycle Challenge Project: Final Report*, Hucknall, Cleary Hughes Associates.

CLEARY, J and MCCLINTOCK, H (2000a), 'The Nottingham cycle-friendly employers' project: lessons for encouraging cycle commuting', *Local Environment* 5 (2), 217–22.

CLEARY, J and MCCLINTOCK, H (2000b), 'Evaluation of the Cycle Challenge Project: a case study of the Nottingham cycle-friendly employers' project', *Transport Policy* 8, 117–25.

DETR (Department of the Environment, Transport and the Regions) (1998), *A New Deal for Transport: Better for Everyone – the Government's White Paper on Integrated Transport Policy*, Cm 3950, London, The Stationery Office, June.

DOT (Department of Transport) (1996), *The National Cycling Strategy*, London, Department of Transport.

DTLR (Department for Transport, Local Government and the Regions) (2001), *Traffic Advisory Leaflet 9/01: The Nottingham Cycle Friendly Employers' Project*, London, DTLR, September.

MCCLINTOCK, H (1992), *The Bicycle and City Traffic: Principles and Practice*, London, Belhaven Press.

MCCLINTOCK, H (1996), 'Cycle facilities and cyclists' safety: experience from Greater Nottingham and lessons for future cycling provision', *Transport Policy* 3 (1/2), January/April.

MCCLINTOCK, H and CLEARY, J (1993), 'English urban cycle route network experiments: the experience of the Greater Nottingham network', *Town Planning Review* 64 (2), 169–92.

MCCLINTOCK, H, CLEARY, J and CHATFIELD, I (1992), 'Nottingham', in H McClintock (ed) *The Bicycle and City Traffic: Principles and Practice*, London, Belhaven Press.

MCCLINTOCK, H and SHACKLOCK, V (1996), 'Alternative transport plans: encouraging the role of employers in changing staff commuter travel modes', *Town Planning Review* 67 (4), October.

NOTTINGHAM CITY COUNCIL (1999), *Cycling and Walking Strategy*, Nottingham, Nottingham City Development Department.

NOTTINGHAM CITY COUNCIL AND NOTTINGHAMSHIRE COUNTY COUNCIL (2000), *Local Transport Plan for Greater Nottingham: Full Plan 2001/02 – 2005/06*, Nottingham, Nottingham City and Nottinghamshire County Council.

NOTTINGHAM CITY COUNCIL AND NOTTINGHAMSHIRE COUNTY COUNCIL (2001), *Local Transport Plan for Greater Nottingham: Progress Report 2000/01*, Nottingham, Nottingham City and Nottinghamshire County Council.

NOTTINGHAMSHIRE COUNTY COUNCIL (1997), *Local Cycling Strategy for Nottinghamshire*, West Bridgford, Nottingham, Nottinghamshire County Council.

PEDALS (1981), *Bike City Bikeways: Proposals for a Network of Cycle Routes for Greater Nottingham*, West Bridgford, Nottingham, Pedals (Nottingham Cycling Campaign).

RICCI, LL (2000), *Monitoring Progress Towards Sustainable Urban Mobility: Evaluation of Five Car Free Cities Experiences (Strasbourg, Bremen, Barcelona, Nottingham and Turin)*, EUR 19748 EN, Seville, Institute for Prospective Technological Studies (Joint Research Centre of the European Commission), ftp://ftp.jrc.es/pub/EURdoc/eur19748en.pdf.

RUSHCLIFFE BOROUGH COUNCIL (1995), *Rushcliffe Borough Cycling Strategy*, West Bridgford, Nottingham, Rushcliffe Borough Council.

12

An efficient means of transport: experiences with cycling policy in the Netherlands

Ton Welleman, Fietsberaad (Dutch Cycling Council)

12.1 Introduction

The Netherlands has a population of nearly 16 million and some 17 million bicycles. Most bicycles are used regularly, by young people and old people, rich and poor, men and women (see Table 12.1). The bicycle is used because cycling is often the most efficient way to go to school, to the shops and to work. In addition, cycling is an attractive form of recreation and sport. The Dutch government realises that the bicycle plays an important role in the mobility of many Dutch people. Bicycle use is promoted because it provides many more advantages than disadvantages. In 1990, the Ministry of Transport therefore started the Bicycle Master Plan project.

In this chapter the role of policy is particularly emphasised. The reason is that good policy – not just cycling policy, but cycling policy as a part of broader transport and planning policy – is a basic condition to make the most of the chances and opportunities for bicycle use. That is true not only in the Netherlands, but in every country and city.

Why does this make such good policy sense? Because cities and all kinds of activities in those cities are becoming more and more difficult to reach. Besides that we have to be worried about the environment and about quality of life, especially in cities. And, last, we have to realise that we have to be economical with our fossil energy supplies. We all know that car traffic is largely responsible for accessibility problems, serious road accidents, high energy consumption and the

Table 12.1 Division of all trips in the Netherlands by persons among primary modes of transport and distance classes, 2000 (in percentages)

Distance, km	Bicycle	Car driver	Car passenger	Public transport	Walking	Other	Total
0–2.5	15.2	6.7	2.7	0.3	17.4	0.6	42.9
2.5–5	5.7	5.5	2.6	0.5	1.1	0.4	15.8
5–7.5	2.7	4.8	2.5	0.6	0.2	0.4	11.2
7.5–20	2.0	7.9	4.2	1.2	0.1	0.5	15.9
Over 20	0.2	7.2	4.3	2.2	–	0.5	14.4
Total	25.7	32.1	16.3	4.7	18.8	2.4	100

resulting negative effects. Politicians, however, often do not have the courage to say that slowing down the growth of car usage is necessary or, even stronger, that we should reduce the use of cars. Something that is possible is talking about alternatives to the use of the car.

Public transport is always mentioned as the first alternative. One reason for this is that public transport costs a lot of money and because the average politician likes spending a lot of money. And all those people who are professionally involved in public transport would be happy to help them to spend this money. There is one problem, however: the large amount of money that is needed for good public transport is not always available. And whatever amount is available is never enough. The result is that public transport does offer an alternative to the use of the car, but not a sufficient alternative at all in a lot of situations.

What are the other alternatives? Walking is the oldest and cheapest means of transport, which is always available for nearly everyone. For many short trips in cities and villages, walking is by far the most efficient means of transport. Nevertheless you rarely hear politicians and traffic experts talk about a 'pedestrian' policy, let alone promoting it. This seems strange. Another alternative that in many countries has been getting more attention over the past few years is cycling. The reason is that cycling can contribute a lot to solving the problems mentioned above. Dutch people do a lot of cycling. And they continue to use the bicycle, even if they can afford one, increasingly often two, and sometimes even three cars per family. How is this possible?

12.2   Geographical background to cycling in the Netherlands

The Netherlands is a small country in the flat delta of three rivers. The climate is not perfect, but it rains only 7% of the time. It is also a densely populated country. For a long time, the population was spread over many relatively small towns and villages. As a result, a dense road network was built up and many destinations are close by. Amsterdam, the capital, has a population of just 725,000.

Rotterdam, one of the largest harbours in the world, is the second city, with 600,000.

The quality of the public transport system is moderate. The Netherlands has had its own bicycle industry for more than a hundred years, but it has never had a major national car industry. The widespread use of cars only became popular at a relatively late stage.

This rough sketch of the Netherlands shows that the circumstances for cycling are favourable in the Netherlands. But does that account for a high percentage of bicycle use?

12.3   The history of bicycle use and bicycle policy in the Netherlands

Knowledge of history is important, and this is also the case when it comes to transport and traffic. The present situation after all is the result of developments over many decades and the near future has for a large part already been determined by the recent past.

When the Ministry wanted to know why there is more cycling in the Netherlands than in surrounding countries, and why there is more cycling in one city than in others, it realised that research on the history of bicycle use and bicycle policy in a number of those countries was necessary. This research was carried out because we were hoping that we would be able to use the lessons from the past for future policy. On the basis of the results the following statement can be formulated: 'Bicycle policy can be effective, but it does require patience.'

To explain this statement, first some frequently mentioned factors that influence bicycle use in the Netherlands have to be discussed:

- *Topography*. The general impression is that the Dutch cycle a lot because they live in a flat country. This flatness of course plays a part, but apparently it is not the only precondition; people living in flat areas outside the Netherlands cycle much less.
- *Spatial structures, town and country planning*. These are influential. This may explain why there is little cycling in the USA and Australia, but in many European countries the average trip distances are fully comparable to those in the Netherlands. Nearly 70% of all trips in the Netherlands are 7.5 kilometres or shorter.
- *Availability of alternative modes of transport*. Mass motorisation got started quite late in the Netherlands. However, there are now about 400 cars per 1000 people. That comes down to one car for every two persons aged 18 or older. Still, the number of trips of 7.5 kilometres or shorter is the same for cars and bicycles. As for public transport, most cities in the Netherlands are too small for profitable bus and tram traffic to be an efficient alternative to the bicycle. The train as a mode of inter-local transport is hardly competitive with the bicycle.

• *Cultural historical values.* That these play a role is illustrated by the choice of transport mode of Dutch people of Turkish, Moroccan and Surinam origin. They cycle much less and use public transport more often. The differences with people of Dutch origin, however, seem to be decreasing with each generation.

These four factors – topography, spatial structure, available alternatives and cultural historical values – do not sufficiently explain the differences in bicycle use between the Netherlands and other countries and between one Dutch city and another. Neither do they explain why bicycle use during a certain period of time increases considerably in one city while it sharply decreases in another.

Additional answers can be found when looking at the influence of policy. What is policy, however? What is bicycle policy? Policy can be taken as comprising the government's intentions and – more important – actions that go with those intentions.

12.4   Dutch bicycle policy during the twentieth century

Up to the Second World War there is not much positive to report in this respect. Between 1900 and 1940, the number of bicycles grew from 100,000 to four million. In comparison by 1940 there were also 100,000 cars. The government considered all these bicycles primarily as a source of income. A major part of the road plans – which encouraged car traffic – was therefore financed from bicycle taxes. The construction of cycle tracks along some of the national highways could be regarded as bicycle policy. However, this happened mainly to decrease the hindrance that the many cyclists caused for a few car drivers!

After the Second World War, cyclists still dominated the scene, but there was hardly any attention paid to cyclists and their infrastructure. Policy makers were primarily occupied with cars, and the construction and widening of roads. Bicycle traffic was generally expected to be marginalised. The bicycle was old-fashioned, a vehicle for the poor. The car symbolised the future, mobility and freedom!

But – and this is crucial – cycling *was* recognised as a mode of transport that is also part of life, as a mode of transport that also uses and may use public space, as a mode of transport that other traffic participants have to take into account. In the Netherlands, transport policy meant a pro-car policy, but in general not an anti-bicycle policy. That was quite wise: at that time there were after all hardly any alternative modes of transport available for most of the Dutch. Mass motorisation did not start until about 1960, and the role of urban public transport was minimal even then (see Fig. 12.1).

The attitude from the 1950s and 1960s – pro-car, but not anti-bicycle – facilitated the turnaround that took place in the 1970s when the rapidly growing car monster started to bite its own tail. The annual total number of traffic casualties increased very rapidly. Traffic congestion occurred more and more and the space that parked cars were occupying formed an increasing problem in the cities.

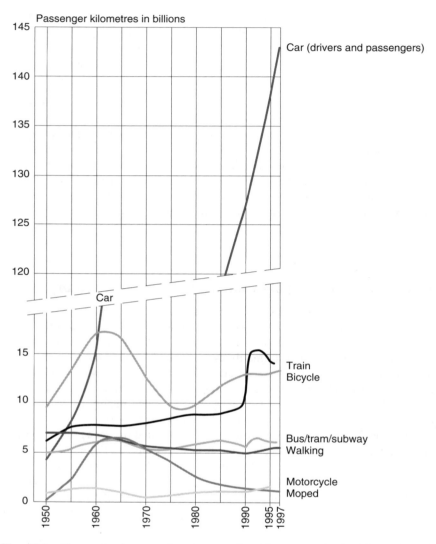

Fig. 12.1   Change in the use of various modes of transport in the Netherlands in passenger kilometres, 1950–1997.

Awareness of environmental issues was growing, there was more and more attention to healthy exercise and with the oil crisis in 1973 people rediscovered the bicycle as an efficient mode of transport. A decrease in bicycle use changed into an increase. And in the cities, policy makers realised that the bicycle might contribute to solving the traffic problems that had arisen.

Central government supported this process in the form of subsidies for the construction and improvement of bicycle facilities by the 540 municipal and 12

Fig. 12.2 Contemporary city street scene. Source: Klats publiciteit, Delft.

provincial authorities. As a result, the total length of bicycle paths increased from 9300 km to 16,100 km between 1978 and 1988, a 73% increase.

Central government also financed pilot projects. Initially this concerned high quality cycle routes, and at a later stage a complete network of routes in the town of Delft (see Figs. 12.2 and 12.3). Evaluations, however, showed that although a good infrastructure for bicycle traffic is a basic condition, it hardly leads on its own to an increase in bicycle use, at least not in the Netherlands. Bicycle policy should comprise more than the construction of infrastructure.

A bicycle-friendly attitude among policy makers in the cities and in central government is a necessary condition for a good bicycle policy. In *all* the plans they make, they should also consider the interests of cyclists. It is a permanent task for cyclists' organisations and their local branches to keep all those policy makers on the alert! Besides enthusiasm and knowledge, this mainly requires a lot of patience and constantly looking out for new social developments. The political discussion on this subject should lead to a different perception in society and ultimately also among policy makers.

Those policy makers will subsequently have to develop new policies and implement new plans. All of this requires a lot of time, among other things because there are so many parties involved. Once the required measures have finally been implemented, it often takes years, if not decades, for the effects to become noticeable. The Netherlands, for instance, is still reaping the rewards of the recognition of the bicycle as a normal mode of transport in the fifties and sixties and of investments made in the seventies and eighties. Bicycle policy can be effective, but it does require patience.

12.5   1990–1999: the Bicycle Master Plan period

In 1990, central government developed a new transport policy. Bicycle policy – expressed in the Bicycle Master Plan of the Ministry of Transport – formed an inextricable part of this. It was a logical continuation of developments in previous years, logical, but apparently not taken for granted. If the Dutch Cyclists' Union had not actively participated in all kinds of social discussions, if they had not lobbied with their relevant knowledge among politicians and had not kept policy makers at the Ministry on the alert, there would not have been any new bicycle policy at all! Fortunately the Cyclists' Union did their work properly and the Bicycle Master Plan project group was able to get to work, with clear targets and a decent budget.

The universal objective of the Ministry's bicycle policy, the main task of the project group, was formulated as follows: 'Promoting bicycle use while simultaneously increasing the safety of cyclists and increasing the appeal of cycling as a mode of transport.' This task was worked out in clear targets for 1995 and 2010. From the outset, an integrated framework was chosen. Bicycle policy was regarded as an inextricable part of transport policy as a whole, and cycling as a mode of transport among other modes. Cycling and bicycle policy were expressly not conceived as objectives in themselves but rather as a means of contributing to solving traffic and transport problems and creating better alternatives to car use.

Mobility was conceived as the degree to which individuals, companies and organisations are capable of taking part in economic and social activities. This was to be seen primarily in terms of the numbers and types of trips, and not the numbers of kilometres travelled. The question at the centre of decision making was therefore: 'Which mode of transport is the most efficient for which type of trip?' Instances in which the bicycle is the obvious choice should be met with facilities that encourage bicycle use. Instances in which the car causes too much nuisance should be met with physical, financial or regulatory measures for reducing car use. At the same time, the spatial planning of functions and activities, and the trip distances that people's participation in these functions and activities entail, must be taken into account. The shorter the trip distance, the greater the possibilities for stimulating bicycle use.

The activities of the Bicycle Master Plan project group consisted of three main elements, as follows.

12.5.1   Creating the right conditions

This means that continuous work will have to be done to provide a safe and high quality infrastructure for cyclists. The primary aim is to reduce the number of places where cyclists and fast moving vehicles meet. This means separating the traffic, on the roads as well as at junctions. Where separation is not possible or desirable, the speed of vehicles will have to be drastically reduced.

To promote infrastructural improvements, the government in 1990 reintroduced a scheme for subsidising municipalities and provinces up to 50% of the

Fig. 12.3 Cyclists are separated from fast-moving vehicles. Source: Klats publiciteit, Delft.

cost of not only constructing but also of upgrading bicycle paths, bicycle bridges and underpasses, junctions, crossings and bicycle parking facilities. So it is not only a matter of more bicycle facilities; the quality of the facilities is equally important.

12.5.2   Developing arguments and instruments

Knowledge, arguments and instruments were developed by carrying out numerous research, pilot and model projects. Arguments are required to convince national and local politicians and other decision makers of the need to co-operate. In addition, instruments are required for implementing policy. In other words government agencies, interest groups, public transport companies and so on must not only *want* to co-operate, but also *be able* to actually take measures.

12.5.3   Promoting bicycle use

Dutch citizens know what a bicycle is and how to use it. So the promotion of bicycle use does not require billboards, but instead adequate, suitable facilities. We therefore did not communicate directly with the general public, but concentrated on informing and influencing bodies concerned with the implementation of bicycle policy. Among them, municipalities formed the most important target group of the Bicycle Master Plan project group by far. Considering their responsibilities and potential, they are in the best position to pursue effective bicycle policy.

The direct outcomes of the project group's activities mainly concern 112 research, trial and model projects that have contributed to a great deal of knowledge and experience. All these projects and activities have cost the Ministry of Transport 40 person-years and about €14.5 million (£9 million).

In addition to this, in the period between 1990 and 1997 about €176 million (£109 million) in government subsidy was given to local authorities. To be clear, the total amount spent on bicycle traffic infrastructure during this period in the Netherlands is estimated at €930 million (£577 million). As a result, the total length of cycle paths has increased by more than 4000 km, an increase of 25% compared to 1988. This does not seem to be very much, but it is especially important that a lot of attention was given to the quality and the safety of all bicycle facilities, including the existing ones.

The large number of projects that have been carried out offered excellent opportunities for intensive communication with all kinds of target groups. The knowledge and experience gained from the projects were distributed via many dozens of reports, manuals, brochures, video films, articles in professional journals, workshops and the project group's own quarterly magazine. A great number of lectures and interviews were given, at home and abroad. In technical institutes, attention was given to bicycle transport and numerous meetings were organised. In combination with frequent oral consultations and personal contacts between the project group and its target groups, all these communication activities led to a certain public familiarity with the Bicycle Master Plan. And this in turn resulted in the political attention for bicycle traffic that is so necessary, but that even in a bicycle country like the Netherlands is often negligible compared with the attention for car traffic and public transport.

So far, so good. But has anything changed in the policy of organisations that could more directly influence bicycle use? Has anything changed within local authorities, county councils, public transport companies, interest groups, consultancies, ministries and industry? Basically, it seems that it has.

- The existence of a bicycle policy at the Ministry of Transport, supported by government subsidy, has put bicycle transport more squarely on many agendas. The objectives of the Bicycle Master Plan have been adopted in nearly all transport plans at local and regional level. Unlike 10 years ago, cycling is now receiving plenty of attention again. Government bicycle policy therefore has a clear role.
- A growing number of companies and government organisations are drawing up transport plans to deal with their employees' commuting habits, in which the bicycle often plays the most important role.

- The knowledge and experience that has been accumulated appears to fill a need, not only of local and regional authorities, but also of public transport companies, consultancies and industry. The quality of facilities for cyclists and for cycle parking is much better now than 10 years ago.
- When asked which alternative they would choose if they were to drive less often, the top response by motorists is the bicycle. For motorists, the bicycle was and is the most important alternative to the car for short distances. In this respect very little has changed in 10, 20, 30 years. Nevertheless, it is good to see that the bicycle has been able to maintain its position, despite increasing car ownership and car use.

Numerous activities, much experience gained, a lot learnt, bicycle traffic more squarely on the agenda . . . is all of this a reason to be satisfied? By no means! Some steps have been taken in the right direction, but it is all about effects. This concerns the question whether the objectives of the national government's bicycle policy have come any closer. Most of these objectives have been developed into distinct targets for the years 1995 and 2010. And although policies are often only effective in the long term, a half-time score can still be interesting. This is now discussed in relation to the main targets for bicycle use, cyclists' safety, the combination of public transport and bicycle, and bicycle parking facilities and theft prevention.

12.6   Bicycle use

Target: 30% more kilometres by bicycle in 2010 than in 1986. The reality is, however, that after an increase of 30% in the 1980s, bicycle use has stabilised since 1990 (see Fig. 12.4). This is the result of two opposing developments:

- On the one hand, we notice that the average length of trips has slightly increased. So the 'bicycle market' has shrunk somewhat.
- On the other hand, this 'market' is still very substantial: in 2000, 70% of all trips were still no longer than 7.5 kilometres. In this market, the bicycle has increased its share slightly between 1980 and 1995, particularly in cities, where use of the car has become less attractive, an important factor being the parking policies. Since 1995 the bicycle has not made any further gains in relation to the car. The booming economy has undoubtedly played a role in this.

12.7   Cyclists' safety

Target: 15% fewer cyclist fatalities in 1995 than in 1986 and 50% fewer in 2010 (see Fig. 12.5). The interim goal for 1995 was met. And this positive development is continuing.

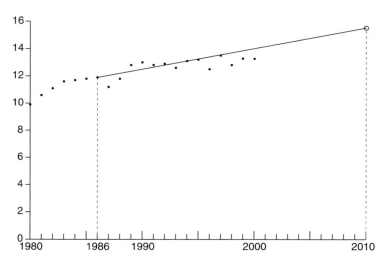

Fig. 12.4   Kilometres by bicycle: annual scores 1980–2000 and the target for 2010 (in billions).

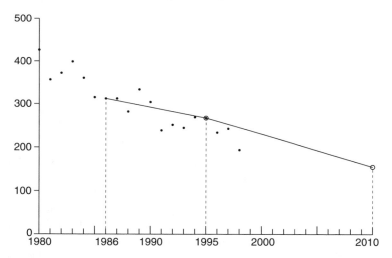

Fig. 12.5   Cyclist fatalities: annual scores 1980–2000 and the target for 2010.

Target: 10% fewer injured cyclists in 1995 than in 1986 and 40% fewer in 2010. The official figures for the total of cyclists admitted to hospital with injuries saw a fall of 27% between 1986 and 1995 (see Fig. 12.6). The interim goal for 1995 was therefore easily exceeded. In reality however, the development is less favourable. Records based on police data are not complete. This incompleteness is greater for cyclists with slight injuries than for those with serious injuries.

The conclusion about cyclists' safety is that the annual number of incidents, 70% of which involve collisions with cars, resulting in the death or serious injury

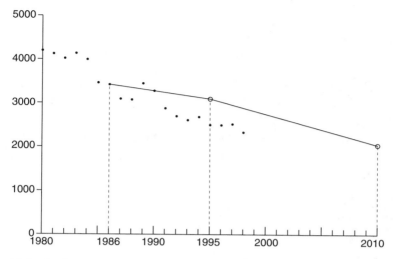

Fig. 12.6   Cyclists admitted to hospital with injuries: annual scores 1980–2000: intermediate objective for 1995 and the target for 2010.

of cyclists has developed favourably. Less favourable is the development of accidents resulting in less serious injuries, which represent by far the largest number of accidents. Of these accidents, only one-quarter involve collisions with cars.

12.7.1   Sustainable safety

The traffic safety policy in the Netherlands aims at a sustainable and safe traffic and transport system. The elements involved in this system are people, vehicles and infrastructure. People are human, which is why they often make mistakes: usually unintentionally, sometimes on purpose. A sustainable, safe traffic and transport system to a certain extent accepts human mistakes. A traffic accident is in that case considered a system error. Such a system should, however, function in such a manner that the risk of major errors, such as fatal accidents, is extremely limited.

The authorities can mainly influence the functioning of the system by providing a safe infrastructure. A truly safe infrastructure will almost automatically lead to the desired form of behaviour of traffic participants. This means that undesired or incorrect behaviour is not or is hardly possible.

These arguments may sound high flown, theoretical and perhaps even unrealistic but they are not. To prove to you that we have been using this line of reasoning for a very long time, both in the Netherlands and in other countries, here is an example.

In the early days there were a lot of fatal head-on collisions on motorways due to the fact that people crossed the central reservation. To try to avoid this you could place signs to alert car drivers that they are not allowed to cross the central reservation. This is a cheap solution which places the responsibility with the

Fig. 12.7   Small differences in mass and speed. Source: Klats publiciteit, Delft.

traffic participants but which does not work, of course. The proper solution would be to install a physical barrier, a crash barrier that makes it impossible to cross the central reservation. Such a measure requires a substantial investment, which, however, pays for itself quite quickly. In this way the authorities assume responsibility while the traffic participant is deprived of a form of behaviour that could have disastrous results. That is sustainable safety!

The sustainable safety philosophy can easily be translated to bicycle traffic. As indicated earlier, serious accidents involving cyclists are largely related to collisions with motor vehicles. This is the result of the differences in mass and speed. There is not much that can be done about the differences in mass, but there is something that can be done about the differences in speed. Sustainable safety with regard to bicycle policy can therefore be translated into three simple rules:

- Encounters of high speed car traffic and cyclists should be avoided. This means separation: in terms of place (such as separate bicycle paths), time (such as with traffic lights) or accessibility (no cyclists on motorways and closing off shopping areas to car traffic (see Fig. 12.7)).
- If separation of high speed car traffic and cyclists is not possible, there is only one solution left: drastically reducing the speed of car traffic, through any measure whatsoever.
- Simplifying traffic situations. The more complex the situation, the more problems traffic participants have. Novice traffic participants especially can benefit from this – children who are learning to cycle, young adults who are learning to drive a car or motorcycle. Elderly traffic participants will also benefit

from simple traffic situations. Research has shown that they do not make more mistakes in traffic, but they do need more time to take all kinds of decisions. Especially in complex situations there is simply no time. And that is when elderly people have problems.

12.7.2   Helmets for cyclists

In the Netherlands you see hardly any cyclists wearing a helmet. This is fortunate because otherwise we would have far fewer cyclists. Proof for this can be seen in countries where cyclists are obliged to wear helmets or where the use of a helmet is urgently advised. In those countries you hardly see any cyclists.

In September 2001 even the Dutch Minister of Transport was heard to say that she would not think of obliging cyclists to wear helmets. She expects that a lot of Dutch people would ignore such an obligation. And if people did wear them, she is afraid that this would drastically reduce bicycle use. That is something which no politician in the Netherlands wants, because it would lead to major problems: much more car traffic in the cities, which is hardly possible, and an enormous overloading of public transport. And in addition it would seriously reduce the possibility for many people to participate in all kinds of activities.

And do these helmets provide any real advantages for cyclists? If you fall on your head, yes. But not so much if you are hit by a car driving at 50 or 80 km per hour. So helmets mainly have a positive effect in accidents that are less serious. If people want to wear a helmet there is no problem, of course. Anyone who cycles in the mountains and descends steep slopes at high speeds would be wise to wear a helmet. The same goes for people who cover long distances on a racing bicycle in a group. But making small children who are cycling under the supervision of their parents wear a helmet? In the Netherlands this is almost considered a form of child abuse. You would not make these children wear a helmet when they are going for a walk or playing on the pavement, would you? Making children wear a helmet would probably cause them to avoid using the bicycle as a means of transportation. In the Netherlands at least it would. If only because they do not want to be teased by their friends. The alternative, for cyclists of all ages? A really sustainable, safe traffic and transport system!

12.8   The combination of public transport and bicycle

In the Netherlands there is not much competition between public transport and the bicycle. As an illustration: in 1995, for every trip by public transport in the 0 to 7.5 km distance category there were 18 trips by bicycle. Only in the largest cities is this ratio more favourable for public transport. In the Netherlands the bicycle even reinforces public transport, as an alternative to the car for trips across distances longer than 7.5 km.

The objective of the bicycle policy was to increase the use of public transport by improving pre- and post-transport by bicycle and by better parking facilities

Fig. 12.8   Parking facilities for bicycles. Source: Klats publiciteit, Delft.

for bicycles at train stations and at bus, tram and metro stops (see Fig. 12.8). To what extent this objective has been realised cannot be expressed in figures, not only because there are no reliable statistics, but also because the public transport system has changed dramatically over the past few years. The effects of this are much larger than the effects of activities within the framework of bicycle policy.

It can be concluded, however, that public transport companies are increasingly recognising the supporting role the bicycle can play in transport before transit. Cycling to and from the station or the stop has a favourable effect on the total journey time. This role appears to increase in importance for longer total journey times. The bicycle's share in transport before transit for various public transport modes is as follows:

- 30 to 40% in combination with the train.
- 20 to 30% in combination with express buses (across long distances).
- Up to 20% in combination with the connecting urban and regional bus transport.
- Less than 10% in combination with local urban transport.

Some 10% of all train passengers use a bicycle to get from the station to their final destination. This largely concerns employees, students and school pupils.

An important link in the chain of public transport and bicycle transport is the bicycle parking facilities at stations and at bus stops. At the moment a lot of money is being invested in order to significantly improve this chain, as far as railway stations are concerned.

12.9   Bicycle parking facilities and theft prevention

Target: a substantially lower number of bicycle thefts in 2000 than in 1990. Unfortunately, this target was not reached. The annual number of bicycle thefts has remained at an even level for the past 15 years, at between 600,000 and 700,000 thefts a year.

It is extremely difficult in the Netherlands to find a solution for bicycle theft, one of the main obstacles in promoting the use of bicycles. A good start was made with the improvement of bicycle parking facilities in such a way that they provide better protection against bicycle theft. But we still have a long way to go before the effects of all those improvements become more visible in the streets.

Bicycle locks help prevent theft. The locks that are available are becoming increasingly more effective, but bicycle thieves are also becoming increasingly clever. At the moment there are hopes for a system entailing a micro-chip in the bicycle, for purposes of simplified and accelerated police monitoring. A lot of parties are involved in this, however. This probably means that it will take a long time before a proper solution becomes available and is actually applied.

12.10   Conclusion

Looking back on 10 years of active bicycle policy in central government the effects are still limited, despite much effort. It takes a long time for bicycle policy to have any effect. In addition, bicycle *use* is not only influenced by bicycle *policy*. It may even be the case that bicycle policy has considerably less influence than, for instance, policy with regard to spatial planning and urban development, parking policy and public transport policy. It is very important that in these policy sectors too attention is given to cycling. But even if this is done, it should be realised that there are all kinds of independent social and economic developments, which are hardly affected by policy.

12.11   Plans for future bicycle policy at a national level

In the spring of 2001 the Ministry of Transport presented a new National Transport Plan, which sets out transport policy for the next decades. One important new feature of this plan is decentralisation: responsibilities, tasks and budgets are shifting from central government to provinces, cities and towns. This means that authorities most directly involved with bicycle issues will have more influence in the future. That is a good thing, because local and provincial governments also benefit the most from easily accessible cities and towns that are pleasant places to live in. An increase in bicycle use will contribute to this.

Decentralisation does not mean that central government no longer cares about cyclists. Several examples of activities at a national level can be cited:

- The central government recently approved a Public Safety and Security Programme, which identifies bicycle theft as a major social problem. Several ministries will be involved in fighting it, in co-operation with the bicycle industry and interest groups.
- A second example concerns traffic safety policy. The Ministry of Transport's slogan is 'sustainable safety'. To move closer to a sustainable, safe transportation system, motor traffic is to be concentrated as much as possible on a limited number of traffic arteries. Cyclists will also be able to follow these routes, but on separate paths. All other roads will be treated as residential areas with a speed limit of 30 km/h within built-up areas and 60 km/h outside the town. These measures should make walking and cycling much more appealing and safer. Although the responsibilities for the execution of this policy lie largely at local and regional levels, the central government has set clear conditions with regard to the decentralisation of the relevant budgets.
- Central government is also doing its part to ensure adequate parking for bicycles. For the next seven years, the Ministry of Transport has allocated about €210 million (£130 million) to be spent on improving and expanding parking areas and facilities for bicycles, in and around railway stations.
- In September 2001 the Minister of Transport set up the Cycling Council. It is a small organisation with an annual budget of €450,000 (£280,000) and brings together policy makers and experts from the cycling world, to co-ordinate activities like acquiring knowledge, distributing it and keeping it up to date. Making reliable data available for local, regional and national use is also important for processes like benchmarking, monitoring and policy making.

13

German cycling policy experience

Wolfgang Bohle, Planungsgemeinschaft Verkehr, Hanover

13.1   Introduction

In Germany an average 11% of all trips are made by bicycle. Public transport has a share of 13% while 31% of all trips are made on foot (Bundesministerium für Verkehr, Bau- und Wohnungswesen, 1999).[1] There are large regional differences in cycling. In some towns, e.g. Münster, Bremen, Freiburg or Erlangen, cycling has a share of 20–35% of all trips. Otherwise, due to topography, socio-demographic characteristics, local attitudes or a lack of infrastructure, in some other towns and regions less than 5% of all trips are made by bike. Compared with other European countries,[2] in the towns mentioned above cycling has a very high share whereas the average German share reflects the European average.

In the 1980s and 1990s, the bicycle enjoyed a renaissance. The share of cycle traffic since the 1970s has increased by more than 50%. Particularly, more well-educated people of higher social status use bikes. Many local authorities and nearly all federal states promote cycling. For a long period, they concentrated on the infrastructure and on traffic safety activities. More recently other spheres of activity to promote cycling have become more important.

In co-operation with the federal states and municipalities, NGOs and experts, the German Ministry of Transport has produced a National Cycle Strategy (Nationaler Radverkehrsplan) up to 2002 focusing on three elements:

- the infrastructure for cycling, parking and combining the cycle and public transport;
- cycle-related services, offered by the private sector and sometimes initiated by the state or the municipalities;
- communication and campaigns in order to promote cycle-oriented attitudes and knowledge.

Besides the states, the municipalities and the cycle-related NGOs, the Nationaler Radverkehrsplan is also aimed at winning over more private sectors – such as employers or shop owners – to the promotion of cycling.

The Radverkehrsplan aims to increase cycling as well as to improve traffic safety for cyclists. Although the number of cycling fatalities since the 1980s has decreased by about 30% and the number of seriously injured cyclists also slightly decreased, the accident risk for cyclists in relation to travel time is still higher than for car users or pedestrians. By comparison with the accident risk in some other European countries too, improvements of safety for cyclists remain on the political agenda.

The following sections briefly discuss cycle strategies and the experiences of two municipalities, Kiel and Troisdorf. Troisdorf represents a concentrated project-oriented strategy, Kiel – more typical of German municipalities – a continuous promotion of cycling, integrating both infrastructure and communication as strategies. In addition, three examples of systematic cycle strategies will be discussed: a continuous co-operation of municipalities, a systematic approach to combine cycling and public transport and two private enterprises promoting cycling. After describing recent cycle parking regulations in broad outline, some German experiences of relevance for other countries are discussed.[3]

13.2   Promotion of cycling by municipalities and federal states

13.2.1   Municipality of Kiel

Kiel, the capital city of the federal state of Schleswig-Holstein in the northern part of Germany, has a population of about 235,000 and is situated on the Baltic Sea. For about 15 years the municipality of Kiel has promoted cycling both in terms of infrastructure and communication measures.

A cycle officer in the municipal administration works as a contact person both for the public and for colleagues within the administration. The officer has, *inter alia*, to ensure a good technical standard for the cycle infrastructure in planning and maintenance of roads.

A cycle forum serves as an advisory board for the municipal council. In the forum the local NGOs, the police, members of the municipal council and the administration departments dealing with the bike are all represented. Meetings of the forum are held six times a year.

Cycle routes as an important element of the cycle network connect the centres of different parts of the town, the central business district (CBD) and the university. The route network – totalling 105 km in length – consists of minor roads and of high standard cycle paths or cycle lanes marked on major roads. Due to financial restrictions, cycle lanes have become increasingly important. By 2000, about 30 km of the route network were completed. With regard to a concentration of quality defects and of spatial competition between cars and bikes, priority for further completion lies on route sections improving the accessibility of the CBD.

Fig. 13.1   A new bridge for pedestrians and cyclists across a bay of the Baltic Sea.

There is special emphasis on minor roads which are – because of their lower volume of motor traffic – more attractive for the users. With regard to safety and attractiveness, there are for example good experiences with a 'cycle-road' (Fahrradstraße). Here the carriageway is signed as a cycle route that may be used by cars travelling at the same speed as cyclists. Shortly after completion of one of the cycle-roads, the volume of cyclists had increased by about 30% and now exceeds the volume of car traffic.

The western and eastern parts of Kiel are connected by a newly constructed bridge for cyclists and pedestrians crossing a bay of the Baltic Sea (Fig. 13.1). Compared to the road network for motor traffic, important connections for non-motorised traffic are shortened by about 1 kilometre.

In addition to the cycle routes, nearly every road with a through traffic function for motor traffic in Kiel has segregated cycle paths or lanes. There are large differences in quality, so the standard has to be improved continuously. With the exception of the through roads, the whole city is covered by zones with a speed limit of 30 km/h or by pedestrian areas. In combination with continuous speed controls and safety campaigns, the speed limits serve to encourage changed behaviour by drivers.

As of 2001, the municipality offered more than 2500 public cycle parking facilities in the CBD. In the district centres and at important public transport stops, additional parking facilities are available. The rack used as a standard parking facility offers good support and allows the user to lock the frame and both wheels of the bicycle. It is suitable for child bicycles too.

During the construction or redevelopment of buildings located at cyclists' destinations, the developer is obliged to establish parking facilities for bikes. At existing buildings, the municipality offers technical support to owners and house users for implementing new parking facilities in public spaces.

In addition the municipality plans to establish a cycle station at the central station (see section 13.2.3).

The infrastructure facilities for cyclists are supported by continuous communication. The municipality regularly publishes information leaflets about safety aspects and infrastructure facilities. Free brochures serve to influence attitudes towards cycling and to promote changed behaviour by road users in general. In co-operation with NGOs, with the university and even local shop owners the positive individual effects of cycling are pointed out.

Due to strong financial restrictions, the marketing activities are regularly supported by insurance companies or other private sponsors. Special emphasis is given to the continuity of communication. There are no short term campaigns but the identified audience for information and marketing activities can be rather well targeted.

In summary, the municipality of Kiel continuously tries to support cycling both by infrastructure and by communication measures. Although the bicycle does not always have priority on the political agenda and – as is typical of many German towns – finances for bicycle measures often compete with other local investments, cycle promotion has had positive effects.

For example, the data available show a growing share of the bicycle in traffic. In 1988 there were only 6% of clients in the CBD getting to the shops by bike (Bundesarbeitsgemeinschaft der Mittel- und Großbetriebe des Einzelhandels, 1989).[4] In the late 1990s the general share of cycle traffic reached 14%. When going shopping or on errands such as trips to the bank or doctor, about 20% of the people of Kiel now choose to cycle.[5]

The traffic safety of cyclists too has been improved. In a period of three years, from 1995 to 1997, an average of 2.3 cyclists per 10,000 people per year were killed or seriously injured in accidents. This number decreased in the period 1998–2000 to 2 cyclists per 10,000 people per year. In spite of increasing cycling, the risk of being injured seriously also decreased in comparison with other German towns (see section 13.2.4).

In the 1980s there was a weak bicycle culture in Kiel. In spite of regular debates about the traffic behaviour of (some) cyclists more positive attitudes towards cycling now predominate: about 85% of the adults in Kiel in general like cycling (Flade, Lohmann and Bohle, 2001).

13.2.2 Municipality of Troisdorf

With a population of about 70,000, Troisdorf – located about 40 km south-east of Cologne – is an example of a medium-sized city that has succeeded in increasing cycling significantly without decreasing walking or public transport. With rather a small bicycle tradition, Troisdorf from 1989 to 1996 became a demonstration project within the programme 'bicycle in communities' of the federal state of North-Rhine-Westphalia. About €13 million (£8 million) were invested by the state and by the municipality.

As well as cycle infrastructure – especially cycle lanes and traffic calming measures – bike parking facilities were established together with bike-and-ride facilities, direction signing and so on. In addition there was a continuous campaign to support public awareness of the individual advantages of the bicycle. Some surveys were carried out at the beginning and at the end of the project in order to discover its results. Within the municipal administration, a project board steered the measures. The municipal council established a special committee for the project. The communication campaign consisted of continuous information for the people living in Troisdorf, a traffic award, information for teachers, a calendar of 'bicycle culture', some bicycle festivals and fairs, an information centre, an international conference, excursions and public exhibitions.

As a result, the municipality succeeded in increasing the share of cycling in traffic from 16% to 21%. Before the project, 56% of all trips were made by car. This share went down to 51%. For trips within a distance of 5–10 km, the share by bike increased from 5 to 16%. At the beginning of the project, 87% of the people felt unsafe when cycling in Troisdorf. In 1996, this proportion had been reduced to 44%.

13.2.3 Cycle stations

Since 1995, the federal state of North-Rhine-Westphalia has run a programme called '100 cycle stations'. The aim is to improve the combination of bikes and public transport, to rejuvenate railway stations with regard to architecture and urban functions and to offer new jobs. By 2001 about 50 cycle stations had been completed at railway stations.

Cycle stations now regularly provide parking facilities protected from rain and theft, cycle hire and technical support or repair shops. Additionally, there are further local facilities at the station such as ticket selling or cycle courier services. Most of the stations provide between 100 and 400 bike parking places. However, the capacity of the largest station in Münster totals about 3000 spaces.

There are both public and private partners responsible for the programme and the stations:

- The state of North-Rhine-Westphalia supports the construction of the stations financially.
- An agency of the Allgemeiner Deutscher Fahrrad-Club (ADFC, a German cycle-related NGO) manages the development and regularly organises an information exchange for the operators of the stations.
- The local authorities establish the cycle stations and co-operate with the operators.
- Local operators of the stations include cycle shops and social organisations. Many stations offer jobs within the framework of social or work-qualifying

programmes. These jobs – and hereby the running of many stations – are supported by the state of North-Rhine-Westphalia too.

• According to a contract with the state of North-Rhine-Westphalia the German railway company (Deutsche Bahn AG) makes the buildings available to the stations for 10 years without rent.

The bike parking fee and the charge for hiring a bike are the same at each station that belongs to the programme. A hired bicycle can be returned at every other station too. Public relations are managed centrally by the agency developing the stations.

In comparison to conventional parking facilities at railway stations, the cycle stations enlarge the radius of potential customers for the station. More than 30% of the customers cycle a larger distance than 2.5 km to or from the station. The stations have even succeeded in winning new customers for public transport. Shortly after establishing the stations, 15% of the customers who previously used to take their car for the whole trip now combine going by bike and public transport.

By 2000 40 people had secure jobs in 30 cycle stations. More than 200 people worked within the framework of a qualifying programme (Ministerium für Wirtschaft und Mittelstand, Energie und Verkehr Nordrhein-Westfalen, 2001).

Due to the limited duration of jobs in the qualifying programme and the heavy running expenses of bike parking facilities controlled by the station staff, there are now some discussions about automatic entrance controls. Some of the stations in North-Rhine-Westphalia already combine entrance control by staff with an automatic entrance control system for the night period and with video surveillance. On the other hand, some new parking facilities at railway stations – for example Celle in the Hanover region, Lower Saxony – are accessible exclusively by an automatic entrance control system. They often provide less service but they are accepted by the customers and running expenses can be reduced.

13.2.4   Co-operation of municipalities
In the federal state of North-Rhine-Westphalia, 29 municipalities are members of a working group Cyclist Cities (Arbeitsgemeinschaft Fahrradfreundliche Städte und Gemeinden in NRW). Since 1993, the working group serves:

• to integrate and to standardise different finance instruments of the state and of the municipalities for cycle traffic;
• to establish a continuous exchange of information and experiences among the municipalities; and
• to support an accelerated realisation of local cycle concepts.

Members of the group include municipalities and regions of different size and geographical and social structures and with different starting points with regard to cycle infrastructure and to the modal split. In this way, the Cyclist Cities may develop best practice solutions for many other municipalities and regions. The long term aim of the state of North-Rhine-Westphalia is to reach an average share of 25% of the bike in urban traffic.

The municipalities involved in the group establish different facilities for cyclists, such as:

- an adapted infrastructure both on main roads and minor roads;
- a continuous infrastructure at both signalised and non-signalised junctions, as well as roundabouts;
- measures for changed behaviour by car traffic;
- parking facilities;
- the combination of bikes and public transport;
- leisure offers for cyclists;
- direction signposting for cyclists;
- measures for the reduction of cycle accidents;
- communication and public relations for cycling in co-operation with NGOs, shop owners and enterprises; and
- support of bicycle services as cycle courier services.[6]

In Germany – with the exception of the municipalities in the federal state of Schleswig-Holstein (northern Germany) – only the municipalities of the group in North-Rhine-Westphalia get financial support from the state for each bicycle-related measure. As a partner of the group the state regularly organises or supports bicycle conferences and publications. In combination with internal discussions, forums and a new service on the worldwide web they serve to inform the municipalities of the group and to inform other municipalities. Additionally, the state develops uniform elements of communication for the Cyclist Cities.

For the municipalities, the group has become an informative and motivating forum. It supports the role of the bicycle on the political agenda, both at the local level and at the federal level. At present, there are discussions on a more formalised and more professional structure for the group, on a more effective evaluation of its activities and on enlarged financial priorities for members with regard to the subsidy instruments of the state.

The activities of the municipalities have already shown good results in terms of the development of cycling and also positive tendencies with regard to cyclists' safety in traffic. At first, the municipalities that became members of the group in 1988 had more accidents with seriously injured cyclists. This was caused by the increasing number of cyclists and, over the first three years, an insufficiently improved infrastructure. Then, subsequently, the accidents involving seriously injured cyclists went down to the general level for North-Rhine-Westphalia (Fig. 13.2).

On the other hand, the share of cycling in these Cyclist Cities increased above the average of all cities in North-Rhine-Westphalia, so the risk of injury per individual cyclist decreased (Fig. 13.3). This corresponds to the development in the towns that only later became members of the working group, with regard to the total number of injured cyclists, i.e. including slightly injured cyclists.

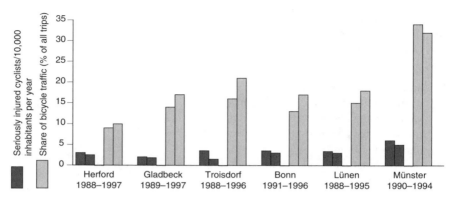

Fig. 13.2   Share of bicycle traffic and seriously injured cyclists per 10,000 inhabitants.

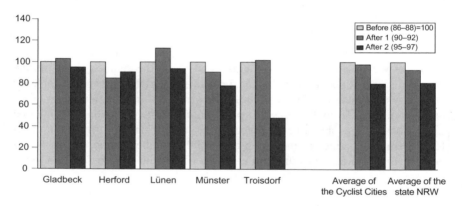

Fig. 13.3   Number of seriously injured cyclists in Cyclist Cities since 1988.

As shown by a recent Danish investigation too (Jensen, 1998), the accident risk for cyclists decreases in the North-Rhine-Westphalian towns when the number of cyclists increases (Fig. 13.4) (Alrutz *et al*, 2000). A large number of cyclists may create more awareness of their presence by other road users. In addition, in the towns with more cyclists the cycle networks tend to be better developed.

13.2.5   Building regulations and bicycle parking facilities
The building regulations of the federal states provide support for the building of cycle parking spaces during the construction of residential buildings or the carrying out of major redevelopments and also for buildings which are important destinations for cyclists. In most federal states the local authorities are authorised to issue corresponding local by-laws. Some federal states stipulate these obligations in the state building regulations.

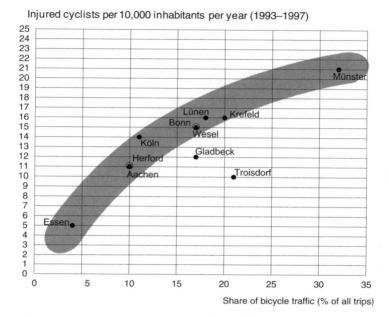

Injured cyclists per 10,000 inhabitants per year (1993–1997)

Share of bicycle traffic (% of all trips)

Fig. 13.4   Number of injured cyclists and share of bicycle traffic in several towns of North-Rhine-Westphalia.

In the first period of the 1990s after these regulations had been introduced, there was widespread uncertainty with regard to their application and to an appropriate number of parking facilities. But now the communities have some positive experiences with the regulations. The number of communities introducing corresponding local by-laws is increasing and in a survey by the German Ministry of Transport, most of the municipalities involved regularly took the regulations into account when giving planning permission. However, some other communities tend to neglect the regulations due to a general reduction of building regulations (Alrutz and Bohle, 2001).

Some municipalities feared establishing an obstacle to investment when applying the full requirements for cycle parking facilities. This fear particularly existed in the period of introduction of the regulations. But when applying existing regulations too, some municipalities reduced the required number of bike parking facilities to a much lower capacity than needed. On the other hand, the municipalities already applying a local by-law since the early 1990s had no negative experiences with regard to investment.

With an increasing number of municipalities introducing local regulations or applying the federal regulations, the technical quality of bike parking facilities has got better. For public parking facilities established by municipalities, a rack allowing the user to lock the frame and a wheel of the bike has become the standard solution. Private building owners also often use this type of parking facility.

The survey mentioned above worked out an adequate number of parking facilities for different buildings' uses. Some results are:

- For office use purposes and office-type service use purposes demand is determined by the proportion of employees who use bikes and by the number of customers using bikes.
- For retail trade use the following differentiation can be made:
  - small shops for daily requirements,
  - retail outlets under $1200\,m^2$ in area selling, e.g., clothes or books,
  - retail trade supermarkets under $1200\,m^2$ in area and
  - retail trade in larger areas.

Due to their workplace densities, customer frequencies and marked variations in the amount of customers using bikes, they have differing demands for parking spaces.

- Large scale leisure facilities with catchment areas extending beyond local requirements, such as cinema centres, have a considerably lower demand than local-orientated leisure facilities.
- The proportion of people using bicycles among building users can be influenced by
  - the location of a building in the urban area,
  - the provision of motor vehicle parking spaces,
  - the quality of the local public transport connections, and
  - the topography.
- Because of the different requirements placed on the location and the implementation of the parking spaces, the guidance values for the number of parking spaces – as worked out in the survey – derived from this are specified separately for employees and customers.

13.3   Cycling to work: promotion of cycling by enterprises

13.3.1   Lufthansa Maintenance Service

At the Lufthansa Maintenance Service in Hamburg, about 7500 employees maintain and repair aeroplanes of the German aviation enterprise Lufthansa. Due to restrictions in car parking capacity and to some severe accidents among employees with a car, the enterprise and the works council made a commitment to support alternative modes of transport for the employees on their way to work.

In the 1990s, 30 decentralised cycle parking facilities were established. On the large area of the enterprise, they are provided directly at the workplaces. Information on cycling and individualised proposals for attractive routes from home to the enterprise are available as leaflets and on the intranet. In addition a cycle repair service for the employees was established.

Due to the long distances many employees have from home to work, there are additional incentives, e.g. solar-powered vehicles, a car sharing system and attractive public transport offers. Only employees getting to work by bike, with a solar-powered vehicle or sharing a car may enter the enterprise area directly. The parking spaces for cars are outside the area, so car drivers have longer distances to walk to their workplaces.

Just some weeks after establishing the cycling facilities, the number of employees cycling to work increased by 50%. About 10% of the employees now get to work by bike. This share at the Lufthansa Maintenance Service, situated on the outskirts of Hamburg, is twice as high as the average share of bike use at enterprises in a comparable location. In particular the fact that some members of higher management cycle to work may indicate the increased role of the bike. The offers for people sharing a car and for public transport users led to a comparable reduction of car trips to work (Hermann, 1996).

13.3.2   Nordmark Pharmaceutical Enterprise

The Nordmark Pharmaceutical Enterprise in Uetersen (Schleswig-Holstein, northern Germany) has for several years been committed to supporting cycling to work. The following are ways in which it does this:

- For 340 employees, there are 337 covered cycle parking facilities directly at the entrance of the workplace buildings. Parking spaces for employee cars are situated outside the enterprise area at about 400 metres from the building. The enterprise area is accessible only for a very few cars of the enterprise service and for visitors' cars.
- For the employees' use there are 30 service cycles belonging to the enterprise. Often the employees are invited to enterprise excursions by bike.
- Nordmark prepares individualised information for the employees about cycle routes from their home to work. Complementary to a planned information campaign, the medical service of the enterprise also encourages cycling to work by employees.
- As the firm is located in a flat region of northern Germany with strong winds, there are also wind indicators to warn of sudden cross winds.
- As a special reward, the enterprise plans to establish a cycle lottery. There is also co-operation with the municipality of Uetersen which serves to improve the cycle infrastructure on the employees' routes.
- Special showers and cabins for changing their clothes are available for the employees.
- For staff who do not know how to cycle, cycling lessons will be held. For those preferring to cycle to work accompanied by their colleagues a cycle contact service is organised. A repair service too is under preparation.

As a result of the common activities of the enterprise and the employees, nearly 45% of the employees cycle to work.[7] In the Schleswig-Holstein region and even in Germany this represents one of the highest shares of cycling in work traffic.

13.4   Comment: relevant German cycling policy experiences

One of the most important experiences of German municipalities is that systematic cycle policies may at the same time both increase cycling and improve cyclists' safety in traffic. Up to a share of about 35% for cycling – as represented in German towns – a well-developed cycle network both encourages cycling and reduces the individual accident risk.

According to German experiences, cycle policy has to include both infrastructure facilities and services for cyclists and communication, for the following reasons:

*   Car drivers very often do not notice new cycle infrastructure. These facilities have to be publicised in order to encourage cycling.
*   Many Germans have generally positive attitudes towards cycling. But the positive individual effects of cycling – health, time saving, promoting a feeling of well-being – are less known. Particularly with regard to the short distances of less than 5 km typical of most car trips, the individual advantages of cycling for some people stimulate cycling better than infrastructure alone.
*   Communication and service initiatives often reach (potential) cyclists directly. Service offers by employers or by colleagues, for example, are made within important social groups for the individual – and therefore may stimulate cycling effectively.

On the other hand, cycle policies orientated exclusively to communication measures would not succeed. For most cyclists, subjective feelings of traffic safety rank very highly. Particularly on roads with higher volumes of motor traffic or with higher speeds, this feeling primarily is given by a spatial separation from car traffic and therefore by a well-developed infrastructure.

With regard to the activities discussed above, some further experiences may be deduced:

*   Municipal cycle policies may be organised as a short term concentrated project or with a more long term commitment. Both may increase cycling and improve cyclists' safety. But a clear responsibility within the municipal administration and financial continuity are also vital. A political committee with members of the town council, of the administration and of NGOs too helps to promote cycle policy effectively.
*   The establishment of cycle infrastructure should be accompanied by measures to change the behaviour of car users. Connections for cyclists between different parts of towns may sometimes be concentrated on through routes without car traffic, but local cycling in towns and the access to buildings and cycle destinations normally have to be organised on networks also used by car traffic. Here, traffic safety too has to be promoted.

- Cycling even in suburban areas or in the outskirts of towns may become an important mode of transport. In combination with for example attractive public transport and a stimulation of shared use of cars, bicycle promotion may reduce motorised commuter traffic in suburban areas.
- Moreover, the combination of the bike and public transport particularly in suburban areas may be an attractive alternative to car use. Safe parking facilities at railway stations win over new customers both for cycling and to public transport.
- There are positive experiences with a supra-local responsibility for some elements combining cycling and public transport. Cycle parking facilities at railway stations often compete with more profitable uses of the station building. A supra-local institution – such as the federal states in Germany – often may more easily succeed in negotiations with the railway operators or the station owner. On the other hand, for the user partly standardised services at different stations are more reliable than different services.
- Marked cycle lanes on the carriageway for many users are – due to the proximity of motor traffic – less attractive than cycle paths next to pedestrian paths. On the other hand, cycle lanes – which cost the municipalities less – are well accepted by users and stimulate cycling too. A specific attractive element is cycle routes on minor roads. When following parallel to major roads or connecting important origins and destinations the routes may even win over people new to cycling.
- There are positive experiences with regulations for cycle parking facilities when building houses or changing the use of buildings. They are accepted by most developers, and the number – and even the quality – of cycle parking facilities may in this way be increased.

13.5   Notes

1 People more than 6 years old. The following data refer to different surveys summarised in Bundesministerium für Verkehr, Bau- und Wohnungswesen, 1999.
2 Only very limited current data are available for the central and eastern European countries.
3 The contribution focuses on western Germany. In eastern Germany, many municipalities also promote cycling, but very little data on the results are so far available.
4 Residential population of Kiel (Bundesarbeitsgemeinschaft der Mittel- und Großbetriebe des Einzelhandels, 1989).
5 Data from the municipality of Kiel. The recent share mentioned above covers the whole town area. There are some cycling surveys not completed as of 2001, so data focused on the CBD are not yet available.
6 The activities and experiences of Troisdorf, a member of the working group, are described in section 13.2.2.
7 Information from Nordmark Enterprise within the framework of the competition 'GewinnFaktorFahrrad', Verkehrsclub Deutschland, LV Schleswig Holstein, and Ministerium für Wirtschaft, Technologie und Verkehr Schleswig Holstein, 2001.

13.6   References

ALRUTZ, D and BOHLE, W (2001), *Bedarf für Fahrradabstellplätze bei unterschiedlichen Grundstücksnutzungen.* Berichte der Bundesanstalt für Straßenwesen. Heft V 79. Bergisch Gladbach, Bundesanstalt für Straßenwesen.

ALRUTZ, D et al (2000), *Begleitforschung Fahrradfreundliche Städte und Gemeinden NRW. Maßnahmen- und Wirksamkeitsuntersuchung.* Düsseldorf, Ministerium für Wirtschaft und Mittelstand, Energie und Verkehr NRW.

BUNDESARBEITSGEMEINSCHAFT DER MITTEL- UND GROßBETRIEBE DES EINZELHANDELS (1989), *Untersuchung Kundenverkehr 1988*, Köln, Bundesarbeitsgemeinschaft der Mittel- und Großbetriebe des Einzelhandels.

BUNDESMINISTERIUM FÜR VERKEHR, BAU- UND WOHNUNGSWESEN (1999), *Erster Bericht über die Situation des Radverkehrs in der Bundesrepublik Deutschland.* Bonn, Bundesministerium für Verkehr, Bau- und Wohnungswesen.

FLADE, A, LOHMANN, G and BOHLE, W (2001), *Stadtbericht Kiel, Forschungsvorhaben Einflussgrößen und Motive der Fahrradnutzung im Alltagsverkehr*, Darmstadt, Bundesministerium für Bildung und Forschung.

HERMANN, H (1996), *Das betriebliche Verkehrskonzept der Lufthansa-Werk Hamburg.* In: *Berufsverkehr als Managementaufgabe. Arbeitsberichte zur Verkehrssicherheit der Berufsgenossenschaft für Gesundheitsdienst und Wohlfahrtspflege*, Hamburg, Berufsgenossenschaft für Gesundheitsdienst und Wohlfahrtspflege.

JENSEN, SU (1998), *DUMAS, Safety of pedestrians and Two-wheelers.* Note no 51. København, Vejdirektoratet.

MINISTERIUM FÜR WIRTSCHAFT UND MITTELSTAND, ENERGIE UND VERKEHR NORDRHEIN-WESTFALEN (2001), *100 Fahrradstationen in Nordrhein-Westfalen. Ein Landesprogramm mit Zukunft. Bilanz, Chancen, Perspektiven.* Düsseldorf, Ministerium für Wirtschaft und Mittelstand, Energie und Verkehr NRW.

Chapter 14

Urban cycling in Denmark

Thomas Krag, Consultant, Copenhagen

14.1   Introduction

Cycling in Denmark is a common activity, especially in urban areas, with average figures such as those shown in Table 14.1.[1] A factor contributing to the high level of cycling is the relatively low level of car ownership in Denmark. On average there are 347 passenger cars per 1000 people.[2] Car ownership in Denmark is therefore still low compared to its neighbouring countries, the main reason being the rather high taxation on cars. The difference is, however, diminishing due to a general increase in Danish incomes, which has resulted in a significant increase in the number of cars in Denmark over the last 15 years.

14.2   Decrease in cycling – with exceptions

In parallel to the increase in cars there has – with some exceptions – been a decrease in the use of bicycles since the mid-1980s. Several studies indicate that the main reason for this is that Danes are normal, lazy people: once the car is there, it is very easy and tempting to use it, and the bicycle is left at home even on short trips, where it would serve as a quick and healthy alternative. Furthermore, the general perception of what is safe and responsible behaviour has made many people transport their children to kindergartens and schools by car.

There are, however, interesting exceptions to the trend. In the central areas of Copenhagen, for example, there has been a more than 100% increase in cycle traffic over the last 25 years, the increase being greatest over the last 10 years, where a significant increase in car ownership and use has also been found. In

Table 14.1   Modal choice in Danish cities

	No of trips per person per average day	Percentage share
Walking	0.54	18
Cycling	0.66	22
Car	1.52	52
Public transport	0.23	8
Total	2.95	100

Source: Statistics Denmark, 1997.[1]

Odense as well cycling has increased. Car ownership in the central municipalities of Copenhagen and in Odense is 223 and 303 per 1000, respectively.

The positive trend for cycling can be attributed to the efforts to promote cycling carried out by the municipalities in question. The municipalities of Copenhagen and Odense have both been active in the area, by providing infrastructure and by various campaign initiatives. Odense was appointed 'The Cycling City of Denmark' in 1997 and granted financial support from the Danish government to carry out various experiments over a five year period in the field of promoting safe cycling.

Positive measures for cycling cannot, however, in themselves explain the positive development of cycling in Copenhagen and Odense. Other important factors are car restraint measures such as car parking charges, car parking restrictions and restrictions on car access to the city centre. Rush hour congestion, where streets are still passable by cyclists (especially where cycle tracks are available), as well as slow or deficient public transport alternatives, also contribute to the popularity of the bicycle. In terms of time and economy the bicycle is highly competitive in the city centres, and is therefore used.

14.3   History

Up until 1950 the history of cycling in Danish cities is probably not so very different from most other cities of Europe. The bicycle was used widely, serving as the main means of transport. Then came a spreading out of the cities and a big increase in the use of cars, and the bicycle lost ground. In the 1960s the road network was greatly expanded and existing cycle tracks often removed in order to create space for cars. In cities whole blocks and even quarters were demolished to give space for new, wide roads.

What distinguishes Denmark from many other European countries, with Holland being the main exception, is that cycling never disappeared as a normal travel mode. The decrease was big, but the bicycle was still visible in urban traffic even when at its lowest. The cycling trend changed around 1975. A combination of the energy crisis and increased awareness of health and environment issues resulted in an increase in cycling. The increase was associated with big bicycle demonstrations and demands to the authorities to pay more attention to cycling.

Since then many new cycle tracks along roads have been built, and the tendency to remove existing cycle facilities was halted. In total numbers cycling probably peaked in 1985 and has since then experienced a decrease on the national level, but with at least some urban areas providing an exception to this general trend.

14.4   National Cycling Strategy

Car traffic has, with a few intermediate declines, been constantly increasing and still is. Traffic plans have pointed out the need for a reduction of CO_2 emissions, if not a reduction of car traffic as such, which during the 1990s led to the development of a Danish National Cycling Strategy.

The strategy is new in the sense that it does not only deal with existing cycling, to be secured by the provision of cycle tracks and other cycling facilities. The aim of the strategy is actually to create new bicycle traffic substituting urban car trips by walking and cycling trips. Making cycling conditions safer is still regarded as a main means of doing this, but also the importance of campaigns and the need for balanced development in cities are considered. The strategy is entitled 'Promoting safe bicycle traffic'.[3]

14.5   Cycle tracks and paths

There has been a long tradition in Denmark for building cycle tracks. A genuine Danish urban cycle track is situated between the roadside and the pavement and is separated from both by a kerb and a difference in level of 10–15 centimetres, exactly in the way that the usual separation between the road and the pavement is brought about (see Fig. 14.1). It has been proposed that the tradition originates from bridle paths, where a kerb was necessary to keep the loose soil for the horses in place.[4]

The reason for making cycle tracks was in the beginning to increase the comfort of cycling, giving an alternative to the normal street surface, which consisted of cobblestones. Later, in the 1930s, cycle tracks were constructed in order to organise the different means of transport on the streets. One could also say that this was done to get the cyclists out of the way, so that cars could go faster. Older pictures from Copenhagen show how cars could be trapped within a flood of cyclists.

Since 1950 the main reason for making cycle tracks has been to make safe conditions for cyclists. So far as comfort is concerned, the cyclist is today quite

Fig. 14.1   A typical Danish cycle track. Source: Thomas Krag.

often worse off with cycle tracks compared to using the road, because of too little attention being paid to the need for cycle track maintenance. Another challenge is represented by the extended use of granite for redesign of urban streets. Often the cyclists' need for even surfaces is forgotten in the strive for aesthetic solutions.

In general, however, cycle tracks serve the needs of cyclists very well, and are usually the backbone of the cycle network of Danish cities, enabling cyclists to use the major urban streets safely and comfortably. Cycle tracks are also very popular. Among planners as well as ordinary people they are believed to be a precondition for cycling. It is even generally believed that a full separation of motorised and non-motorised road users is the ideal situation. Apparently the obvious way forward, such a concept was developed under the headline 'The SCAFT principle' in the late 1960s and employed to a great extent for the construction of new urban agglomerations in Denmark from the 1970s onwards. So far the results have not been better than before. Fully separated path systems are fine for children going to and from school (see Fig. 14.2). Usually, however, the paths are felt to be socially unsafe and confusing with lots of detours compared to the road system, resulting in people walking and cycling along the roads where this was not intended. This is very similar to the experience of Milton Keynes in the UK.

Planners today tend to believe more in integration of cars and bikes where car speeds can be kept at a moderate level and, where car speeds are higher, separation with cycle tracks.

Fig. 14.2   The morning rush outside a typical school in Copenhagen. Source: Thomas Krag.

14.6   Traffic safety

Cyclists are generally safe on cycle tracks along roads. Side roads and bigger intersections, however, pose significant safety problems, not to mention roundabouts. The traffic rules in Denmark generally treat cyclists like all other vehicles, meaning that a car coming from a low priority road should give way to cyclists on the more important road. This also applies when a cycle track is present, and means that motorists turning right at junctions should also give way to cyclists going straight ahead. All Danish roundabouts have priority for the traffic circulating on the roundabout, meaning that those entering should also give way to cyclists circulating on the roundabout.

Research has shown that the safety problems at junctions and roundabouts generally are associated with motorists failing to watch out for – and give way to – cyclists. The urban traffic situation is very complex, and a 'new' element such as the cyclist is difficult to take into account. This was especially demonstrated in connection with some of the cycle tracks constructed in the 1970s and 1980s. It was a major surprise when such cycle tracks, built to make cyclists safer, actually resulted in an increased number of accidents at junctions.

The lessons learned are that it is very important to give all classes of road user the best possible opportunities to watch out for each other:

- There should never be a verge between the road and the cycle track close to the junction.

- For at least the last 30 metres the road and cycle track should be directly alongside each other, if possible with no difference in level.
- The stop line for cars should preferably be five metres behind the one for cyclists so motorists will not overlook cyclists when both are waiting at red lights.
- The extension of the cycle track should, at least at big junctions, be clearly marked with a distinctive colour and bicycle symbols repeated through the junctions.

Another important lesson is that the safety of cyclists improves as the number of cyclists increases. This is closely linked to the attention paid by drivers – if there are many cyclists, car drivers will more easily remember to look out for them.

More recently, retarded stop lines for drivers and blue cycle lanes in the bigger junctions have become widely used. Quite often one will also find that a cycle lane some 30 metres before the intersection ends in a combined lane for cyclists and for cars turning right. Cyclists do not necessarily feel much safer as a result of this measure, originally introduced to increase capacity, but it tends to be quite safe in terms of accidents between cars and bikes.

Roundabouts still represent a serious challenge so far as the safety of cyclists is concerned. Roundabouts with two parallel car lanes are especially problematic. In a number of cases the cycle track has been removed from such roundabouts, and cyclists have instead been obliged to give way at all entry and exit lanes, meaning that the passage through the junction becomes very annoying and time consuming. A much better, but far from cheap, solution is to separate cyclists and motorists in two levels at such roundabouts. A number of such roundabouts are found in Denmark.

Some 12% of the total road network in Denmark is equipped with cycle tracks. Taking into account that a large part of the network consists of residential roads and rural roads with low traffic volumes, this means that a major part of the most important urban streets have cycle tracks. This is despite the fact that the construction of an urban cycle track along both sides of an existing street is rather expensive, costing DKK 3–6 million (€0.4–0.8 million or £0.25–0.5 million) per kilometre.

It has indeed been shown that is possible to both increase cycling and improve cyclists' safety. Both in Copenhagen and Odense, for example, an increase in cycling has been brought about with a corresponding decrease in the number of accidents involving cyclists. This is probably partly due to specific safety measures, but it is also due to the fact that an intensive level of cycling automatically makes motorists more aware of cyclists.

A Collection of Cycle Concepts, including a lot of references, has been published as part of the Danish cycling strategy.[5]

14.7   The national cycling city of Denmark

Odense is situated in the middle of Denmark on the island of Funen (Fyn) and is, with a population of 184,000 in the municipality, the third largest city in

Denmark. The municipality covers 304 km². Odense and its suburbs have several cycle paths in green surroundings, some of them built on disused railways. North-east and south-west cycle routes were established through the city centre as part of an experimental scheme in the mid-1980s. Significant car restraint measures – pedestrianisation and streets redesigned to be used exclusively by cyclists and public transport – were also brought about in the city centre between 1985 and 1988. Shortly before this, on the other hand, a large and today much regretted road widening project, requiring the demolition of major parts of an old quarter, was carried out.

Odense is, among other things, famous for its Safe Routes to School projects. For decades a systematic redesign of roads and paths based on surveys of pupils has taken place.

A number of ideas for the promotion of cycling and cycle safety are currently being tried out in a four year state-supported project in Odense, with a total budget of DKK 20 million (€2.7 million or £1.7 million) including 50% support from the state. The aim of the project is to treat all aspects of the trip. No less than 50 different areas of action have been identified under five headings: the user, the vehicle, the home, the road and the destination.

Users and potential users are addressed through various campaigns. Children in schools and kindergartens are the target for some of the campaigns, designed to reach the parents as well. Another campaign given a special Odense angle is the country-wide bicycle to work campaign. A map of Odense showing facilities for cyclists has been distributed to every household in the city. The same is true for newsletters about cycling.

Efforts are being made to raise the standard of the bicycles used. Activities comprise exhibitions of attractive bicycles and arrangements to supply social workers, who typically cycle to their clients, with new, high quality bikes. The cycle parking centre at the station mentioned below can also be regarded as part of these activities. Special bikes can be borrowed at the municipality, giving the people of Odense the possibility to test them out. Another project is devoted to the use of trailers and trailerbikes for transporting children to and from kinder-gartens and schools.

Several activities aimed at improvements of roads, cycle tracks and cycle paths are being carried out. Some are directed towards safety and traffic calming, others at speed and comfort. A really innovative one has been the installation of a row of small light posts along a cycle track helping cyclists to avoid stopping at a red light – the world's first green wave for cyclists (see Fig. 14.3). To raise aware-ness about the quality of the cycle track surface among the staff responsible for road maintenance, bikes have been introduced to be used for surveying roads and tracks.

Cycle parking is another area where several activities have been undertaken. New, alternatively designed and weather protected cycle parking racks have been installed in the city centre (see Fig. 14.4). At the station a big cycle shop and a lot of parking facilities have been installed. A system has been developed enabling the user to lock the bike to the spot using the built-in cycle lock (virtually all Danish bikes are equipped with such a lock). Part of the parking takes place in

Fig. 14.3   The world's first green wave for cyclists, installed in Odense to help cyclists pace their journeys to reach traffic lights when they are on green. Source: Municipality of Odense.

a room in the basement. There is easy access to the room by ramps, and special care has been taken to give it a pleasant appearance. This part of the bike parking area is locked, and users must have a special key, rented on a daily, monthly or annual basis. The need already exceeds the available capacity. In total more than 2000 new parking places have been provided.

More information on the project can be found on the Internet.[6] The management philosophy is also included in *Collection of Cycle Concepts*.

14.8   Copenhagen – city for cyclists

Copenhagen, the capital of Denmark, is situated on the coast in the eastern part of the country. There are some 499,000 people in the municipality and around 1.8 million in the Greater Copenhagen area on the eastern part of the island of Sealand. The size of the municipality is 89.6 km². When figures for the municipality of Frederiksberg, situated within and surrounded on all sides by Copenhagen, are added in the total area becomes 98.3 km² and the population 590,000. The figures mentioned in the following section, however, relate solely to Copenhagen.

Fig. 14.4 New covered cycle parking area at night in Odense, a Danish city with a particularly strong commitment to cyclists. Source: Municipality of Odense.

Cycle tracks have been built since the beginning of the twentieth century. A specific cycle track plan did not, however, first appear until 1980. The municipality claims that it never removed any cycle tracks, as several other Danish municipalities did.

Many central offices and businesses are situated in Copenhagen, and many people commute into the city centre. Up till the early 1990s Copenhagen was exceptional in being a capital city but still allowing an almost free flow of cars on the streets. Since then the traffic situation has become international in the sense that roads are now very congested in the rush hours, with correspondingly slow car and bus traffic.

The bicycle traffic in Copenhagen is also intense, and has been increasing in the central parts of the city since about 1975. The flow of cyclists increases from the city boundary towards the centre. On the busiest street 20,000 cyclists per day pass by in either direction. The bicycle was used as a means of transport to work by 34% of the employees in 2000. A ring around the city centre was crossed by 160,000 cyclists and 315,000 motor vehicles on a week day the same year. Total cycling within the municipality is estimated to cover 0.96 million km per day, compared with 4.43 million km from motor traffic.

The backbone of the cycle traffic network is the cycle tracks. Cycle tracks are found along almost all major streets, the width usually being at least 2.1 metres. There are 307 km of cycle tracks and 10 km of cycle lanes in Copenhagen, which

Fig. 14.5   A well-functioning cycle lane in Copenhagen. Source: Thomas Krag.

means that more than 42% of the total road network of 378 km is equipped with cycle facilities on one or both sides. On top of this there are 43 km of green cycle routes.

In order to make fast progress towards having all major roads supplied with cycle tracks, cycle lanes have been established as an intermediate solution along the roads where cycle tracks are planned but not yet built. In contrast to the cycle tracks, which are separated by a kerb and a difference in level, the only separation between cycle lanes and the motor traffic is a white line. Cycle lanes have been shown to be quite effective, in spite of not giving the same degree of safety as cycle tracks. They are to a surprisingly high degree respected by motorists (see Fig. 14.5).

As a supplement to the cycle track network a plan for a network of high level cycle routes has been developed. The routes will be designed to enable the cyclists to go fast with as few stops as possible. Altogether the length of the planned route network is 100 km. Part of the network will make use of already existing roads or paths; the remaining 62 km has to be built from scratch. The network will be constructed to a high standard and the implementation of the plan is estimated to cost about DKK 500 million (€70 million or £40 million).

The municipality is using a special tool – the so-called Bicycle Account – to follow the development of the plans and monitor the situation. Hard figures about the route network and the number of accidents as well as figures about the user assessment of a number of issues are included in the Bicycle Account. Other issues covered are the cycling and traffic policy of Copenhagen. The Bicycle Account is produced every two years and has been published since 1995 (see Fig. 14.6).

User surveys have revealed that 60% of city cyclists are female, and that 48% find Copenhagen 'good' or 'excellent' for cycling. The satisfaction with the maintenance of cycle tracks and streets is much smaller, which has made the municipality increase the activities – and the budget – for cycle track maintenance. DKK 9.1 million (€1.2 million or £0.75 million) was set aside for maintenance in 2000. The key figures from the most recently published Bicycle Account (2000) are shown in Table 14.2. The Bicycle Account can be found on the Internet and is also available in print.[7] Other publications describe the bicycle strategies of Copenhagen in more detail.[8]

14.9   The City Bike – a positive publicity project

Copenhagen has become widely famous due to its City Bike project. Even before the launch of the project the rumour had spread all over the world that you could find free bikes to ride in Copenhagen. The basic idea of the City Bike is to have a fleet of public bikes which can be used by everybody on payment of a small deposit (see Fig. 14.7). The bikes and associated maintenance activities are paid for by advertisements on the City Bikes and special racks have been made for them. The racks are equipped with deposit/lock units like the ones used for trolleys in supermarkets. This is a very simple system indeed.

The idea has received much support and generated much enthusiasm since it was first launched in 1989. Several attempts to find private investors were made, but only with a significant input of public money and practical support could the project be realised in 1995. The municipality of Copenhagen has supplied the project with parking racks as well as the necessary space for them all over the city centre at no cost.

There have been many difficult experiences since then. Vandalism and theft of the bikes have been a major challenge. Today a very rugged bicycle design as well as a well-functioning maintenance system have been developed. Rehabilitees from the Reva Centre in Copenhagen look after most of the repairs. Four vans are used, enabling a large part of the repairs to be done on the spot.

The City Bikes are made with a special frame and make use of non-standard bicycle parts to reduce the risk of theft of single parts. The wheels are made in a plastic mould. The tyres are made of massive rubber, i.e. puncture-free but hard to ride. The City Bikes carry no lights, no lampholders and no luggage carriers. It is no great pleasure to take a ride on a City Bike and you do not feel like doing a longer trip.

The deposit needed to take a City Bike is a single coin of DKK 20 (€2.70 or £1.70). You are allowed to use the City Bikes within a certain zone in the city centre. Outside this zone it is forbidden to ride them, the penalty for doing so being as high as DKK 1000 (€135 or £80). The fleet of bikes is probably

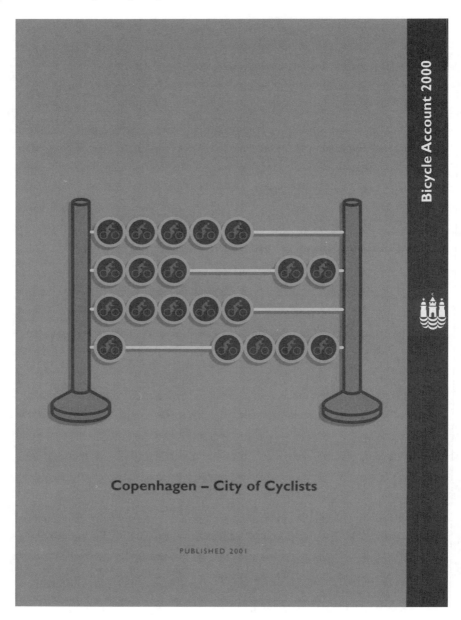

Fig. 14.6   The Copenhagen Bicycle Account.

Table 14.2 Statistics from *Bicycle Account 2000*

Cycle traffic	0.96 m km
Percentage who cycle to work (1999)	34%
Motor traffic	4.43 m km
Cycle tracks	307 km
Cycle lanes	10 km
Green cycle paths	43 km
Cycle track maintenance allocations	DKK 9.1 m/ €1.2m/£0.7m
Serious cyclist casualties (1999, corrected)	211
Signal junctions with cyclist priority	28%
Bicycle messengers, annual trips	877,800

Fig. 14.7 Two City Bikes by the harbour front in Copenhagen. Source: Soren Svendsen.

somewhat below 1500. Despite high public interest it has been difficult to sell the advertising space.

An investigation has shown that the bikes are used primarily by tourists and young men, most trips lasting just 5–10 minutes. It seems that less than 10% of the fleet is available in the racks, meaning that a good many must have left City Bikes elsewhere than in the designated racks. Compared to the number of people cycling on private bicycles, the City Bike adds virtually nothing to the cycle traffic of Copenhagen. The positive publicity of the City Bikes has, however, been limitless, and the idea has inspired several cities and countries to install similar systems. More information on the project can be found on the Internet.[9]

14.10   Notes

1 STATISTICS DENMARK (1997), *General Transport Survey: Municipalities with Cities over 10,000 inhabitants selected*, Statistics Denmark.
2 STATISTICS DENMARK (2000).
3 The Danish Cycling Strategy exists in English and can be accessed from the Ministry of Transport, trm@trm.dk.
4 DANSK CYKLIST FORBUND (1975), *Dansk Cyklist Forbund: 75 år,* publication for the Danish Cyclist Federation's 75 year anniversary, Copenhagen, Dansk Cyklist Forbund.
5 JENSEN, SU (ed) (2000), *Collection of Cycle Concepts*, Copenhagen, Danish Road Directorate, http://www.vd.dk/wimpdoc.asp?page=document&objno=17291.
6 The project concerning Odense, the cycling city of Denmark, has a website at http://www.cyclecity.dk.
7 The Bicycle Account can be found at http://www.vejpark.kk.dk/cykelregnskab/frames.html, or ordered (also in English) from vejpark@btf.kk.dk.
8 For example, THOMAS, M (1997), *Copenhagen City of Cyclists*, Copenhagen, Municipality of Copenhagen.
9 Information about the Copenhagen City Bike system can be found at http://www.bycyklen.dk.

15

Traffic calming on the national road network to improve cycling conditions in small towns in Poland: the case of Kobylnica Slupska on National Road 21

Andrzej Zalewski, Lodz University of Technology, Poland

15.1   Introduction

Among the principal road safety problems in Poland are the dangers to cyclists and other vulnerable road users and also road crossings in small towns. There are very dangerous traffic conditions for cyclists on through national roads with high volumes of motor traffic. One of the measures for improving cycling conditions and reducing dangers is traffic calming on these roads. This chapter discusses the implementation of a traffic calming scheme on National Road 21 in Kobylnica Slupska as a means of improving cycling conditions in a suburban area. This project was implemented between August 1994 and June 1995. It was the first realisation of traffic calming through a small town in Poland on the national road network. Six years after implementation the results are consistent with the original expectations. Traffic safety and conditions for cycling have improved significantly.

15.2   National context of bicycle use in Poland

During the 1990s a big change in the role of the bicycle as a transportation and recreational mode in Poland was observed. In the 1980s the bicycle was perceived as an element of anti-socialist activities and cycle infrastructure was not implemented. However, the bike always remained popular for recreation. Today, with a new age the attitude to the bicycle has changed. It is in vogue as a mode of recreation and in small and medium-sized towns as a daily means of transport. Local authorities take the bike into consideration as an element of urban and transportation policies and there are local programmes for the development of cycle infrastructure. They calculate that a pro-bike policy will help them win votes in local elections. These changes of attitude to cycling and infrastructure are the results of the opening of borders and frequent contacts and collaboration with western European local authorities, as well as being the result of the transformation of the democratisation process.

The extent of bicycle use in Polish towns depends on the size of the town. The bicycle as a transportation mode is most popular in small and medium-sized towns and in the suburban areas of conurbations, where accessibility to public transport is limited. In Poland, a bicycle is the most common household means of transport. According to the author's estimates, the number of bicycles per 1000 population is about 300 (Central Statistical Office, 2001). Mopeds represent only 5% of this number. Nearly 50% of bikes are of modern design, equipped with a minimum of 10 gears and 26-inch wheels. They are foreign products or assembled in Poland. The cost of a medium quality model mountain bike is the equivalent of one average month's industrial salary, which is too high for many people. There have also been big changes in the level of motorisation. Individual motorisation grew rapidly from 112 cars per 1000 people in 1990 to 258 cars per 1000 people in 2000 (Central Statistical Office, 2001).

The lack of physical protection against collisions discourages cycle use, especially as far as collisions with cars are concerned. Cyclists more often than drivers do not obey traffic regulations, and they are less experienced in traffic than drivers. However, drivers too do not respect traffic laws affecting cyclists. Streets and roads are very dangerous, because speeds are high and exceed the 60 km/h limit in urban areas. Drivers frequently use the footways as a place to park and block the passage of cyclists and pedestrians. It must be noted that in Poland parking cars on the footway is permitted! Data on traffic safety show that in 2000 a total of 692 cyclists were killed in accidents and about 7000 injured. The numbers killed amount to 11% of all people killed in road accidents, and 7000 cyclists injured amounts to 10% of all the injured in Poland (Zielińska, 2001).

There is also a great threat of cycle theft and much danger to cyclists' personal security. In spite of rigorous penalties, bikes are stolen from parking areas and from closed buildings. Away from the main streets, cyclists are attacked by hooligans and bikes are taken from their owners.

Bypasses for through traffic, which heavy vehicles use in great numbers, have not been built in all towns, with the result that there are heavy flows of through traffic in towns. Riding a bicycle in heavy traffic is not pleasant because a cyclist has to breathe in car exhaust fumes.

Numerous improvements in public transport have negatively influenced cycle use for local urban trips. The number of cycle networks and their length in Polish towns and conurbations are very low, but in recent years have been expanded to a total of about 1000 km (separate paths, shared paths for pedestrians and cyclists, on-street cycle lanes).

National transportation policy is very passive and regards the bike as only a minor means of transportation. Cycling represents a significant element of transportation policy only in a few towns where cycle paths exist or where strong pro-ecological attitudes prevail. Promotional actions are infrequent and only occasionally result in increased cycling. Changes in the official attitudes of government and local authorities are in many cases very slow. In some cases changes in attitude to cycle infrastructure problems are apparent. However, many local authorities and many traffic managers still do not feel the necessity or importance of improving cycling conditions in urban traffic or think that it is very simple to achieve. When municipalities' budgets are tight, local authorities face a problem in selecting new investments. They choose investments they regard as most indispensable. For this reason implementation plans are often postponed. This situation is not accepted by pro-bike campaign groups or generally by ecological organisations. They are very active particularly in big towns and conurbations. They organise many bicycle demonstrations and events, but their requirements are acted on only very slowly.

Polish traffic law includes an optional bicycle licence for children and young people between the ages of 9 and 18.

Whether there is a tradition of using bikes (a very important element of treating a bike as a means of transportation in Polish towns) or not depends on the region. The Bicycle Peace Race and Tour de Pologne have for many years aroused interest, especially among young people. As official surveys show, riding a bicycle is the most popular form of active recreation, after gardening for pleasure, and taking long walks.

Generally it can be stated that use of the bicycle in Polish towns varies and fluctuates from 0.04 to 0.42 trips per person per day. A bicycle in Polish climatic conditions is a seasonal means of transport. The distribution of temperature during a year results in the maximum use of bikes in the period between May and September. The month of considerable growth of bicycle use is April and the decline occurs in October.

15.3   Case of Kobylnica

15.3.1   Outline of the pre-project stage

The location of the experimental traffic calming scheme on the national road network is shown in Fig. 15.1. The traffic problems on the road through Kobylnica suburban village (2500 population) are the result of the following facts: traffic 5500 ADT (average daily traffic) vehicles/day in 1993, 7700 ADT vehicles/day in 1995 and 8500 ADT vehicles/day in 2000. These figures were recorded in the central zone of Kobylnica with its public buildings including a school, church, cultural centre, shops and public services. In the summer period, the traffic increased by nearly 30%, because of the tourist character of the road. National Road 21 in Kobylnica central area has a width of 7.0–7.5 metres with, in addition, a parking lane of 2.5 metres width. The footways (minimum 2.0

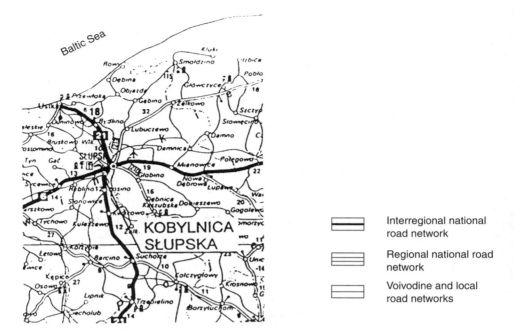

Fig. 15.1   Location of traffic calmed section in Kobylnica Slupska on National Road 21.

metres width) are situated on both sides of the road. The surface of this road is constructed in asphalt concrete and was in a good state before reconstruction.

From the north, this route leads traffic to and from Slupsk – the former Voivodine capital (to 1999). The distance between the centres of Kobylnica and Slupsk is nearly 2 kilometres. This highway fulfils a very important role in the national road network, because it is the only road connection of this class between central Poland and the central Polish coast (Slupsk, Ustka). It also has to provide heavy vehicle access to Ustka port. In Kobylnica centre, Road 21 crosses Voivodine Road 39177, at a complex junction. This road is a connection with another important route – National Road 6, Szczecin–Gdansk. In October 1993 heavy through traffic was moved here in order to relieve Slupsk centre. This category of traffic amounted to 17% of all traffic to and from Slupsk.

Furthermore the main (Główna) street (part of Road 21) runs by a school and a church, a zone where many pedestrians cross. Going in the direction of Slupsk, Road 21 crosses a local street, which is the transportation axis of the new residential zone built in recent years. The left turn in this intersection is very difficult from main street to local street, because it is very difficult to have to give priority to vehicles from the south.

The capacity of the junctions after the implementation of traffic calming and after changes of traffic organisation remains assured in spite of additional increased through traffic. The connection with the Voivodine road, which is

reduced in importance, works on an acceptable level for national roads in suburban areas (i.e. the second level of traffic service).

There is quite a high level of cycling – 60 bikes per hour at the analysed junction. It has a local and access character and a recreational function in the direction of the Landscape Park of Slupia river valley to the south of Kobylnica.

On the section discussed above, according to data from the regional police
office, during the years 1988–1992, 29 accidents were recorded, in which 11
pedestrians were killed or injured. Increased heavy traffic, with a long braking
distance, causes a further danger, especially for pedestrians and cyclists.

The other factor which makes conditions worse for traffic safety at the main
junction is poor visibility. Cars which turn left from the Voivodine road, observations show, force their right of way. The limited visibility for the cars passing
from the south makes for difficulties in reaction time (reduction of speed or
pulling up to make way for cars arriving from the Voivodine road).

The heavy traffic along the street on which the main public service buildings
are situated degrades the living environment and causes the disintegration of
social attitudes. It is necessary to note that according to American research (Spitz,
1982), streets are friendly when traffic flow does not exceed 1200 vehicles per
day, i.e. 120 vehicles per hour and 2 vehicles per minute. In the case of the two
streets mentioned above, these values are exceeded many times, and this causes
many difficulties.

Noise and vibration are most unpleasant for local residents. The estimated
existing intense noise level (70 dB) before traffic calming exceeded the admis-
sible level (45 dB) for suburban zones. Buildings situated along the main street
(National Road 21) were threatened by vibration.

15.3.2   Aims of the traffic calming project

In the Urban Plan for Kobylnica Slupska are plans for a National Road 21 bypass
of the village from the east, with connections to the street network of Slupsk also
planned. In practice, there has been no progress on this solution, primarily for
economic reasons.

Therefore, taking into consideration conclusions from analyses of the local
conditions and the impossibility of bypass construction, the AZ-Plan design
team suggested applying a traffic calming scheme to the main street through
Kobylnica following examples from other countries. In the opinion of AZ-Plan,
this solution would give priority to surrounding buildings and the living environment over the movement of traffic, while maintaining capacity, improving
landscape values and reducing adverse environmental impacts.

15.3.3   Outline of the scheme

Traffic calming in the central zone of Kobylnica Slupska on a length of about
1.4 km of the National Road shown below in Fig. 15.2 was approved by order of

Kobylnica Commune Council. The scheme was designed in two stages: the preliminary project (Zalewski and Cielecki, 1993) and the technical project (Zalewski et al, 1994). In the detailed projects mentioned above use was made of the traditional type of design of this section prepared by Bikotex-Nord Design Office from Gdansk. Experience gained by the author during a long collaboration with CETUR (Centre Etudes des Transports Urbains in Bagneux near Paris) and during training in ISIS Design Office in Lyon in 1992 was very useful in the elaboration of these projects.

There were two stages of liaison with important agencies: the former Regional Directory of Public Roads in Koszalin, the former Transportation and Construction Department of Slupsk Voivodine Office and the Road Traffic Department of the former Voivodine Police Office.

There was consultation on the project with the employer, i.e. the council and councillors of Kobylnica Commune. In the author's opinion, lectures on the implementation of the traffic calming idea for small towns and villages on national transit roads, presented at special seminars in Warsaw, Koszalin and Kobylnica, were very important elements in the procedure for design and realisation of the project. During these meetings, officials at different levels could be convinced of the value of the traffic calming idea and of the case for the realisation of the pilot section in Kobylnica.

The traffic calming scheme in Kobylnica Slupska is designed according to the following principles:

- Aims of traffic calming:
  - improvement of traffic safety conditions,
  - improvement of cycling conditions,
  - improvement of conditions for pedestrians and the disabled,
  - reduction of environmental impacts,
  - improvement of road traffic conditions at two main junctions,
  - improvement of public transport services,
  - improvement of aesthetic values and the layout of streets.

- Means of traffic calming:
  - introduction of 40 km/h speed limits and adaptation of street geometry to encourage compliance with this limit,
  - designing of shared pedestrian/cycle paths or footways or other forms of bicycle roads,
  - relocation of bus stops in bays.

In the preliminary project (Zalewski and Cielecki, 1993) the following traffic calming means were considered:

- 40 km/h speed limits;
- speed reduction at junctions from 60 km/h to 40 km/h by optical braking (red thermoplastic perpendicular road markings) and gateway effects (gate elements with enlarged signposts limiting speeds to 40 km/h, situated outside the shoulders or in the sidewalks);
- realignment of carriageway on right sections;

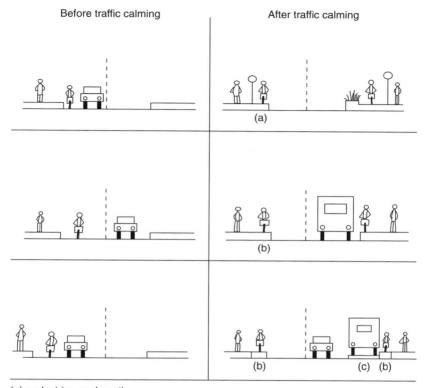

Before traffic calming   After traffic calming

(a)

(b)

(b)

(b)   (c) (b)

(a) pedestrian-cycle path
(b) cycle lane
(c) bus bay

Fig. 15.2 Variants of transformations of selected sections of National Road 21 in
Kobylnica Slupska before and after traffic calming in stage of preliminary project.

- reduction of the width of the lane from 3.5 metres to 3.0 metres;
- channelisation islands (central refuges) in the junctions or in the sections
  between junctions;
- resurfacing in places where point calming of traffic or separate left turn lanes
  are necessary;
- parking bays changed to longitudinal parking along part of the modernised
  street section;
- road markings.

In the technical building project (Zalewski et al, 1994) the following solutions,
illustrated in Figs. 15.2 and 15.3, were used:

- width of traffic lane 3.25 metres instead of 3.0 metres as initially proposed;
- shared cycle and pedestrian path (minimum width of 2.5 metres) along the
  road, with differentiation of surface colour by concrete bricks (pedestrians are
  situated between cyclists and the fence);

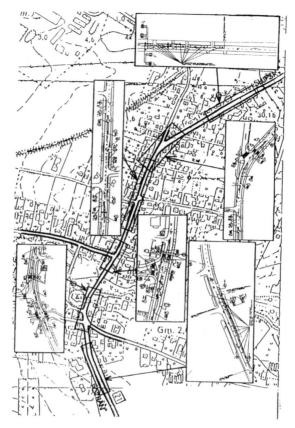

Fig. 15.3   Detailed elements of the traffic calming scheme in Kobylnica Slupska.

- limitation of traffic speed by signs to prohibit speeds of more than 40 km/h; signs are located above intersections to remind drivers of the speed limit;
- standardisation of junction types and central refuge islands on the sections between crossroads;
- relocation of signs outside the footways taking account of the width of the shared pedestrian–cycle paths;
- preservation of existing lights on pedestrian crossing near the school and church zone.

Furthermore, having regard to economics, the range of traffic calming means used and the extent of changes of street geometry were minimised. The surfaces of the channelisation islands, pedestrian–cycle paths, parking lanes and bus stop bays are designed in concrete bricks in different colours.

15.3.4   Rebuilding stage

The rebuilding of National Road 21 was carried out from August 1994 to June 1995 with a break during the winter. Special difficulties appeared during the reconstruction of the new bridge. It required moving traffic onto part of the recon-structed section. Speeds of traffic decreased to 30–40 km/h during the period of construction works. The rebuilding of National Road 21 was a specific invest-ment with a design that differed from traditional layouts. Many corrections of line of kerb, new construction of channelisation islands, longitudinal parking along part of the modernised street section, as well as bus stop bays required many small works and higher precision in execution than in typical roadworks. During implementation the design of the gate elements was successfully changed.

15.3.5   Post-rebuilding stage

Traffic calming investment allows for a range of experimental solutions of this type in Poland and from this point of view the evaluation was a very important element of adapting the process of traffic calming to Polish conditions.

After one month, one year and six years of use, the traffic calming project through Kobylnica was shown to be satisfactory in terms of its traffic, spatial and social aspects. This is confirmed by the results of the survey and the opinions of local residents, road users and the road and traffic management institutions, as well as the media (Zalewski, 1995, 1996 and 2001). Photographs of the imple-mented solutions are presented in Fig. 15.4 (a) and (b).

Fig. 15.4 (a)   Traffic calming in Kobylnica Slupska: differentiated surface for cyclists (red) and pedestrians (grey) on the shared cycle–pedestrian path.

Fig. 15.4 (b)   Traffic calming in Kobylnica Slupska: separate lane for left turn.

Transportation aspect
The implementation of 40 km/h speed limits by the adaptation of street geometry with growth of traffic density has meant that the following aims have been
achieved:

- Average speeds through the village were reduced from nearly 70 km/h before
  to 48 km/h in 1995 and to 44 km/h in 2001. The average speed level for 85%
  of traffic (V85%) was 50 km/h in 1995 and 48 km/h in 2001 depending on the
  part of Kobylnica; this result means that traffic is more even in character, and
  that the reduction of average speed is effective and has the potential to
  improve safety levels through Kobylnica.

- Average journey time through Kobylnica increased from 24 seconds to 41 seconds.
- Average waiting time at the principal junction in the village was 17 seconds. This means that despite increasing traffic volumes, there was a minor improvement in traffic conditions, because average waiting time at this give-way junction did not decrease very much.
- The safety level for all road users was improved significantly; within one year of the reconstruction fatal accidents disappeared and only one injury accident was recorded. The improvement of conditions for the safety of pedestrians and cyclists is a result of the transformation of the use of street space and the segregation of cyclists outside the carriageway. Shared bike–pedestrian paths and crossings at intersections were created and new road signs and markings were introduced, as well as adapting the street geometry, with different colour surfaces for different road users.
- The geometry of the implemented solution obliged drivers, especially heavy vehicle drivers, to reduce their speeds and to be more careful driving, helped by the channelisation of traffic flow in the junctions and in the sections between junctions.
- The formation of ruts has been observed on the carriageway in the sections between the channelisation islands and kerbs, where there are narrowings. This means that the surface in these places must be made more durable in future traffic calming projects (for example using concrete or paving stones).
- The traffic calming scheme has withstood its winter tests well. The cycle paths then change their functions. During the winter they are used for storing snow. During the winter of 1995/1996, cycling practically died out because of the long lasting and hard frosts at this time.

Socio-economic and spatial aspects
The results of the implementation of the traffic calming scheme in terms of its socio-economic and spatial aspects are as follows:

- The aesthetic image of the main street was improved significantly; it gained a more modern and urban appearance.
- Social and neighbour contacts were increased.
- The transformation of the street appearance and new aesthetic aspects stimulated local initiatives: summer gardens were planted in front of a bar and a shop; many local residents renovated the fronts of their houses and fences.
- The residents of Kobylnica Commune petitioned the local authorities to continue construction of the shared cycle/pedestrian path and crossings at junctions in the direction of the neighbouring Landscape Park of Slupia river and to reconstruct National Road 21 through neighbouring villages with blackspots.

The cost of the traffic calming project in Kobylnica (with reconstruction of the roadway throughout its length and construction of a new bridge) was €270,000 per km ($US250,050) and was nearly 20% more expensive than a traditional solution; the authorities of Kobylnica Commune contributed 20% of the costs. This is very important, because the reconstructed road is managed by the Regional Division of the General Directory of National Roads and the Commune does not have to pay for this road network.

Environmental aspects
Adverse environmental impacts were reduced. Surveys of noise and air pollution before and after reconstruction were carried out at the same points and in the same conditions by the local division of the National Environmental Service and by the Voivodine division of the Sanitary Service from Slupsk. They showed that the level of noise had not changed, but that air quality had improved because the speed of traffic was reduced. In the designers' opinion, the improvement of traffic conditions at the principal junctions as well as bays for bus stops and car parking had a significant influence on improving air quality.

Results of the surveys of local residents and drivers
Surveys were undertaken by the AZ-Plan design team one month, one year and six years after implementation, among adult residents and drivers passing through Kobylnica Slupska (90–110 interviews in each group of road users). The results of the three surveys (Zalewski, 1995, 1996, 2001) were similar and they can be summarised as follows:

- The average value of the whole traffic calming project on a point scale from 1 to 10 in two opinion surveys of local residents was nearly 7 points, while 'drivers' evaluated the scheme higher, at 7.3 points (Zalewski, 1995).
- In the range of evaluation of different project elements, the most frequently mentioned by local residents were:
  - 40 km/h speed limits,
  - shared cycle–pedestrian paths (divided by surface colour),
  - signs and road markings;

The most frequently mentioned elements by drivers were:

  - 40 km/h speed limits,
  - channelisation islands at the junctions or in the sections between junctions.
- The optical braking and gateway effects were very seldom mentioned.
- Residents and drivers gave primary ranking to 'improvement of traffic safety conditions' and 'reduction of speed' as the aims of traffic calming. They are interdependent aspects.
- In answer to the question 'What did you like about the project?', the two groups of respondents were unanimous: 'aesthetic appearance' (i.e. the reciprocal composition of elements of a street)
- It is very difficult to determine the most important response to the question: 'What did you dislike about the project?' The most common response was 'too narrow lanes of traffic and channelisation islands', but that element was mentioned by only 8% of respondents.

- Residents and (passing) drivers had the same opinions in response to the question: 'Where else could traffic calming solutions through small towns and villages be implemented?' They gave the names of various small towns and villages on National Road 21 to central and south Poland. Respondents claimed that traffic calming schemes should be applied on the sections of national roads that are dangerous.
- Only 25% of drivers said that they had come across similar schemes abroad in countries such as Germany, France, Holland or Denmark, i.e. the countries where Polish travellers go relatively frequently.

It is difficult to draw precise conclusions from the survey because of the nature of the survey sample. Respondents were chosen very randomly, with different levels of road traffic knowledge (for example: some respondents claimed that traffic calming improved safety conditions, but would be better if traffic lanes were wider, i.e. two contradictory elements!).

Opinions among institutions responsible for road and traffic management, the former Voivodine Council, the Voivodine architect, the head of the Voivodine division of the National Police and independent experts, confirm that implementation of traffic calming along National Road 21 in Kobylnica Slupska is a satisfactory project that has made positive changes in traffic and the urban environment.

The case of Kobylnica was the first traffic calming scheme in Poland on a road through a small town. Monitoring of this case is therefore necessary and should be analysed comprehensively and systematically.

15.3.6 Conclusions and comments

The satisfactory results of the implementation of traffic calming through Kobylnica would have been impossible without the co-operative efforts of the AZ-Plan design team, the local authorities of Kobylnica, the Regional and General Directory of Public Roads, funders and representatives from the safety and traffic management institutions. The local authorities, councillors and residents of Kobylnica played a particularly important role. Their contributions were: the political will to implement the scheme; providing 20% of the costs of implementation; and creating a pro-traffic calmed atmosphere in their own Commune.

The formation of a new official attitude in local society to an innovative solution and pride in its achievement were crucial to the success of the project. It means in this situation 'we can have something that others haven't'. The Kobylnica experience confirms similar experiences in this field elsewhere in Europe (AIPCR, 1991; CETUR, 1992; CERTU, 1994).

15.4 Final conclusions

The traffic calming project on National Road 21 in Kobylnica Slupska was the first complex design and implementation of traffic calming on a through route of the national road network in Poland. The traffic, spatial and environmental aspects of the project and a range of legal, physical and organisational analyses of the traffic calming measures applied, together with the results of implementation, have been presented in this chapter. This example shows that traffic calming is a very effective means of improving traffic safety and cycling conditions simultaneously. In this case the focus was on the implementation of safety measures on national roads through small towns and villages in Poland. After six years of operation, the results of the traffic calming scheme through Kobylnica are satisfactory. The project passed its examination favourably. The design assumptions of the traffic calming were confirmed in practice. The reconstruction of the main street in Kobylnica brought additional benefits to the Commune. The street changed its appearance completely. A former through road was converted into a friendly and living street. Fatal accidents stopped. Traffic safety improved as a result of the project. Kobylnica as the entrance gate to Slupsk conurbation gained a modern and attractive image. Conditions for cyclists improved by transferring cyclists from the carriageway to shared cycle–pedestrian paths. Reducing the average speed from 70 km/h to 44 km/h has been beneficial, though the legal limit of 40 km/h is still often exceeded. It can be supposed that future further increases in traffic will help force down speeds a few kilometres per hour lower. This will be very advisable.

15.5   References

AIPCR (1991), *Circulation de transit dans les Petites Agglomérations, Sécurité et Environnement: Rapport 04.003B,* Comité Technique AIPCR de Routes Interurbaines et de la Route en Milieu Urbain, Route Spécial 1991.

CENTRAL STATISTICAL OFFICE (2001), *Concise Statistical Yearbook of Poland,* Central Statistical Office, Year XLIV, Warsaw.

CERTU (1994), *Ville plus sûre quartiers sans accidents,* Lyon, Réalisations Évaluations.

CETUR (1992), *Ville plus sûre quartiers sans accidents: Savoir-faire et techniques,* Bagneux, CETUR.

SPITZ, S (1982), 'How much too much (traffic)', *ITE Journal,* May.

ZALEWSKI, A (1995, 1996 and 2001), *Technical Analyses of Road Traffic Conditions After Reconstruction of National Road 21 Through Kobylnica Slupska, Parts I, II and III,* AZ-Plan, Warsaw, unpublished.

ZALEWSKI, A and CIELECKI, A (1993), *Preliminary Project of Traffic Calming Through National Road 21 in Kobylnica Slupska,* AZ-Plan, Warsaw, unpublished.

ZALEWSKI, A et al (1994), *Technical Building Project of Traffic Calming Through National Road 21 in Kobylnica Slupska,* AZ-Plan, Warsaw, unpublished.

ZIELIŃSKA, A (2001), 'Voivodines in the lens of statistics', *Traffic Road Safety (BRD),* Research Review of Automobile Transportation Institute, Warsaw, March, s. 2, 30–1.

16

Padua: a decade to become a cycle city

Marcello Mamoli, Istituto Universitario di Architettura di Venezia, Italy

16.1   Introduction

For almost a decade, Padua (Pàdova in Italian) has been a cycle-friendly town well known in Italy and abroad. If nowadays it has become a real cycle city, this is the outcome of more than 10 years of continuous effort which acknowledged the widespread natural community interest in cycling and inspired some coherent policies in planning and adapting urban spaces for it. This activity took the form of several simultaneous and mutually reinforcing measures.

The social dimension must be considered first. While such cycle-friendly conditions as 'easy' geography, ideal urban size and relatively mild climate, enjoyed by Padua, are generally found in Italy, in the country as a whole these features have not been in the past, nor are still, sufficient to provide a substantial cyclist share in local traffic. Neither are they sufficient to create spontaneous cycling flows along many recreational routes, which are also attractive and potentially available. If this was indeed the case, cycle cities would be common in Italy and not confined to a handful of exceptions. There would certainly be more in total, and they would be more regularly located along the peninsula.

The difference for Padua is one of culture. Padua might be compared with the outstanding Dutch examples, as well as Italian urban centres of different size or university towns, where students, either resident or commuters, often help in creating or supporting the local cycling atmosphere to an above average level.

A cycling culture means spreading consciousness of local cycling traditions, in sport or leisure as well as in everyday utility trips, and means maintaining these habits or, at least, a willingness to come back to them, even occasionally. A cycle-minded culture today means directing new attention and consciousness to the environmental quality of our lifestyle, to the conservation of natural resources, heritage centres and landscapes, which are the main features of our living context. It means also changing our behaviour, both as individuals and collectively. This is the background that has enabled the community of Padua to cultivate its potential character as a cycle city.

16.2   The development of a cycling strategy

Ten years ago Padua's natural inclination towards cycling gained serious political interest for the first time, which gradually involved the municipal authorities, the provincial administration and the Department of Public Works of the regional government. Since then there has never been a huge concentrated effort, but rather constant step-by-step activity by different independent agents, fortunately operating with similar attitudes in fields of non-conflicting authority.

As in many cycle cities in Europe, the municipality had the hardest job, which was to cope with the general upgrading of the densely urbanised central areas and with the traffic and parking crisis. Today, like 10 years ago, more than 55% of the resident population (about 235,000) make two or more trips daily in urban areas, mainly to or from the central business district (CBD). To this must be added the high share of urban traffic accounted for by thousands of commuters from all over the region, travelling to the inner urban core of Padua (see Fig. 16.1).

The street layout of the city is mainly of medieval origin, reflecting the underlying Roman schemes, and despite modern marginal changes, is not suitable for expansion or for radical change. The streets cannot cope with the volumes of car traffic, but it would be hopeless, in the current way of life, to wait for a spontaneous shift to walking or public transport by those who use their cars intensively.

At the end of the 1980s Padua was still a middle-sized town[1] where the total usual number of non-motorised trips, consisting of pedestrians (32%) and cyclists (10.5%), together amounting to 42.5% of the general modal split, was still so important as to equal the proportion of drivers (44%), with very few car passengers (only 10% of private car trips). Public transport's share of the modal split was about 13%.[2]

This means that in the general modal split there was already a healthy balance between non-motorists and motorists, and that, in view of the still increasing private motorisation, it was necessary to support this positive statistic by upgrading the conditions for the 'green' modes of transportation by means of the provision of infrastructure. A dedicated city-wide network modelled on other European experiences was likely to fit this purpose and was intended to remove the main barriers to a growth of green modes' share in traffic.

The first step was the formulation of the Cycle- and Walkways Plan (1986–1989),[3] to provide the city with environmental opportunities for the huge mass of non-motorist commuters, i.e. pedestrians and cyclists, who until then had received almost no attention or funding. The Plan was started as an advisory feasibility study to provide the municipality with an overall view of possible

Fig. 16.1   Typical cycling conditions in Padua centre at the end of the 1980s before the Cycle- and Walkways Plan.

infrastructure for non-car users, offering greater safety, safe trip continuity, more utility connections and amenity.

The planners drew on successful experiences with cycle networks in northern Europe, in places such as Delft, Erlangen, Munster and Odense, which at the time were newly established or still ongoing. The intention was to build up a network of specialised cycle- and walkways mainly out of the existing material resources. The idea was not only to look at case studies where the network was obtained by means of a clever reshaping of current street sections or junctions, but also to involve other kinds of spaces to form the general scheme. These included pedestrian areas, parks, river or canal embankments, disused railways, fringe areas and so on, so that it was likely to be easier to find a solution to the widespread problems of continuity and fragmentation of the network. The study's general approach was thus designed to bring together single features and resources into a whole organic scheme, adapted to various local opportunities.

16.3   Key requirements for the cycling network

After the study of the most reputed cycle city models, the team decided that the whole network should fulfil the following requirements:[4]

- *City-wide size*, to provide all significant parts of the town with protected or independent facilities and to allow door-to-door trips within the whole municipal area.

- *Higher safety standards*, aiming to remove one of the most common objections to the use of the bike or walking in the city, i.e. the fear of accidents to vulnerable users on car-dominated streets. This is a matter not only of provisions looking safe to a specialist's eye, but also of ensuring that each solution was easy for children or the elderly to understand and use.
- *Continuity* of the cycling and walking facilities, mainly through finding links at critical points or overcoming barriers, to allow all significant connections and create new ones.
- A *systematic network concept* based on a clear outline structure, (inner) physical and functional articulation and a clear hierarchy of the different parts, so that it would be easy for everyone to achieve a clear mental image of the cyclists' and pedestrians' city, which needs not necessarily to be like that of motorists, but is meant to have its own design as regards spacing, functions, construction, amenity, etc.
- A *multi-purpose approach*, allowing for the fact that the same trip could be made at different times of the day, of the week or of the year, for different reasons and for more than one purpose (e.g. trips to work, but also shopping and social trips). Cyclists' and pedestrians' journeys never can fit rigid origin – destination patterns and their freedom of movement must be supported and increased.
- A *multifold pattern combination*, which means that the network design is open to various suitable mixtures of possible patterns, according to the character of each site or city district, to fit in appropriately.

This approach was generally conceived to achieve sound infrastructural performance in matters such as safety, directness, comfort and amenity all over the network, looking at the best practice examples, but also having no prior design standard to be strictly adhered to. This has made all technical steps flexible from the planning phase up to construction.

The methodical procedures employed were also a departure from the usual demand/facility design path. Indeed as soon as possible a first scheme was drawn up including all available and potential resources. This was then submitted to a systematic grid of tests, to check how well it resulted in the surveyed and desired uses, and to assess its potential to generate new cycling and pedestrian activity as a consequence of better environmental conditions for non-car trips. This procedure was repeated several times, each with a series of adaptations, in a sort of a loop shifting from the proposed solution to analytical controls and back again, until a general balanced outcome was achieved in the definitive Network Plan.

The ladder of analytical controls provided a set of corresponding general proof-maps, showing the suitability of the chosen pattern to the purposes and the goals of the planned network. This ladder was mainly made of the following components:

- *General street and highways structure and hierarchy*, which also showed a few random cycle lanes already existing or planned as resources, and displayed the restraints, i.e. the parts of the scheme where no cycling could be allowed such as motorways or other roads used as such.

- *Morphology of the infrastructure network*, with critical evaluation of the distances between adjoining streets or highways, and analysis of the net junctions, with relevance to their position in the general network and their patterns.
- *Location of main public and private services* with reference to their level (neighbourhood, town, district, or regional level) and related catchment areas.
- *Traffic conditions* on the current sections of each route, evaluation of flows and volumes, modal split, accident density, length of queues, parking conditions.
- *Heritage townscape or landscape*, environmental or natural features, protected areas, and suggested upgrading opportunities alongside the cycle- and walkways.

The resulting General Scheme proved that about 300 kilometres of protected or independent cycle- and walkways were suitable for implementation in the whole urban area, more than about 90% of which could be made out of existing paths of various types, and only about 10% of which would need new construction or equivalent restructuring. These consisted of a few particular but very necessary connections, for continuity of existing paths beyond physical barriers, plus several extensions, largely sited within the new planned housing or park developments.

This meant also that a cycle- and walkways network with an overall length of a quarter of the town-owned street network was suitable for making safer, free from car access, ready for a widespread general environmental upgrading, for the network users as well as for the adjacent settlements.

The general Network Plan was supported by three special detailed schemes, of which one was designed for the entire Old Town, within the sixteenth century Venetian fortifications, providing not only the spaces reserved for non-motorised users but also suggestions for a general reorganisation of the motor traffic patterns and parking facilities, since at that time no general traffic planning was yet enforced in the central area. These latter suggestions, although not requested, appeared to the planning team a necessary device to demonstrate that cycle- and walkways, once introduced in the existing traffic situation, would not interrupt car traffic flows nor car parking in the most critical area of the inner city. On the contrary, by means of this detailed scheme it was possible to show that the number of car parking places could remain substantially the same and that the traffic circulation patterns could still have more than one satisfactory upgrading solution, with possible short term and mid-term gradual evolutions from the status quo to new schemes.

This helped also to ensure that more specialised space for pedestrians and cyclists would be available within the most attractive places of the town, and this developed into a denser network covering all the points of interest of the city core. All the centre of Padua was meant to be available mainly to pedestrians

Fig. 16.2   Cycling conditions today after the construction of cycleways to end inside the central area.

and cyclists, but without thereby becoming an impenetrable 'island', as in other Italian cities of smaller size. How was that to be achieved? By means of a critical evaluation of all the resources in the area, where it was quite natural to give back to pedestrians and cyclists only the main market squares and most of the related web of narrow streets or lanes dating from Roman or medieval times.

This eased a consequent ordered reshaping of the adjoining spaces under a non-sectoral, but generalist site planning and design approach, where priority was given to urban public spaces and resident motorist convenience came second, given the fact that several huge central parking facilities were planned in and around the central area, within walking distance of any destination.

The other detailed schemes were intended to allow the creation and the appropriate mobility of two suburban natural parks: one to the east, adjoining the new industrial area of Roncajette, and the second in the west near the airport, named Basso Isonzo, by a famous old inn. Both areas are quite close to the urban core, but have difficult connections to the city centre because of canals, railways or motorways. The cycle- and walkways network was to overcome those barriers, creating new links to the parks and through them to allow new alternative mobility lines from the outskirts to the city, bypassing some very busy junctions (see Fig. 16.2). Figure 16.3 shows a map of the municipal cycleways in 2001.

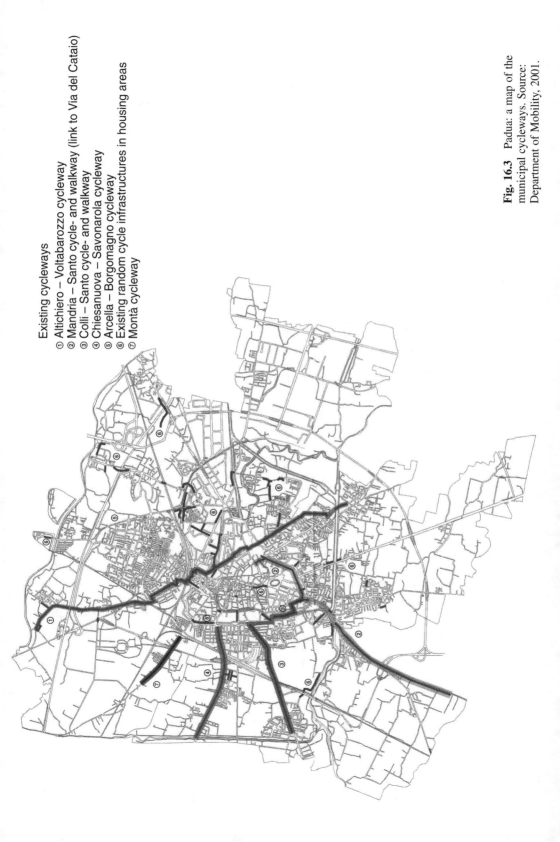

Existing cycleways

① Altichiero – Voltabarozzo cycleway
② Mandria – Santo cycle- and walkway (link to Via del Cataio)
③ Colli – Santo cycle- and walkway
④ Chiesanuova – Savonarola cycleway
⑤ Arcella – Borgomagno cycleway
⑥ Existing random cycle infrastructures in housing areas
⑦ Montà cycleway

Fig. 16.3 Padua: a map of the municipal cycleways. Source: Department of Mobility, 2001.

16.4   Implementation

The general Network Plan was also accompanied by two step-by-step implementation schemes, encompassing either parts of the general network or parts
chosen from the detailed studies.

The first implementation stage regarding the Phase A short term operational
scheme aimed to create, in a period of about three years, the essential backbone
of the network. The idea was to provide, as soon as possible, a selected coherent
part of the web, which would fulfil the same requirements as those of the whole
scheme and would demonstrate what the final scheme would be like. The aim
was also to connect and fashion into the new and coherent functional scheme a
handful of cycleways, either already existing or under construction during the
network design period.

At the same time a series of through connections, forming an 'X' shaped
pattern, would be developed running from north to south, across all districts of
the town, overcoming several barriers and involving rail and bus stations, the
main university areas, many schools, hospitals, parks and housing areas.

This first phase, although restricted to a part of the system, was designed to
create at an early stage the character of the new townscape generated by the cycle-
and walkways policy. In this manner the more urgent needs of the community
could find an appropriate and definitive answer. The total costs were estimated at
!1.59 million.

Phase B, the second implementation stage, concentrated again on a selection
of parts of the general network; it was shaped in a rough 'Y' pattern and was conceived as complementary to the achievements of Phase A. Obviously it was meant
as the natural extension of the network previously achieved, coping with the basic
needs of all the housing neighbourhoods, mainly with regard to routes to schools,
more connections to the CBD, to the university and to general and local service
areas. This was a task to be started in the third year of the implementation programme and to be ended in the sixth year, again requiring an investment of !1.59
million.

The rest of the network scheme, to be completed in 10 years from the beginning of the project, was conceived mainly as comprehensively covering the urban
area and a fuller provision of infrastructure facilities. The remaining investment
requirement was estimated at !2.12 million.

This planning and the corresponding financing took account of the design
resources of municipal planning offices and the ordinary budget allocations of
about !1.06 million per year for walkways maintenance. The planners' and officers' suggestion was that about 50% of this ordinary sum could be used to enable
construction of cycle- and walkways belonging to the planned network, and also
to fix roads needing this kind of attention. At that time no national or regional
subsidies were available.

The general plan underwent wide discussion with the public, interest groups
and so on, and finally received the unanimous approval of the City Council at the

beginning of 1989. The city soon started a new general campaign to promote urban cycling, and in the spring a huge cycle rally was organised to attract more of the population back to bike use in urban areas. This initiative attracted more than 5000 participants who literally 'took over' the city for hours in a festival-like atmosphere. One of the local newspapers sponsored the rally and the Network Plan, so that all participants and thousands of readers were given a poster-sized copy of it, which made it well known to everybody. Even now letters are still received recalling some of the expected features of the plan. In the following years these kinds of rallies were repeated successfully by the municipality with the help of associations and groups.

The city of Padua, as a founder member of the AICC (Italian Cycle City Association), was an example for other cities, providing administrative or technical advice for other campaigns of this kind. Padua was chosen to be a case study in the two Velo-City conferences in Milan and Nottingham and also featured in several national meetings and publications.[5]

Other public institutions such as the provincial and the regional governments started to take an interest in cycle infrastructure, including reusing suburban or countryside canal or river embankments, very attractive and interesting routes that lay outside the jurisdiction of the municipality.[6]

In the climate of growing attention to matters relating to environmental quality of life and transport, the Ministry of Urban Affairs in 1991,[7] following lobbying by the AICC, showed a new awareness and successfully steered through an Act of Parliament, subsidising demonstration cycling and walking infrastructure in all main towns. This was followed by a parallel Veneto Regional Act[8] to subsidise cycleways as well, open to all municipalities. To make it easier to get the necessary know-how the local corporations also received a special design handbook,[9] which soon received national interest.

In Padua these financing measures made it possible in 1992 to start the design of the first phase operative schemes, and to submit them for funding to national and regional selections. Both applications were successful so that in 1993 investments of about !0.53 million[10] were subsidised with !0.265 million given by the regional fund, out of a programme of !1.855 million. The construction work financed from this fund started in 1996 and ended in 1998. In 1996 the national fund was also allocated and co-financed, with a sum of !0.265 million, an investment of about !0.927 million, allowing building to start in 1998 and to finish in 2000.

16.5   The current situation

After more than 10 years of work the situation is that from the constructional and functional point of view things have gone broadly as planned, since adaptations made here and there on site have all been only marginal. On the other hand, implementation is certainly late with regard to the original phasing. This is partly a consequence of nationwide events and influences, and partly a matter of local factors.

Following the Maastricht Treaties national policies imposed severe austerity on public investments of all kinds. For Padua's cycle- and walkways this meant relying just on schemes subsidised by the state or by the region, and getting almost nothing more from the ordinary budget, which was greatly reduced. Secondly it must be added that a mixture of sudden political changes and the usual bureaucratic slowness delayed the allocations and all decision making outside the municipality's authority. Thirdly the building of the cycle- and walkways network was started in a period of general transition and radical change in the field of public works, and it somehow managed to continue in a context of very limited public building activity.

The analysis of the first cycle- and walkways schemes, which simulated three main scenarios for transport, highlighted the need for comprehensive traffic and transport planning and a need to assist the 'green' modes. This clearly called for a general reference framework on the part of the municipality. A new awareness of this need grew quickly among politicians and officials, generating new concern for the city's public transport and for policies to restrict car access to the inner city. A special Mobility Department[11] was created within the municipality and comprehensive plans and investments were started, aiming to obtain national subsidies for metropolitan transport modernisation, with funds mainly devoted to light rail, metro and tramways. The general Mobility Plans supported the need for all possible integration of the two networks (cycle- and walkways plus bus and tramways), but also gave absolute priority to investment in public transport. In the last decade public transport was greatly improved and restructured with the newest bus vehicles and a better structure of routes and services.

In terms of modal share, the private car is at substantially the same level as 10 years ago (43% of traffic). Cycling is about 12%, maintaining the achievement of the first campaigns, but pedestrians now account for only 18% of trips. Their loss is accounted for by the surprising growth of public transport. Any conclusions about these figures need to be stated with care. It should be pointed out that in 10 years (1991–2001) car ownership in Padua increased by about 20% and so did general car traffic. However, according to the Mobility Department, this has not affected all parts of the city in the same way. More car traffic is reported in the outskirts, but not in the central area. Here more people are daily attracted by bus or by bike, and cycling is a current mode of mobility within the central areas (the city wall extends 11 kilometres!) for residents and commuters. Again individual habits have increased the range of trips, which are on average longer. This might at least partly explain the decrease in walking, the increase of bus use and the only modest increase in cycling.

So in general we can say that the essential Phase A scheme, in spite of some setbacks, was able to achieve its original objectives, not only to prevent cycling declining, but even increasing it significantly. In times of austerity, it provided essential links and bridged gaps in the network. It was also able to upgrade marginal areas at almost no cost, revealing in this way its second, but no less important, identity. The network indeed is conceived by generalist planners (and not by transport specialists) as a tool for upgrading urban spaces, physically and culturally. The municipal Parks and Gardens Department understood this aspect and has sustained it in recent years.

The cycle- and walkways network proved to be a simple device to rediscover the complexity of the original features of local town and landscape, and provide enough space and environmental conditions for enjoying the view of them, freed from the dominance of the car.

If a broader and significant part of the central area can soon be changed according to the Network Plan, the municipality will still be able to increase the number of pedestrians in the city. These are rarely just strollers but more often customers and audiences for cultural events, who give life to the inner city. If a significant part of Phase B was now to be implemented, despite the delay, and the Basso Isonzo park was at last to become a reality, after decades of promises, there will surely be an overall positive impact, responding to public expectations about the quality of the environment. A mature cycle city like Padua deserves new perspectives and developments.

16.6   Notes

1 The population was at the 1981 Census about 235,000. In the 1991 Census it was a little lower, i.e. about 225,900. There were about the same number in 2001.
2 Census 1981 and up to date estimates by the municipal Mobility Department.
3 This Plan was made by a team co-ordinated by Marcello Mamoli, Paolo Stella, Giovanni Abrami, Chiara Stefani and others on behalf of the Planning Department of the municipality.
4 M MAMOLI, 'Progettare una rete ciclabile', in L. Gelsomino (ed) *Recupero Edilizio* 4, Edizioni Ente Fiera di Bologna, 1985.
5 M MAMOLI, 'Starting a cycleways network: The case of Padua', in *VII Velo-City Conference Papers: The Civilised City Responses to New Transport Priorities*, Nottingham, Nottinghamshire County Council, 1993. See also: M MAMOLI, 'Pedonalità e ciclabilità nel centro storico. Il caso di Padova', in A.U. – Rivista dell' Arredo Urbano (INASA, Roma): n.32-bis; M.MAMOLI, 'Padova: un sistema integrato di percorsi pedonali e ciclabili', in Urbanistica Informazioni (INU (Istituto Nazionale di Urbanistica), Roma), Dossier n. 1/90 – 'Una città in bici', eds W Ameli and F Conte.
6 The provincial government and the Department of Environment and Parks started several studies using the embankments of the rivers Brenta (north), Bacchiglione (south) and the canals Battaglia and Vigenzone linked in one scheme called Via del Cataio also linking the town southwards to the Regional Park of Euganean Hills (Colli Eugànei). The Regional Office of Public Works, at that time with a branch office in Este, studied a scheme to connect this city to the adjoining Park of Euganean Hills. This is connected to the Via del Cataio routes.
7 The Minister at the time was Mr Tognoli who promoted the Act. In the following government the Act was formally signed by his successor Mr Conte. The Act is currently known as Legge 208/1991. A new Act now in force is known as Legge 366/99 by Mr Galletti, MP.
8 This Act combines proposals by Mrs Sartori, Regional Councillor for Transport, Mr Pupillo, President of the Regional Council and Mr Valpiana, Councillor; it is known as Legge Regionale 29.12.1991 n. 39.

9 M MAMOLI, 'Manuale per la progettazione di itinerari ed attrezzature ciclabili', Regione Veneto, Assessorato alla Viabilità e ai Trasporti, Segreteria Generale per il Territorio, Venezia, 1992.

10 !530,000 is the equivalent of about 1 billion liros in the former Italian currency. !265,000 corresponds to 0.5 billon liros and !900,000 to 1.7 billion liros.

11 The up to date map of the implemented network and mobility data are printed by courtesy of the Mobility Department.

17

US bicycle planning

Andy Clarke, Association of Pedestrian and Bicycle Professionals, Washington DC

17.1 Introduction

The 1990s were a period of extraordinary change in the US bicycle field. At the start of the decade, federal spending on bicycle facilities averaged a few million dollars annually (see Fig. 17.1); the US Department of Transportation (DOT) had no staff working full time on bicycle issues and no real strategy for dealing with those issues; fewer than 10 states and 25 cities had a bicycle co-ordinator position, and none had an equivalent pedestrian staff person. The 1990 Pro Bike Conference in Washington, DC was attended by less than 250 people. Little bicycle-related research had been done since the end of the 1970s.

As the new millennium began, average federal spending on bicycle and pedestrian facilities exceeded $300 million (€335 million) annually. The US DOT has staff working on bicycle and pedestrian issues in most of its agencies (Office of the Secretary of Transportation, Federal Highway Administration (FHWA), National Highway Traffic Safety Administration, Federal Transit Administration and the Federal Railroad Administration). In addition, the DOT has set ambitious national goals for bicycling and walking (US DOT, 1999), has completed a multi-year bicycle and pedestrian research programme, has elevated bicycle and pedestrian safety to priority status, and has published a wide range of related technical and promotional literature.

Every state department of transportation is required by law to have a bicycle and pedestrian co-ordinator, and dozens of cities and counties have full time bicycle and pedestrian staff. Moreover, all the major professional associations (American Planning Association, Institute of Transportation Engineers, Transportation Research Board (TRB), American Society of Civil Engineers) have bicycle or pedestrian technical committees, or both; more than 500 people

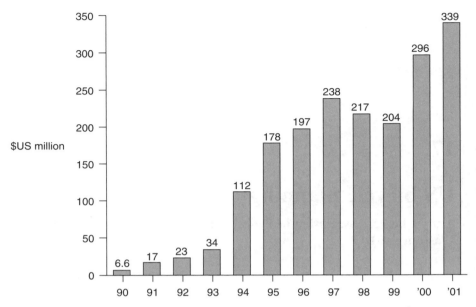

Fig. 17.1 US bicycle planning: federal funds spent on bicycling and walking. Source: Federal Highway Administration.

attended the 2000 Pro Bike/Pro Walk conference; and an association exclusively for bicycle and pedestrian professionals – the Association of Pedestrian and Bicycle Professionals – has been formed.

Public opinion and attitudes towards bicycling have undergone an equally dramatic transition. Known by the unglamorous nickname of the 'forgotten modes' in the 1980s, bicycling and walking have evolved to become emblems of a high quality of life. Major advertisers sell products ranging from breakfast cereals to financial services by using images of bicycling; bicycle sales have boomed throughout the decade; professional football stars promote bicycle safety during the Super Bowl; homebuyers want access to walking and bicycling trails more than golf courses and swimming pools; and 'liveable communities' designed to accommodate these modes are a major new focus of the architecture, landscape architecture, planning, health promotion and transportation professions.

The unprecedented growth of interest in bicycling has been matched by greater opportunities to improve access and safety for bicyclists through funding and planning, as well as a rapid increase in the expertise and professionalism of the growing number of people involved in the field. There has also been a concomitant increase in the need for information that can help translate the desire to improve conditions for bicycling into tangible programmes and projects aimed at reducing injuries and fatalities among bicyclists, and at expanding use of the mode for everyday as well as recreational travel. Some bicycling statistics are:

- 2 miles (3.2 kilometres): average length of bicycle trip (Nationwide Personal Transportation Survey, 1995);
- 39.6%: trips by all modes that are 2 miles (3.2 km) or less (Nationwide Personal Transportation Survey, 1995);
- 467,000: number of people bicycling to work in 1990 (US Census, 1990);
- 567,000: number of people bicycling to work in 2000 (US Census, 2000);
- 42 million: number of adults who rode a bike during the previous 30 days (US DOT, 2001);
- 690: bicyclists killed in traffic crashes in 2000 (National Highway Traffic Safety Administration, 2000);
- 38%: adult bicycle riders regularly wearing a helmet, August 1998 (Consumer Products Safety Commission, 2000);
- 69%: children under 16 regularly wearing a helmet, August 1998 (Consumer Products Safety Commission, 2000);
- 16.1 million: bicycles sold in the United States in 1998 (Bicycle Retailer and Industry News, 2000);
- $5.6 billion (€6.3 billion): approximate retail value of US bicycle market in 1998 (Bicycle Retailer and Industry News, 2000).

17.2   Future prospects

At the dawn of the third millennium, the future for bicycling as a component of the transportation system looks bright. Bicycling is a healthy, clean, efficient and economical means of transport that offers the rider speed and flexibility over short distances and can be accommodated with relatively little space. Bicycling is accessible to a great many people who cannot drive, especially the young. The activity is popular and is capable of generating significant economic returns to communities that choose to promote appealing riding conditions and services. In addition, improved prospects for bicycling are resulting from developments in such key areas as updated technical information, renewed national policy support and heightened awareness.

17.2.1   Updated technical information

The American Association of State Highway and Transportation Officials has published a new *Guide to the Development of Bicycle Facilities* (AASHTO, 1999) that significantly expands the amount of technical information available to planners and designers of bicycle facilities.

Part IX of the *Manual of Uniform Traffic Control Devices* (US DOT, 2000), which addresses signs, markings and signals for bicyclists, has been overhauled with the co-operation of a newly formed bicycle task force; the National Committee on Uniform Traffic Control Devices recently welcomed its first bicycling organisation – the League of American Bicyclists – as a voting member.

The National Committee on Uniform Traffic Laws and Ordinances has also established a bicycling task force and is updating relevant sections of the Uniform Vehicle Code to reflect the evolving status of the bicycle.

17.2.2   Renewed national policy support

The Transportation Equity Act for the 21st Century (TEA-21) expanded opportunities to integrate bicycling projects and programmes into the planning and funding of transportation infrastructure, including making bicycle improvements eligible for the hazard elimination (safety construction) programme.

TEA-21 also required the US DOT to study various approaches to accommodating bicycle travel. The resulting report endorses the policy approach of establishing bicycle and pedestrian ways in new construction and reconstruction projects unless doing so would be contrary to public safety, excessively costly, or unnecessary because of sparse population.

The FHWA has issued guidance on the bicycle-related provisions of TEA-21 whereby 'every transportation agency [is expected] to make accommodation for bicycling . . . a routine part of their planning, design, construction, operations, and maintenance activities.' The FHWA has also renewed its commitment to the goals emerging from the National Bicycling and Walking Study of doubling the proportion of trips made by foot and bicycle in the United States to 15% and simultaneously reducing fatalities and injuries to these users by 10%.

17.2.3   Heightened awareness

Bicycling continues to slowly permeate American culture. In 2000, First Lady Hillary Clinton and US DOT Secretary Rodney Slater championed the Millennium Trails initiative to celebrate and promote long distance trails, including several bicycle trails. The successive Tour de France victories of Lance Armstrong captured the hearts of many Americans and opened their eyes to bicycling. Growing concern over obesity, diabetes and heart disease attributed to a lack of physical activity has just begun to turn people on to bicycling and other low impact physical activities.

17.2.4   Action at the local level

Another hugely encouraging sign is the willingness of many of the largest US cities to promote bicycling with major infrastructure and promotion programmes. Cities such as Los Angeles, New York, Houston, Philadelphia, Phoenix, Washington DC and San Francisco are all engaged in striping extensive bike lane networks, developing long distance trails, and signing and mapping bicycle routes. Leading the charge is the city of Chicago.

17.2.5   Chicago case study

In May 1992, the newly appointed Chicago mayor's Bicycle Task Force conceived a modest seven page plan with great ambition: to 'make Chicago more
bicycle friendly'. Almost 10 years and $30 million (€34 million) later the city
has added more than 100 miles (160 kilometres) of bikeway and 6600 bike
parking racks, and has significantly upgraded the centerpiece Lakefront Trail.
Together with the Chicagoland Bicycle Federation (CBF), the city has also initiated a series of innovative projects to improve safety, education, enforcement and
promotion – so much so that the mayor, Richard Daley, now wants Chicago to
be the *most* bicycle-friendly city in the United States.

In pursuit of that target, the city has plans to add another 100 miles (160
kilometres) of bike lane, 150 miles (240 km) of bike route and 2000 more bike
racks, and is spending almost $7 million (€7.8 million) on two critical bridges
carrying the Lakefront Trail through the downtown area.

The numbers are impressive. For years, Randy Neufeld, Executive Director of
the CBF, has used his unrivalled knowledge of local, regional and state transportation funding programmes to secure funding for the city's projects and mil-
lions of dollars worth of regional bicycle infrastructure projects. For example,
bicycling routinely gets around 7% of the area's Congestion Mitigation and Air
Quality funds, and this rose to 10% in the most recent round of grants in 2001 –
more than twice the national average.

The CBF has also plugged away at local transit agencies for years and is now
reaping the rewards. The Chicago Transit Authority (CTA), which runs the city
buses and heavy rail system, has opened its trains to bicycles outside of rush
hours and is expanding the number of bus routes that have bike racks. Frank
Kreusi, President of the CTA and an enthusiastic bicyclist, wants to do more and
points to their support of Bike Chicago events and the 6000-rider Boulevard
Lakefront Tour as proof of their commitment. The CTA can also claim to be the
only transit agency that offers a cycling jersey as part of its gift catalogue!

The PACE suburban bus service has completed a successful bike rack experiment and has committed to ordering all future buses to be equipped with a
Sportswork rack and intends to have its entire fleet with racks by the end of 2001
(see also Chapter 8, section 8.12). Even METRA, the regional commuter rail
provider, has announced plans for a limited trial of bikes on board their trains.

In the region as a whole, bicycle and pedestrian planning is growing fast. Two
of the six suburban counties now have dedicated bicycle staff in those two counties, DuPage and Kane, and most communities are expected to have developed
bicycle plans in the next two years. In Kane county, all roadway projects now
include bicycle and pedestrian consideration. The Chicago Area Transportation
Study (the metropolitan planning organisation for the area) is just embarking on
a regional bicycle and pedestrian plan.

What may ultimately set the city apart from other communities, however, is
the string of creative initiatives that go way beyond mere concrete, asphalt and
steel. For example, the city has distributed 110,000 copies of *Safe Bicycling in*

Chicago and has published the guide in Spanish and Polish as well as English. The state of Illinois has published a statewide version. The guide has been adapted for children and a shorter 'quick reference' edition has appeared. See http://www.cityofchicago.org/Transportation/Bikes/safety.htm. Each year the city produces a bike map showing the existing and proposed routes and more than 70,000 copies of that have been circulated, not including the copies distributed courtesy of the *Chicago Sun Times* as a special section of their Sunday paper.

In 1999 the CBF worked with the city to launch a Shopping by Bike programme, a pilot programme to encourage bicycle riding to local shopping centres. They have produced a bike to work guide and bicycle parking guide for businesses. Education and promotion initiatives include training for taxi and bus drivers and a wide selection of 'Bike School' courses for cyclists of all ages and abilities.

Bike Month in Chicago is celebrated in late May and early June and is coordinated entirely by the mayor's Office of Special Events. As a consequence, the celebration has been able to expand from a nine day event with 45 activities and 10,000 participants to a month-long series of more than 100 events attracting 25,000 cyclists. Last year, the CBF's annual Boulevard Lakefront Tour concluded Bike Chicago 2001 with more than 6000 riders exploring the city's neighbourhoods and lakefront.

In the spring of 2001, a team of four cycling ambassadors began roaming the streets and trails of Chicago providing advice, encouragement, information, safety tips and assistance to fellow cyclists or potential riders. Mayor Daley's bicycling ambassadors (a moniker shared only with the mayor's graffiti removal squad) are based on a similar programme that has been running for several years in Toronto. In 2002, the city hopes to get enough Governor's Highway Traffic Safety Programme funding for 10 ambassadors.

Ben Gomberg, the city's Bicycle Programme Co-ordinator, has organised fact-finding missions for bicycle programme staff to Toronto, Philadelphia and Portland, Oregon, to learn how they do things. He has contracted with CBF and local consulting firms to research bike network needs and opportunities, and to assist with the bike parking programme.

Not everything is perfect in Chicago. The CBF still bemoans the lack of parking at, and access to CTA stations; the availability of indoor parking in downtown is an issue; the reconstruction of Wacker Drive (a monumental project in the heart of the city) didn't adequately consider bicycle access issues; the suburban counties and Illinois DOT continue to build a hostile environment for cycling; riders are still getting killed on city streets; and the car is still king.

Over the next 10 years, however, any city wanting to claim the title of 'Best City for Bicycling' in the United States is going to have to contend with Chicago. For more information:

- City of Chicago bicycle programme,
  http://www.cityofchicago.org/Transportation/Bikes/bicycle.htm;
- Chicagoland Bicycle Federation, http://www.biketraffic.org/;

- Chicago Transit Authority, http://www.transitchicago.com/store/giftx.html;
- League of Illinois Bicyclists, http://www.bikelib.org/index.html;
- Chicago Area Transportation Study,
  http://www.catsmpo.com/bikeped/index.htm.

17.3   Obstacles

All of the developments outlined above have played their part in creating a more bicycle-friendly infrastructure at the local level, where increasing numbers of cities and metropolitan planning organisations are adopting and implementing bicycle plans, hiring bicycle programme staff and encouraging bicycling as a means of transport. However, it is still too soon to declare the bicycle the vehicle of the third millennium. Bicycling remains a relatively minor means of travel in all but a handful of US communities and faces many significant obstacles before it can break through as a mainstream transportation mode.

17.3.1   Bicycling is largely invisible

Few of the basic planning tools used by state and local agencies (e.g. planning models and forecasting tools) incorporate bicycling data because such data are sparse. The US census data that do exist and are available by census tract focus exclusively on the journey to work. Journeys to work by all modes account for just one in five of all trips and an even smaller proportion of bicycling trips.

Data from the Nationwide Personal Transportation Survey are readily available only at the national level, not the state and local levels, and few communities have gathered their own data on bicycle use and patterns. As a consequence, only limited information is available on levels of bicycling, bicyclists' habits and preferences, and their exposure to danger. Although efforts to develop forecasting tools and exposure measures for bicycling are under way in some communities, this lack of basic information is hindering progress.

The US DOT has identified six areas in which better documentation of bicycling and walking is necessary (US DOT, 1999):

- development of a method for accurately recording bicycle and pedestrian trips;
- development of a method for measuring and tracking bicyclist and pedestrian exposure rates;
- measurement and tracking of bicycle helmet use rates;
- capture of expenditure information for bicycle and pedestrian projects and programmes;
- improvement of the 2000 Nationwide Personal Transportation Survey instrument to generate better data on bicycle and pedestrian activity, and detailed analysis of the results; and
- the impact of bicycle and pedestrian investments on air quality, public health and other quality of life indicators.

17.3.2   Responsibility for bicycling issues is split

One of the most positive aspects of bicycling is the role it can play in so many aspects of everyday life. However, the fact that bicycling relates to transportation, recreation, health and safety, environment, energy, community development and tourism can also be detrimental as no one agency or body takes responsibility for the creation of an overall bicycle-friendly environment.

17.3.3   Bicycling is a local solution

Most bicycle trips are quite short and are affected most directly by very local traffic conditions. However, transportation planning and funding decisions are made largely at the regional or state level (even though they reflect a composite of local issues), and the decision makers have tended to overlook small, local improvements to the street environment that would make bicycling safer, more convenient and more appealing.

17.3.4   Urban sprawl makes bicycling less viable

The seemingly inexorable flight of people and jobs from urban areas to suburban and exurban areas poses a significant threat to the long term viability of bicycling as a means of travel. Suburban development has typically increased individual trip distances for bicycling and failed to accommodate bicycles in roadway design. In addition, the dispersion of destinations characteristic of suburban areas has increased the need for trip chaining and reduced the likelihood that bicycling will be an option for many trips. While individual destinations within a suburban area may all be within a manageable distance from home or work, the combined distance of a multi-destination trip may make travel by bicycle unrealistic for the casual cyclist.

17.3.5   Technical issues remain unresolved

Although there has been tremendous progress in the planning and design of roadways to accommodate bicyclists, there are still significant knowledge gaps in such areas as the accommodation of bicyclists at intersections; the safety of bicyclists at roundabouts; and the use of coloured pavement (carriageway) markings, bicycle signal heads and other bicycle specific traffic control devices. The results of decades of experience from European and other countries (e.g. Denmark, Germany, the Netherlands, Japan, Australia, China) remain untried in the United States because of a lack of access to foreign language research and an unwillingness to accept 'other people's' solutions.

17.3.6   Bicycling suffers a credibility problem

For too many Americans, bicycling already appears to be an unlikely option for everyday travel. The obstacles described above, especially sprawl, threaten to make the mode even less accessible and desirable, even though a majority of Americans say they want to bicycle more and would ride more often if conditions were better (Parkwood Research Associates, 1991). For bicycling to thrive, it must be perceived as a practical, feasible alternative to driving.

17.4   New directions

Bicycling in the United States has the potential to play a significantly greater role in everyday transportation than is currently the case. More than one-quarter of all trips in the United States are still less than 1 mile (1.6 km), and almost one-half are 3 miles (5 kilometres) or less – a 15 to 20 minute bike ride. Thus the potential is there. The goals of the National Bicycling and Walking Study could be met if every adult in the United States walked or bicycled for just three of the trips made each week that are currently made by automobile. Indeed, there are some hopeful signs that this potential can be realised.

17.4.1   Integrated roadway design

In states such as Florida and Oregon and many of the communities within those states, virtually all new and improved roads automatically incorporate accommodations for bicyclists. In urban areas, arterial and collector streets routinely feature striped bicycle lanes, residential roads are managed to keep vehicle speeds more compatible with bicycling, and facilities are in place so that bicycles can be parked securely at trip's end. Where appropriate, trail and greenway corridors are used to supplement on-road opportunities for bicycling. It appears that this approach, endorsed by the US DOT bicycle and pedestrian design study mandated by TEA-21, will become the standard for good highway design early in the new millennium.

17.4.2   Focus on speed and safety

Bicyclists are over-represented in crash statistics, yet safety improvements for bicyclists have historically received little funding or attention. Bicyclists and pedestrians together account for 14% of traffic fatalities, only 7% of trips, and less than 1% of safety-related construction funds. As this discrepancy becomes apparent, and with safety guiding the actions of many federal, state and local agencies, the new millennium will see a much greater emphasis on protecting these most vulnerable road users. Awareness of the critical role of speed in the incidence and severity of crashes will grow, prompting greater interest in bicycle-compatible traffic calming.

17.4.3   Education for all ages, all modes, all professions

Planners and traffic engineers are increasingly being taught how to plan and design transportation facilities with the bicyclist in mind as part of their graduate and undergraduate courses. In this new century, comprehensive bicycle education programmes for elementary school children, their parents and their teachers will start to be included in the regular school curriculum, following the lead taken by school districts in Texas and Southern California.

17.4.4 Technology transfer

The need for up-to-date, practical and accessible information on all aspects of bicycle-friendly communities is growing along with the number of people in the field. Fortunately, TEA-21 established a national Pedestrian and Bicycle Technical Information Centre to collect, synthesise, and disseminate information on engineering, encouragement, enforcement and education strategies related to bicycling and walking. This centre has become a critical resource for professionals (www.bicyclinginfo.org).

17.4.5 Globalisation

In June 2000, the international bicycling community met in Amsterdam, the Netherlands, for the second World Bicycle Conference (Velo-Mondial 2000). This conference provided a unique opportunity for a significant cross-section of US practitioners to visit a developed nation where bicycling has been integrated into the transportation system and where nearly one-third of all trips are made by bicycle: a record turnout of almost 100 US delegates made the trip. The conference also fostered a much greater awareness and appreciation of the work done in other nations to research bicycle-related issues, thereby encouraging a greater investment in the translation and use of foreign language research in the field.

17.4.6 New beginning in bicycle research

The year 1999 marked the culmination of the FHWA's five year bicycle research programme, publication of a set of research problem statements by the TRB Committee on Bicycling, unprecedented interest in bicycle-related topics at the state and local levels, and improved funding opportunities for research at the national and state levels. As the twenty-first century gets under way, there is a compelling need to establish a national strategic research programme in bicycle transportation in the United States to address the large gaps in knowledge identified above.

17.5 Conclusion

A final requirement if bicycle travel is to reach its full potential in the third millennium is the very twentieth century activity of marketing. Although Americans say they want to ride bicycles more for both personal and work travel, few believe bicycling is a real option for them today. While planners and engineers are more likely than ever to consider bicycling as they plan and design the transportation system for the next generation, special lanes and other facilities for bicyclists too rarely leave the drawing board. Politicians and policy makers include bicycling in the rhetoric of sustainable communities and healthy cities, but few turn those policy statements into action.

In the first decade of the twenty-first century, bicycling must be bolstered by more research, continued funding, education initiatives and other activities described above. But it must also be elevated in the minds of the American public, among the professional transportation community, and in the realm of public policy to the point where people believe bicycling can really make a difference, and that it is a worthwhile investment in the future.

17.6   References

AMERICAN ASSOCIATION OF STATE HIGHWAY AND TRANSPORTATION OFFICIALS (AASHTO) (1999), *Guide to the Development of Bicycle Facilities*, Washington DC, AASHTO, *http://www.aashto.org*.

BICYCLE RETAILER AND INDUSTRY NEWS (2000), *StatPak*, Santa Fe.

CONSUMER PRODUCTS SAFETY COMMISSION (2000), Washington DC.

NATIONAL HIGHWAY TRAFFIC SAFETY ADMINISTRATION (2000), *Traffic Safety Facts*, Washington DC.

NATIONWIDE PERSONAL TRANSPORTATION SURVEY (1995), Washington DC, US DOT.

PARKWOOD RESEARCH ASSOCIATES (1991), *Pathways for People*, Emmaus, Pennsylvania, Rodale Press.

US CENSUS (1990), Washington DC.

US CENSUS (2000), *Census 2000 Supplementary Survey Summary Data Tables, 2000*, Washington DC.

US DOT (US Department of Transportation) (1999), *National Bicycling and Walking Study: Five-Year Status Report*, Washington DC, US DOT.

US DOT (US Department of Transportation Federal Highway Administration) (2000a), *Manual of Uniform Traffic Control Devices*, Washington DC, FHWA.

US DOT (US Department of Transportation Federal Highway Administration) Bureau of Transportation Statistics (2000b), *Omnibus Survey*, Washington DC, http://www.bts.gov.

18

Increasing cycling through 'soft' measures (TravelSmart) – Perth, Western Australia

Colin Ashton-Graham, Government of Western Australia, with Gary John, Bruce James, Werner Brög and Helen Grey-Smith*

18.1   Introduction: the Western Australia situation

Perth, Western Australia, is one of the most isolated cities in the world; a metropolitan population of 1.4 million lives in an area bounded by marginal desert on one side and the Indian Ocean on the other. The Commonwealth of Australia government in Canberra sets fuel duties, collects income taxes and provides strategic funding for projects such as roads. The Western Australia state government provides a range of regional transport infrastructure and services (including public transport). A single tier of local governments (30 in the metropolitan area) provide local road management and development control within a state planning framework.

Perth has low development density and high levels of road provision and yet has basic mobility patterns similar to many European cities. More than 40% of all trips made are less than 3 kilometres in length and almost 60% of trips are less than 5 kilometres. A major difference between Perth and European cities is the very high level of car use. Recent data (Socialdata, 2001) shows that mode shares in Perth, expressed as main mode only per trip, are:

- car as driver 58%;
- car as passenger 21%;

* Note: the views expressed in this chapter are those of the authors alone and do not represent the policy of the Government of Western Australia.

- walking 12%;
- public transport 6%;
- cycling 3%.

This mode distribution is partly due to high levels of car ownership and infrastructure giving the car a time advantage over other modes. A plentiful supply of free or cheap car parking and very low fuel costs add to the attraction of the car for even the shortest trips. The typical distance distribution of car trips is (Socialdata, 2000a):

- 8% less than 1 km;
- 30% less than 3 km;
- 46% less than 5 km;
- 67% less than 10 km.
  (above figures are cumulative)

The recent trend in travel patterns has been towards a small increase in travel distance per person to undertake the same number of activities and trips as were carried out 15 years previously. A significant trend has been in mode choice away from cycling, walking and public transport and towards the car as driver. These factors, coupled with a strong population growth, have caused traffic levels (total kilometres consumed) to rise sharply.

This case study of the TravelSmart Individualised Marketing programme will illustrate that, even in an environment where the car has competitive advantage, simple soft policy interventions can radically change (and sustain) mode choice in favour of walking, cycling and public transport. Further, TravelSmart in Perth has demonstrated a behaviour change approach to be cost-effective in socio-economic terms and to maximise efficient use of existing, and expensive, infrastructure and services.

18.2   Perth policy context

18.2.1   Metropolitan Transport Strategy (MTS)

In light of the unacceptable social and environmental consequences of significant traffic increases the Western Australian government has developed a suite of policy documents and implementation strategies to manage travel demand.

The Metropolitan Transport Strategy (Department of Transport *et al*, 1995) for Perth was published in 1995 and remains the foundation document for transport and land use planning. The MTS has established a set of target outcomes for modal share, seeking to reduce the share of trips by car as driver whilst not constraining access to places.

The Strategy is a trend breach approach, designed to limit car traffic growth and reverse the decline in the use of alternatives to the car. Against the 2029 trend scenario the aim is to significantly increase mode shares for cycling and public transport and to maintain the relative share for walking.

18.2.2   Bike Ahead

A specific modal group for cycling was established in the mid-1980s and developed to a 12-strong staff unit by the late 1990s. This group (Bikewest) was responsible for strategy, promotion, audit and encouragement of cycling in partnership with all tiers of government and with the non-government sector. A bicycle strategy was established in 1985 (Departments of Local Government and Transport, 1985) and updated under the title Bike Ahead in 1996 (Department of Transport, 1996a) to encompass the established strands of education, enforcement, encouragement and engineering.

Implementation of the cycling strategy has included the delivery of a strategic cycle network (1996 to 2001) and a marketing campaign (Cycle Instead, October 1999).

18.2.3   TravelSmart 10 year plan

Travel demand management was established as an implementation response to the MTS challenge. A small unit within the Transport Department of Western Australia (now the Department for Planning and Infrastructure) was established to develop programmes to intervene in mode choice. This travel demand management unit looked to European best practice to seek out transferable interventions with a strong emphasis on voluntary behaviour change. A series of pilot programmes in dialogue marketing, schools, workplaces and local government interventions were implemented under the TravelSmart brand. Basic research and close monitoring of the impacts of these interventions demonstrated the effectiveness of the TravelSmart approach and resulted in the adoption of a TravelSmart 10 year plan in 1999 (Department of Transport, 1999).

18.3   Infrastructure approach (Perth Bicycle Network)

The development of a strategic bicycle network plan (Department of Transport, 1996b) was undertaken by Bikewest in 1995. A process of public planning workshops was utilised to ensure the relevance of the network to end consumers. Consultants to Bikewest surveyed the identified routes and developed works lists for agreement with local governments.

The Perth Bicycle Network Plan consisted of a three stage, 12 year programme for A\$112 million (€65 million/£40 million) of bicycle routes. The major elements of the network plan are 1400 km of on-street local bicycle routes, 64 km of principal shared paths along the rail and freeway reserves, spot improvements to overcome major barriers to cycle access and regional recreational paths along the river and coastal foreshore areas. The proposed network density aims to establish a strategic network grid of 500 m spacing with priority access to the major employment centres.

Implementation of the Perth Bicycle Network Stage 1 took place between 1997 and 2001 at a cost of A\$24 million (€13.9 million or £8.7 million) with

Stage 2 commencing in 2002 at a further A$20 million (€11.6 million or £7.3 million). Annual screenline surveys have recorded significant increases in cycle use at the spot improvement sites and along the principal routes. Strong migration of corridor traffic has also been recorded on some local bicycle routes.

The most recent annual monitoring report (ARRB, 2001) shows that cycle numbers at the central business district cordon count increased 19.8% between 1996 and 2001. Increases at 23 network-based screenline sites across the metropolitan area were 90.7% more cycle traffic between 1999 and 2001.

18.4   Non-built approach (TravelSmart)

An innovative approach to achieving increases in cycling, walking and public transport in Perth has been developed as a result of understanding mobility patterns and taking a 'situational' approach to consider individual trips, rather than systems, as the basis for change. The traditional approach to changing community behaviour, especially in the health promotion area, is social marketing (Brög et al, 1999). However, there is little evidence that traditional social marketing campaigns based on the health promotion model are effective at changing individuals' travel behaviour. Brög argues traditional social marketing focused on target audiences is not appropriate to change travel behaviour on the basis that:

• People's travel decisions are based as much on their environment as their attributes;
• People's misperceptions of cycling and public transport are best improved through direct experience of the modes; and
• People need help to identify which trips can be used by alternative modes, which is different for each household and each household member.

Individualised Marketing, a specialised dialogue marketing technique developed by Socialdata, establishes direct contact with individuals to take them through a process that identifies the real demand for information and motivates them more easily to think about and change their behaviour. It has been successfully applied to more than 75 projects in 13 European cities.

Studies of mobility in Perth reveal that there is strong potential to increase cycle use within the existing infrastructure. This has led to a suite of non-built solutions being developed and tested with local communities. The name given to this programme is TravelSmart. Individualised Marketing is the major TravelSmart programme.

The soft policy approach offered by TravelSmart Individualised Marketing has become a mainstream programme in Western Australia with a A$24 million (€13.9 million or £8.7 million) plan to deliver information, motivation and travel experiences to 650,000 people (one-half of the metropolitan population) by 2010. The first half of this programme has partial funding commitment to 2004.

18.4.1   Understanding mobility – baseline surveys

Between 1997 and 2001 random sample travel diary surveys were conducted in 11 of the 30 local government areas across Perth. The travel diary approach collects full information on mobility including the journey purpose, time of day, mode, travel time and distance. The data is used in conjunction with the Geographical Information System (GIS) to run community learning processes on the travel opportunities, barriers and patterns in each local area.

Further 'in-depth' research has been completed for five local government areas representing a mix of inner, middle and outer metropolitan areas (Socialdata, 2000b). The in-depth approach is to 'reality check' the alternative travel options for every trip recorded in the travel diary surveys. This analysis is supported by follow up interviews identifying the awareness, perception and choice barriers currently preventing individuals from using real alternatives. Considering each person's attitudes and skills individually and then further examining their options for every trip is the 'situational approach' pioneered by Socialdata.

The analysis reveals that a reduction of car use is possible in principle. This in-depth research has identified that for Perth potentially:

• Nearly two-thirds of all trips could be undertaken by environment-friendly modes; or
• Seven out of eight trips could be made by motorised private modes.

These are theoretical potentials, but they give information on the possible shifts between motorised private modes and the alternatives. The current travel patterns in the areas of Perth surveyed show that around 80% of trips are made by car (as driver or passenger) and 20% by the alternatives (walking, cycling and public transport).

The research also shows that almost half of the current car trips are in principle replaceable by the environment-friendly modes:

• around a fifth by public transport;
• about a quarter by cycling; and
• around a fifth by walking.

The pie chart (Fig. 18.1) shows that one-fifth of all trips were made by walking, cycling or public transport (the environment-friendly mode (EFM) segments). Four-fifths are made by car as driver or passenger (the motorised private mode (MPM) segments). The in-depth research shows that for 39% of all trips the car is the only real option (e.g. for long distances, carrying grocery goods or taking passengers) and for 13% of trips the environment-friendly modes are the only option (i.e. a car or licensed driver is not available). This leaves 48% of trips where there are travel choices between the car and the alternatives. Some (7%) are currently chosen to be made by environment-friendly options; most (41%) are made by car. It is these 41% of all trips (around half of car trips) that is the focus of TravelSmart. The systems and services exist for 41% of trips to be changed to walking, cycling or public transport. It is only subjective issues (lack

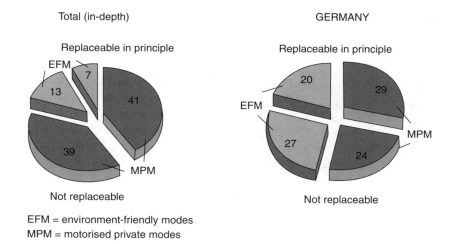

EFM = environment-friendly modes
MPM = motorised private modes

Fig. 18.1   Modal split and possible change of behaviour.

of awareness, misperception, habit) that need to be overcome to achieve real changes.

In summary, these are the potentials for change for the Perth total (in-depth), compared with the situation in Germany. Theoretically, the total share of environment-friendly modes could increase threefold to a mode share of 61% of all trips, provided that all potentials are completely exhausted. But equally the share of car trips could gain 7% of the mode share from the alternatives, raising it to 87% of all trips. These are theoretical limits of the existing system, but they reveal the potential and give information about possible shifts. A number of information and motivational interventions have proven that one-quarter of this potential for environment-friendly mode use is readily realised.

In Germany the environment-friendly mode and car trips are almost equal: 47% of all trips are covered by the alternatives and 53% by the car with potential for change still available.

18.4.2   Potential for cycling

Having established travel patterns and completed the reality checks of car trips, it is possible to examine the potential (those car trips where there are no barriers) for cycling in more detail.

Cycling currently accounts for 3% of trips in the areas studied (travel surveys for 9 of the 30 Perth local government areas) and car trips for 76%. We know that more than half of all car trips are replaceable by walking, cycling or public transport. Some of these car trips may offer more than one alternative (e.g. walking or cycling to the local café/shop). Cycling is measured to have the potential to replace 26% of all car trips or almost half of those that are in principle

Car trips with
no constraints,
bicycle
available (26%)

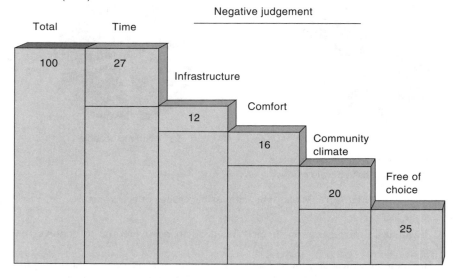

Fig. 18.2   Potential for cycling: total (in-depth) trips within Perth.

replaceable. Such a shift would represent almost a 700% increase in cycle use. Interestingly, this is a similar potential to that identified by a UK cycling model (Rowell and Fergusson, 1991). Clearly these are theoretical maxima and are subject to barriers to achieving real change. The major barriers are discussed in the Perth potentials report (Socialdata, 2000b). Figure 18.2 illustrates these barriers.

There are 26% of all trips currently undertaken by car where there are no constraints against cycling, and a bicycle is available. Of the potential bicycle trips, 27% were not made because of the amount of time it would take, and for 12% of the trips, it was because of the lack of bicycle infrastructure. In 16% of the trips, lack of comfort (car emissions, safety risk, clothing) was an important reason for not cycling, and for 20% of the trips there was a negative view in the community about cycling (that is, a negative community climate).

The remaining 25% of this potential (5% of all trips) are trips where there are no constraints requiring the car to be used, a bicycle available and no negative time impacts or judgements by the individual. These trips are described as 'free of choice'. Soft measures can motivate people to exercise this 'free choice' to cycle and can inform people of their (unjustified) negative judgements. A doubling or trebling of cycle use in Perth should be achievable with the correct soft policy interventions.

From a 'customer' perspective changing four trips (two return journeys) per week to walking, cycling or public transport is almost enough to double the share of EFM trips in Perth. It is this customer focus that underpins the successful TravelSmart dialogue marketing project. This project is discussed in the next section of this chapter.

18.4.3   TravelSmart concepts

Based upon the mobility research presented, the government of Western Australia developed a suite of projects under the 'TravelSmart' brand and the call to action 'it's how you get there that counts'. TravelSmart is a registered trade mark of the government of Western Australia. The TravelSmart projects are based around the principle of voluntary behaviour change, and encompass corporate partnerships, community development and individual empowerment approaches. TravelSmart has active programmes with local governments, schools, workplaces and households. The dialogue marketing project, conducted with households, has been the standout success.

18.4.4   Individualised Marketing

After extensive reviews of world best practice, Transport WA engaged the services of the international company Socialdata to deliver their proven, and proprietary, product of Individualised Marketing within Perth. Transport WA developed an innovative application of Individualised Marketing to extend its reach to cover cycling and walking in addition to the more usual public transport focus.

18.5   Project history

Between 1997 and 2000 WA set out to test the applicability of Individualised Marketing to the Perth situation and to test the inclusion of cycling and walking in the action.

The first project was a pilot with a small random sample (380 households) in the inner city municipality of the city of South Perth. The project had three distinct stages:

*   a benchmark travel survey of existing behaviour (August 1997);
*   implementation of the Individualised Marketing programme (September–October 1997); and
*   travel surveys to measure travel behaviour after the programme – immediately after (November 1997), one year later (September 1998) and two and a half years later (February 2000).

The first two after surveys also contained separate control groups. This was undertaken to identify any changes that may have occurred due to effects outside of the marketing intervention. The sample size for the 12 month survey was 206 households for the marketing group and 207 households for the control group. The only maintenance marketing function performed was the circulation of a newsletter to participants outlining the results of the marketing intervention.

The third after survey contained only the respondents of the second after survey in the South Perth marketing group. There were no further maintenance marketing activities other than some local newspaper advertising for the commencement of a large scale marketing programme in February 2000 (this did not commence until after the third survey had been completed).

The pilot project achieved an average 10% reduction in car trips across all 380 households approached and including non-participants.

On the strength of the success of the 1997 pilot programme WA embarked on a large scale demonstration project in 2000. The large scale project extended to all 35,000 residents of the city of South Perth at a cost of A$1.3 million (€0.77 million or £0.47 million). The 1997 to 2000 before and control group surveys were utilised as the project baseline. The service delivery phase took place between February and May 2000 and the evaluation travel diary survey was conducted in October/November 2000. Annual follow up surveys are planned to track the sustainability of project changes.

Results from the large scale project exceeded those of the pilot. In September 2001 government commitment to the next stage of the programme of TravelSmart dialogue marketing was announced. Subject to annual budget rounds the programme is being extended to up to 370,000 people, being around one-quarter of the metropolitan population. The method and results of the large scale demonstration project are discussed in the following section.

18.6   Methods

Unlike traditional social marketing, which is based on a segmentation philosophy and targeted information for identified groups, Individualised Marketing does not pre-select. Rather, it assumes everybody to be a potential customer. It is the potential customer, rather than the expert, who determines involvement (including the intensity of information and motivation needs) in the marketing process. The process (shown in Fig. 18.3) therefore involves contacting the entire population in the first instance. Depending on individual responses to six or so simple questions asked over the telephone households are segmented into three main groups. This then leads to a different (individualised) implementation programme for each household.

In the large scale programme about 15,300 out of 17,500 city of South Perth residences were matched with names, addresses and telephone numbers. These households were contacted first by letter and then by telephone (T). Of these, 94% could be reached for the self-selection (segmentation) dialogue (S) and a record of the loosely structured discussion maintained by the interviewer.

After answering the questions households are divided into three groups:

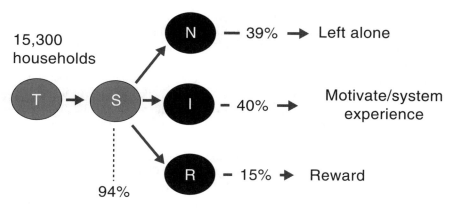

Fig. 18.3   Method diagram.

1   N = not interested; 39% were in this group and these people were left alone. This is important because it saves scarce resources being spent on people who are unlikely to respond to the action.

2   R = regular users of walking, cycling and public transport (15%); these people were given a reward and offered information that in turn led to these people using these modes more.

3   I = interested; 40% were in this group and interested in starting to use alternative modes. These people were given the individually tailored information that they requested and were offered a home visit if it seemed necessary. People having a home visit were sometimes given the opportunity to experience public transport with a 'test ticket'.

In all, 55% of households (8465 households) actively participated in the programme. Almost all of the households that actively participated in the large scale programme requested information by completing and returning a service sheet. Information on public transport was requested by 86% of these households with 73% also requesting information about walking and cycling. This indicates that the vast majority of people interested in public transport are also interested in walking and cycling. This response endorsed the 'situational approach' with people considering a range of travel options for each trip that they make.

The public transport information was localised and personalised to make it as user-friendly as possible for residents. This involved providing convenient wallet size local (bus stop specific) timetables, a local access map/guide with bus routes and bus stops shown, personalised timetables as well as ticket and fare guides and ferry timetables.

The cycling information and incentives consisted of a suite of practical leaflets on aspects of cycling (buying a bike, night riding, children, etc.), a local bicycle route map, a discount voucher for local bike shops (servicing, parts and new bike discounts) and a TravelSmart water bottle.

The walking information was a booklet containing health-related and practical information combined with a motivational walking challenge chart.

All information requests were hand delivered in a TravelSmart shopping bag and doorstep advice offered. Follow up with all households receiving information was conducted and households with additional needs identified. These households were offered visits to counsel them on issues such as how to use public transport ticket machines, restoring their old bike and where to ride. Total customer care and reward characterised the action resulting in 90% approval of the action and hundreds of spontaneous letters of thanks.

During the information, motivation and convincing phases of Individualised Marketing, 7795 households requested around 42,000 items of information, which were personally delivered as 6000 individualised packages. This was followed by around 2600 home visits for walking and cycling and over 600 home visits by bus drivers for public transport. More than 1200 households responded to the questionnaires, with 97% positive about the programme.

18.6.1   Resulting behaviour change

The large scale TravelSmart Individualised Marketing programme, offered to all households in the city of South Perth, was evaluated as a resounding success. It exceeded the projections that were determined possible by the pilot study conducted in 1997, which showed a 10% reduction in car trips.

Random sample surveys of the population were conducted before and after the Individualised Marketing programme to measure its effect. The results from the after survey are expressed across the whole population, not just for those people taking part. The surveys measured a 14% reduction in car trips, with these trips changing to:

- walking (up 35%);
- cycling (up 61%);
- public transport use (up 17%); and
- car sharing (up 9%).

A combination of mode shift and destination substitution (e.g. shopping more locally) resulted in a 16.7% reduction in car travel time. This is assumed to translate into a similar reduction in vehicle kilometres. People spent more time exercising through walking and cycling, and overall they made more trips within the city of South Perth than before.

There was a significant change in the main mode share of trips:

- 8% from car as driver ('before' survey 60%, 'after' survey 52%);
- 2% to car as passenger ('before' survey 20%, 'after' survey 22%);
- 4% to walking ('before' survey 12%, 'after' survey 16%);
- 1% to cycling ('before' survey 2%, 'after' survey 3%);
- 1% to public transport ('before' survey 6%, 'after' survey 7%).

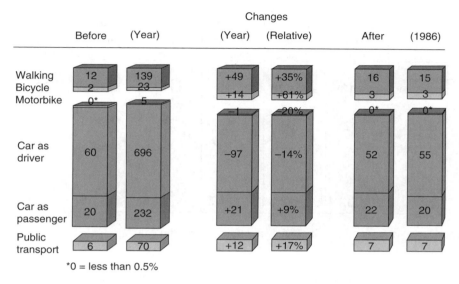

Changes

	Before	(Year)	(Year)	(Relative)	After	(1986)
Walking	12	139	+49	+35%	16	15
Bicycle	2	23	+14	+61%	3	3
Motorbike	0*	5	−1	−20%	0*	0*
Car as driver	60	696	−97	−14%	52	55
Car as passenger	20	232	+21	+9%	22	20
Public transport	6	70	+12	+17%	7	7

*0 = less than 0.5%

Fig. 18.4   South Perth evaluation: large scale application.

The 8% reduction in main mode share of car as driver trips has restored travel patterns to lower car dependence than the 1986 levels of 55%.

The results are presented in more detail in Fig. 18.4. These results show the main mode share (as a percentage) before Individualised Marketing, the travel behaviour changes following the intervention, and a comparison to 1986. The main mode is also expressed as trips per person per year.

Every South Perth resident makes on average 1160 trips per year, the majority of which are made by car, 696 as car as driver and 232 as car as passenger. Trips undertaken by walking, cycling and public transport combined total 232. Following Individualised Marketing, car as driver trips decreased by 97 trips per person per year, and these trips were converted to the environment-friendly modes. Nearly half of the car as driver trips became walking trips (49), around a quarter were undertaken by public transport and cycling (12 and 14) and 21 of the 97 car as driver trips were converted to car as passenger.

This shows that it requires only small changes, two trips a week (for example, walking to the local shops and back) instead of using a car, to achieve significant effects.

18.6.2   Impacts upon cycling

Further analysis (Fig. 18.5) tracks the intensity of the changes among user groups. The change in trips per person per year is presented, as in the summary results, across the whole population. The monitoring method also identifies the share of

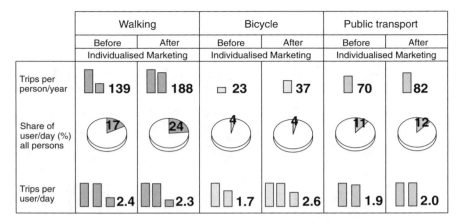

Fig. 18.5   Use of modes: South Perth.

population using any one mode before and after the TravelSmart Individualised Marketing intervention. The number of trips is then expressed relative to the number of users of a mode.

It can be seen that the success of the intervention for walking was to get more people to make walk trips with a small drop in the average number of trips per person (i.e. new walkers made fewer trips than existing walkers). This is a strong result if we can encourage the new walkers to tackle additional trips in the future.

For public transport, the increase in the number of trips is caused by both a number of new public transport users and an increase in the number of trips per user.

For cycling, the substantial increase in trips is dominated by more bicycle use by existing users. The number of new users is small and, due to rounding, does not show up as a significant increase in the cycling population. The impressive result for cycling (61% increase in trips) is tempered by the lack of new users. The project managers are now investigating additional communication methods to strengthen the rate of recruitment of new bicycle users through the Travel-Smart intervention. Information on practical bicycles, bicycle maintenance and health benefits will be added to the existing suite of maps and practical advice.

18.7   Programme evaluation

18.7.1   Financial and socio-economic evaluation

The delivery and evaluation of the Individualised Marketing pilot in 1997 provided the opportunity to undertake a thorough socio-economic and financial evaluation. The results of this assessment of the pilot programme demonstrated a 15 : 1 benefit–cost ratio (Ker and James, 1999). The 15 : 1 estimate is now considered to be an underestimate because it was based upon rapid decay of the behaviour change benefits, a scenario that is now evidenced to be incorrect. Traditionally, transport projects require significant building of roads and infrastructure. They usually have a benefit–cost ratio in the range of 2 : 1 up to 7 : 1.

The favourable assessment of the pilot programme provided a strong case to secure capital works funding from the Western Australian state treasury to deliver the first stage of a large scale programme for the entire population of the city of South Perth (35,000) in 2000.

The delivery and evaluation of the large scale programme together with further evaluation of the pilot programme (to demonstrate sustainability) and in-depth research to determine the potential for delivering Individualised Marketing to other local government authority areas has allowed further development and refinement of the financial evaluation.

The socio-economic costs and benefits expressed as net present value (discounted at 8% over 10 years) from the proposed delivery of the Individualised Marketing programme (650,000 people) is A$1.015 billion (€0.59 million or £0.37 million). This gives a benefit–cost ratio of between 30 : 1 and 37 : 1 depending upon sensitivity tests for a range of factors including congestion, health benefit, road trauma and additional public transport capacity (Ashton-Graham, 2000). The costs and benefits include those incurred by the:

- Overall community – socio-economic value from effects such as air and noise pollution calculated by applying an internationally recognised dollar value that represents the cost that would be incurred in cleaning up or avoiding these pollutions using methods other than Individualised Marketing;
- Government – such as the cost to the government of delivering Individualised Marketing, and the costs and revenues of public transport and health care; and
- Individual – actual dollar savings to people (users) such as savings from lower car running costs.

The socio-economic evaluation does not include public transport fare box revenue. This is because fare box revenue is a financial transaction (transfer payment) only and does not represent a net use of resources (i.e. it is a cost to the user and a benefit to the government and private bus companies).

In financial terms, if we consider additional fare box revenue to the public transport system as a whole (government and private bus companies), the first year rate of return from additional patronage by South Perth residents was 48%. This results in cost recovery, from delivering the Individualised Marketing programme to South Perth (A$1.3 million, €0.76 million or £0.47 million), in a little over two years. Over a 10 year period, the present value of net costs and revenues would be three times the initial investment. These levels of financial return are based on an isolated project requiring no additional bus capacity to meet the additional patronage.

However, the additional fare box revenue from the Individualised Marketing programme does not fully recover additional public transport costs for the government. In Perth, this is primarily because of the contractual arrangements with the private bus operators that require approximately 62% of additional fare box revenue and A\$1.50 (€0.87 or £0.55) per additional km payment to operators (for additional services required during peak periods). As a project the future programme is projected to recover 90% of public transport costs compared to the system average of around 30%. The project improves public transport financial performance and delivers large socio-economic benefits.

18.7.2 Data validation

The travel diary surveys are not the sole data used to evaluate the programme. Corroboration of diary results is carried out against electronic bus ticketing data. An independent audit of the diaries was conducted in 2000 (Goulias, 2000). Further, the reach of the programme (55% active participation) and the reaction (90% approval) add confidence to the measured impacts.

18.8   Next steps

The Department for Planning and Infrastructure Western Australia is delivering TravelSmart dialogue marketing across much of the Perth metropolitan area. Work to refine the information materials and motivation methods is ongoing.

The TravelSmart programme provides both a stand-alone action with significant outcomes and strong value adding to public transport, cycling and walking infrastructure investments. The TravelSmart approach also offers an alternative to some immediate road expansion, and does so at a much lower cost.

18.9   References

ARRB TRANSPORT RESEARCH (2001), *Monitoring usage of PBN Routes 2001*, Perth, Western Australia, Transport WA.

ASHTON-GRAHAM, C (2000), personal communication based upon unpublished treasury documentation (cited in TravelSmart leaflet, 2001, Perth, Western Australia, Transport WA).

BRÖG, W, ERL, E, FUNKE, S and JAMES, B (1999), *Behaviour Change Sustainability from Individualised Marketing*, 24th ATRF Conference, Perth, September.

DEPARTMENTS OF LOCAL GOVERNMENT AND TRANSPORT (1985), *Perth Bike Plan*, Perth, Western Australia, Transport WA.

DEPARTMENT OF TRANSPORT (1996a), *Bike Ahead*, Perth, Western Australia, Transport WA.

DEPARTMENT OF TRANSPORT (1996b), *Perth Bicycle Network Plan*, Perth, Western Australia, Transport WA.

DEPARTMENT OF TRANSPORT (1999), *TravelSmart 10 Year Plan*, Perth, Western Australia, Transport WA.

DEPARTMENT OF TRANSPORT, MAIN ROADS WESTERN AUSTRALIA, MINISTRY FOR PLANNING, FREMANTLE PORT AUTHORITY, WESTRAIL AND METROBUS (1995), *Metropolitan Transport Strategy*, Perth, Western Australia, Transport WA.

GOULIAS, K (2000), *Audit of South Perth Individualised Marketing Evaluation Survey*, Perth, Western Australia, Transport WA.

KER, I and JAMES, B (1999), *Evaluating Behaviour Change in Transport: Benefit Cost Analysis of Individualised Marketing*, Perth, Western Australia, Transport WA.

ROWELL, A and FERGUSSON, M (1991), *Bikes Not Fumes*, Godalming, UK, Cyclists' Touring Club.

SOCIALDATA (2000a), *Mobility Behaviour Melville 2000*, Perth, Western Australia, Transport WA.

SOCIALDATA (2000b), *Potential analysis 'Perth'*, Perth, Western Australia, Transport WA.

SOCIALDATA (2001), *TravelSmart Program – Variation Report 4*, Perth, Western Australia, Transport WA.

Appendix
Sustainable urban travel:
contacts and websites

Contacts

AA (Automobile Association) Foundation for Road Safety Research, Priestley Road, Basingstoke RG24 9NY, Basingstoke, tel. +44 (0) 1256 491925

Access Committee for England, tel. +44 (0) 20 7250 0008

ACEM (European Motorcycle Manufacturers' Association), Brussels, fax +32 (0) 2 230 1683

Advisory Committee on Business and the Environment (ACBE), c/o Department for Trade and Industry (DTI), 4th Floor, 151 Buckingham Palace Road, London SW1 9SS, tel. +44 (0) 20 7215 1984

Allgemeiner Deutscher Fahrradclub (ADFC) (German National Cycling Club), Postfach 107747, D-28077 Bremen, email: kontakt@adfc.de

American Association of State Highway and Transportation Officials (AASHTO), 444 North Capitol Street NW, Suite 249, Washington DC 20001, tel. +1 (0) 202 624 5800, fax +1 (0) 202 624 5806

American Council of the Blind, 115 15th Street NW, Suite 720, Washington DC 2005, tel. +1 (0) 202 467 5081

A Pie (Spanish Pedestrian Association), Madrid, fax +34 (0) 91 5632799

Association for Commuter Transport (ACT), 1 Vernon Mews, Vernon Street, London W14 0RL, tel. +44 (0) 20 7348 1977, fax +44 (0) 20 7349 1988, email: mail@act-uk.com

Association for European Transport (AET). Secretariat: Glenthorne House, Hammersmith Court, London W6 0LG, tel. +44 (0) 20 8741 1516, fax +44 (0) 20 8741 5993, email: info@aetransport.demon.co.uk

Association of Cycle Traders (ACT), 31a High Street, Tunbridge Wells, Kent TN1 1XN, tel. +44 (0) 1892 526081, fax +44 (0) 1892 544278, email: act@cyclesource.co.uk

Association of Train Operating Companies (ATOC), tel. +44 (0) 20 7214 9941

Association Vervoermanagement Nederland (Mobility Management Netherlands), c/o Hans de Vrede, St Jacobsstraat 331, NL-3511 BP Utrecht, tel. +31 (0) 30297 1495, fax +31 (0) 30297 1497, email: bureau@vmnl.nl

Associazione delle Citta italiane per la mobilita sostenible (Association of Italian cities for sustainable mobility), Corso Turati 19/6, I-10128 Torino, tel. +39 (0) 010 552 347115, fax +39 (0) 010 5577852, email: sozzi@fol.it

Associazione Italiana Citta Ciclabili (Italian Association of Cycle Cities), Segreteria Organizzative ANCMA, Via Mauro Macchi 32, I-20124 Milano, tel. +39 (0) 2 6698 1818, fax +39 (0) 2 66982072, email: eicma@galactica.it

A to B Magazine (focusing on integrating cycling and public transport), 19 West Park, Castle Cary BA17 7DB

Austrian Mobility Research (FGM-AMOR), Graz, tel. +43 (0) 316 810451-12

Auto-Free Times (US), PO Box 4377, Arcata, California, CA 95518, tel. +1 (0) 707 826 7775

Bicycle Association (represents UK-based manufacturers and importers of bicycles, components and accessories), Starley House, Eaton Road, Coventry CV1 2FH, tel. +44 (0) 1203 553838, fax +44 (0) 1203 228366

bikerail (consultancy service, supported by the DTLR and the Countryside Agency), 3 Pottery St, London SE1, tel. +44 (0) 20 7252 3296, email: bikerail@pro.net.co.uk

BRAKE, Paper Mews Place, 290 High Street, Dorking RH4 2TU, tel. +44 (0) 1484 559909, email: brake@brake.org.uk

British Cycling Federation (BCF), email: development@bcf.uk.com

British Rickshaw Network (Erica Steinhauer), 40 Cowley Road, Oxford OX4 1HZ, tel. +44 (0) 1865 251620, fax +44 (0) 1865 251134

British Schools Cycling Association (BSCA), email: susanknight@bsca.fsnet.co.uk

Bundesanstalt für Strassenwesen (German Federal Institute for Roads), Brüederstrasse 53, D-51427 Bergisch-Gladbach, tel. +49 (0) 2204 430

Bundesministerium für Verkehr (Austrian Federal Ministry for Transport, Innovation and Technology), Renngasse 5, A-1010 Wien, tel. +43 (0) 1 53464 3105, fax +43 (0) 1 5346 42230

Bundesministerium für Verkehr, Bau- und Wohnungswesen (Federal German Ministry of Traffic and Construction), Postfach 200100, D-53710 Bonn, tel. +49 (0) 228 300-0, fax +49 (0) 228 300-3428 and 300-3429

Cambridge Travel for Work (Cycle-friendly employers) Project, 9 Portugal Place, Cambridge CB5 8AF, tel. +44 (0) 1223 712455, fax +44 (0) 1223 712426

Car-Busters (Magazine and Resource Centre), Kratka 26, 100 00 Praha 10, Czech Republic, tel: +420 (0) 2-781-08-49, fax +420 (0) 2-781-67-27, email: carbusters@ecn.cz

Car Club Publications (publishers of the Car Club Kit, etc.), PO Box 1237, Coventry CV6 3ZB

Car Free Cities Network, Brussels, tel. +32 (0) 2 552 0874, fax +32 (0) 2 552 0889, email: cfc@eurocities.be

CAST (The Centre for Alternative and Sustainable Transport), School of Sciences, Staffordshire University, College Road, Stoke on Trent ST4 2DE, tel. +44 (0) 1782 294111, fax +44 (0) 1782 747167, email: CAST@staffs.ac.uk

Centre d'études sur les reseaux, les transports, l'urbanisme, les constructions publiques (CERTU), 9 rue Juliette Recamie, F-69456 Lyon Cedex 06, tel. +33 (0) 472 745805, fax +33 (0) 472 74 5980, email: inter@certu.fr

Centre for Independent Transport Research in London (CILT), 3rd Floor, University House, 88–94 Wentworth St, London E1 7SA, tel. +44 (0) 20 7247 1302, email: cilt@dial.pipex.com

Centre for Sustainable Transportation, 15 Borden St, Toronto MS5 2M8, tel. +1 (0) 416 923 8839, fax +1 (0) 416 923 6531, email: cstctd@web.net

Child Accident Prevention Trust, 4th Floor, Clerk's Court, 18–20 Farringdon Lane, London EC1R 3AU, tel. +44 (0) 20 7608 3828

Children's Play Council (promoting Home Zones Schemes), 8 Wakley Street, London EC1V 7QE, tel. +44 (0) 20 7843 6016

Cities for Cyclists Project (European Cyclists' Federation), c/o Dansk Cyklist Forbund (Danish Cyclist Federation), Romersgade 5–7, DK-1362 Kobenhavn K, tel. +45 (0) 33 323121, fax +45 (0) 33 327683, email: cfc@dcf.dk

Cleaner Transport Forum, tel. +44 (0) 1273 735802

Cleary Hughes Associates, 51 Wood Lane, Hucknall, Nottingham NG15 6LR, tel. (0115) 964 1869

Club des Villes Cyclables (French National Association of Cycle-friendly Cities), 33 rue du Faubourg-Montmartre, F-75009, tel. +33 (0) 1 156 039214, fax +33 1 (0) 156 039216, email: villes.cyclables@wandaoo.fr

CODATU (Co-operation for the Continuing Development of Urban and Suburban Transportation), Espace Ville, rue Maurice Audin, F-69518 Vaulx en Velin, France, tel. +33 (0) 4 7204 7701, fax +33 (0) 4 7204 7702, email: codatu@entpe.fr

Colibi (Comité de Liaison de Fabricants Européens de Bicyclettes, Boulevard de la Woluwe 46, Box 6, Brussels, B-1200 Belgium, tel. +32 (0) 2 778 6458, fax +32 (0) 2 762 8171

Commission for Integrated Transport (CfIT), 5th Floor, Romney House, Tufton Street, London SW1P 3RA, tel. +44 (0) 20 7944 4101/4813, fax +44 (0) 20 7944 2919, email: cfit@dtlr.gsi.gov.uk

Commission of the European Communities Directorate-General for Transport, Avenue Beaulieu 31, office 5/30, B-1160 Brussels, tel. +32 (0) 2 299 1111, fax +32 (0) 2 299 0895

Community of European Railways (CER), Boulevard de l'impératrice 13, Box 11, B-1000 Brussels, tel. +32 2 525 9080, fax +32 2 512 5231, email: guy.hoedts@cer.be

Confederation of Passenger Transport, London (CPT), 15–19 Kingsway, London WC2N 6UN, tel. +44 (0) 20 7240 3131

CORAC (Swiss Association of Cycling Co-ordinators) (Mr A Staheli), Postfach, CH-4001 Basle, Switzerland

Council for the Protection of Rural England (CPRE), Warwick House, 25 Buckingham Palace Road, London SW1W 0PP, tel. +44 (0) 20 7976 6453, email: cpre@gn.apc.org

Countryside Agency Publications, Wetherby, tel. (0870) 1226 236

CROW (Dutch Centre for Contracts Standardisation), PO Box 37, NL-6710 BA Ede, Netherlands, tel. +31 (0) 8380 20410, fax +31 (0) 8380 21112, email: crow@crow.nl

Cycle Campaigns Network (CCN), c/o 201 Prestbury Road, Cheltenham GL52 3ES, tel. +44 (0) 1242 512881, email: ccn@cyclenetwork.org.uk

Cycle Touring & Countryside Trust, c/o The Beeches, 20 Reddings Road, Moseley, Birmingham B13 8LN, tel. +44 (0) 121 449 9241, fax +44 (0) 121 449 9181

Cycling Project for the North West (CPNW), Grosvenor House, Agecroft Road, Manchester M27 8UW, tel. +44 (0) 161 745 9099

Cyclists' Touring Club publications (CTC), Cotterrell House, 69 Meadrow, Godalming, Surrey GU7 3HS, tel. +44 (0) 870 0060, fax +44 (0) 1483 426994, email: cycling@ctc.org.uk

Danish Council of Road Safety Research, tel. +45 (0) 39 680444

Danish Road Directorate (Vejdirektoratet), Niels Juels Gade 13, Postboks 1569, DK-1020 Kobenhavn K, tel. +45 (0) 33 933338, fax +45 (0) 156335, email: los@vd.dk

Danish Transport Research Institute, Ermelundsvej 101, DK-2820 Gentoft, tel. +45 (0) 39 680444, fax +45 (0) 39 657362, email: dtf@dtf.dk

Department for Transport (replacement in May 2002 for the Department for Transport, Local Government and the Regions (DTLR)), Great Minster House, 76 Marsham street, London SW1P 4DR, tel. +44 (0) 20 7944 3000

DFT (DTLR) Publications Sales Centre, Unit 8, Goldthorpe Industrial Estate, Rotherham S63 9BL, tel. +44 (0) 1709 891318, fax +44 (0) 1709 881673

DFT (DTLR) Free Literature Service, PO Box 236, Wetherby, West Yorkshire LS23 7NB, tel. (0870) 1226 236, fax (0870) 1226 237, email: dtlr@twoten.press.net

DFT (DTLR) Public Inquiries Unit, tel. +44 (0) 20 7890 3000, email: press@dtlr.gov.uk

Department of Health (DOH) (Health Promotions Division), 133–155 Wellington House, Waterloo Road, London SE1 8UG, tel. +44 (0) 20 7972 4453/5

Deutsches Institut für Urbanistik, Postfach 126224, D-10593 Berlin, tel. +49 (0) 30 39001-0, fax +49 (0) 30 3900 1-100

Detour Publications, c/o Transportation Options, 427 Bloor St W, Box 16, Toronto, Ontario, M5S 1X7, Canada

Disabled Persons Transport Advisory Committee (DPTAC), Department for Transport, 1/11 Great Minster House, 76 Marsham St, London SW1P 4DR, fax +44 (0) 20 7890 6102, email: dptac@dtlr.gsi.gov.uk

Ecologica (and the *Journal of World Transport Policy and Practice* – Editor: Professor John Whitelegg), 53 Derwent Road, Lancaster LA1 3ES

Ecoplan, 8/10 rue Joseph Bara, F-75006 Paris, France

Environ (Leicester Environment City), Parkfield, Western Park, Hinckley Road, Leicester LE3 6HX, tel. +44 (0) 116 285 6675

Environmental Change Unit (University of Oxford), 5 South Parks Road, Oxford OX1 3UB

Environmental Law Foundation, Suite 309, 16 Baldwin Gardens, Hatton Square, London EC1N 4RS, tel. +44 (0) 20 7404 1030, fax +44 (0) 20 7404 1032, email: info@elf-net.org.uk

Environmental Transport Association (ETA), 10 Church Street, Weybridge, KT13 8RS, tel. +44 (0) 1932 828822, fax +44 (0) 1932 829015, email: info@eta.co.uk

Environmental and Transport Planning (ETP), Stanford House, 9 South Road, Brighton BN1 6SB, tel. +44 (0) 1273 540955, fax +44 (0) 1273 508791

EPOMM (European Platform on Mobility Management), c/o Eurocities, 18 Square de Meeus, B-1050 Brussels, Belgium, tel. +32 (0) 2 552 0874, fax +32 (0) 2 552 0889, email: info@epommweb.org

European Academy of the European Environment (EAUE), Bismarckallee 46–48, D-14193 Berlin, tel. +49 (0) 30 8959 990, fax +49 (0) 30 8959 9919

European Bicycle Manufacturers Association (EBMA), tel. +33 (0) 1 45 01 9186

European Car-Free Cities, 18 Square de Meeus, B-1050 Brussels, tel. +32 (0) 2 552 0875, fax +32 (0) 2 552 0889, email: cfs@eurocities.be

European Committee of Ministers of Transport (ECMT), fax +33 (0) 1 4524 9742

European Conference of Ministers of Transport/Conférence Européenne des Ministres des Transports, Documentation Centre, 2 rue André Pascal, F-75775 Paris Cedex 16, tel. +33 (2) 1 45 249727, fax +33 (2) 1 45 249742, email: marie-astrid.desoutter@oecd.org

European Cyclists' Federation (ECF), Brussels Office (Contact: Marie Caroline Coppieters, Secretary-General), Rue de Londres 15/3, B-1050 Brussels, Belgium, tel. +32 (0) 2 512 9827, fax +32 (0) 2 511 5224, email: office@ecf.com

(ECF Conference Director: Olly Hatch), 31 Arodene Road, London SW2 3AA, email: oh@velo-city.org

European Environmental Agency (EEA), Copenhagen, tel. +45 (0) 33 367100

European Federation for Transport and the Environment (T&E), Boulevard de Waterloo 34, B-1000, Brussels, Belgium, tel. +32 (0) 2 502 9909, fax +32 2 (0) 502 9908, email: t+e@arcade.is.be

European Federation of Road Traffic Victims (FEVR), PO Box 2080, CH-1222 Geneva 2 Depot, email: fevr@worldcom.ch

European Greenways Association/Association européenne des voies vertes, Gare de Namur, Boîte 27, B-5000 Namur, tel. +32 (0) 81 224256, fax +32 (0) 81 229002, email: aevv.ogwa@gate71.be

European Local Transport Information Service (ELTIS), email: ELTIS@pophost.eunet.be

European Transport Safety Council (ETSC), Rue du Cornet/Hoornstraat 34, B-1040 Brussels, Belgium, tel. +32 (0) 2 230 4106/4004, fax +32 (0) 2 230 4215

Eurostat (European Statistical Office), Luxembourg, tel. +352 (0) 522 51

Euro-velo (proposed European cycle routes being promoted by the European Cyclists' Federation), Brussels, tel. +32 (0) 2 762 2868, fax +32 (0) 2 3003, email: EuroVelo@compuserve.com

Federazione Italiana Amici della Bicicletta, Via Cesariano 11, I-20154 Milano, tel. +39 (0) 2 331 3664, fax +39 (0) 2 331 3663, email: gerry@micanet.it

(FEVR) European Federation of Road Traffic Victims, PO Box 2080, CH-1222 Geneva 2 Depot, email: fevr@worldcom.ch

FGM (Forschungsgesellschaft Mobilität)/AMOR (Austrian Mobility Research), Schoenaugasse 8a/I, A-8010 Graz, tel. +43 (0) 316 810451-21, fax +43 (0) 316 810451-75, email: fgm@fgm-amor.at

Fietserbond enfb (Dutch National Cyclists' Union), Postbus 2828, NL-3500 GV Utrecht, tel. +31 (0) 30 291 8110, fax +31 (0) 0 291 8188

Folding Society (folding bikes), 19 West Park, Castle Cary, Somerset BA7 7DB, tel. +44 (0) 1963 351649

Friends of the Earth (FOE), 25–28 Underwood St, London N1 7JQ, tel. (020) 7490 1555, email: info@foe.co.uk

FOE Publications, 56 Alma Street, Luton, tel. +44 (0) 1582 482297

FUBicy (Fédération française des Usagers de la Bicyclette), 4 rue Brulée, F-67000 Strasbourg, tel. +33 (0) 3 8875 7190, fax +33 (0) 3 8822 5607, email: FUBicy@sdv.fr

Fussverkehr Schweiz (Swiss Pedestrians' Association), Klosbachstrasse 48, CH-8032 Zürich, tel. +41 (0) 1 383 62 40, fax +41 (0) 1 383 97 88, email: info@fussverkehr.ch

Handcycle Association of the UK, 15 Chiltern Way, Sandway, Northwich, Cheshire CW8 2NE, email: tom.doughty@talk21.com

Health Development Agency, Trevelyan House, 30 Great Peter Street, London SW1P 2HW, tel. +44 (0) 20 7413 2637

Health for All Network UK, tel. +44 (0) 151 231 4283, email: ukhfan@livjm.ac.uk

Hedgehog Consortium (integrated transport solutions), 3 Pottery Street, London SE16 4PH, tel. +44 (0) 20 7252 3696, fax +44 (0) 20 7394 7030, email: hedgehog@pro-net.co.uk

Home Zones Scotland Network, c/o Stirling Council Children's Services, tel. +44 (0) 1786 430129

Idevaerkstedet De Frie Fugle (Danish Long-distance cycle routes), Ny Adelgade 5A, DK-1104 Copenhagen K, tel. +45 (0) 33 111175, fax +45 (0) 33 117512, email: friefugl@post8.tele.dk

Institut für Landes- und Stadtentwicklungsforschung (ILS), Deutsche Strasse 5, D-44339 Dortmund, Germany. tel. +49 (0) 231 90510, fax +49 (0) 231 9051155, email: ils.do@t-online.de

Institut Wohnen und Umwelt (Institute of Housing and Environment), Annastrasse 15, D-64285 Darmstadt, tel. +49 (0) 6151 29040, fax +49 (0) 6151 290497, email: iwu-darm-stadt@t-online.de

Institute for European Environmental Policy (IEEP), Dean Bradley House, 52 Horse-ferry Road, London SW1P 2AG, tel. +44 (0) 20 7799 2244, fax +44 (0) 20 7799 2600, email: central@ieeplondon.org.uk

Institute for Transport Studies, University of Agricultural Sciences, Vienna (Universität für Bodenkultur, Wien), A-1190 Wien, Peter Jordan Strasse 82, tel. +43 (0) 1 47654/5301, fax +43 (0) 1 47654/5344, email: sammer@mail.boku.ac.at

Institute for Transportation and Development Policy (US), 115 W 30th Street, Suite 1205, New York NY 10001, tel. +1 (0) 212 629 8033, fax +1 (0) 212 629 8033

Institute of Logistics and Transport (IoLT) (formed on 1 June 1999 from the merger of the Chartered Institute of Transport (CIT) with the Institute of Logistics), 11/12 Buckingham Gate, London SWIE 6LB, tel. +44 (0) 1536 740100, fax +44 (0) 20 7592 3111, email: enquiry@iolt.org.uk

ILT Supply Chain Centre (publications), PO Box 5787, Corby NN17 4XQ, tel. +44 (0) 1536 740100

Institute of Road Safety Officers (IRSO), Pin Point, 1–2 Rosslyn Crescent, Harrow HA1 2SB, tel. (0870) 010 4442, fax (0870) 55 533 7772, email: irso@dbda.co.uk

Institute of Transportation Engineers (ITE), 525 School Street SW, #410, Washington DC 20024-2797, tel. +1 (0) 202 554 8050

Institution of Civil Engineers (ICE), 1 Great George St, London SW1P 3AA, tel. +44 (0) 20 537 3631, fax +44 (0) 20 665 2019

Institution of Highways and Transportation (IHT), 6 Endsleigh Street, London WC1H 0DZ, tel. +44 (0) 20 7387 2525, email: iht@iht.org

International Bicycle Fund, 4887 Columbia Drive South, Seattle WA 98108, USA, tel./fax +1 (0) 206 767 0848, email: ibike@ibike.org

International Mountain Biking Association, PO Box 7578, Boulder, Colorado CO 90306-7578, USA, tel. +1 (0) 303 545 9011, fax +1 (0) 303 545 9026, email: imba@aol.com

International Union of Public Transport/Union Internationale des Transports Publics (UITP), Avenue de l'Uruguay 19, B-1050 Brussels

International Veteran Cycle Association (Ron Sant), 74 Nantwich Road, Middlewich, Cheshire CW10 9HG, tel. +44 (0) 1606 833168

Japan Bicycle Promotion Institute, 9-3, Akasaka, 1-Chome, Minato-ku, Tokyo, Postal code 107-0052, tel. +81 (0) 3 5572 6410, fax +81 (0) 3 5572 6407, email: master@jbpi.or.jp

Joint Mobility Unit (DFT), tel. +44 (0) 20 7387 2233

Landscape Institute, London, tel. +44 (0) 20 7738 9134

Langzaam Verkeer vzw, JP Minckelersstraat 43A, B-3000 Leuven, Belgium, tel. +32 (0) 16 239465, fax +32 (0) 16 290210, email: epomm@langzaamverkeer.be

Light Rail Transit Association (LRTA), PO Box 302, Gloucester GL4 4ZD, or 3 Pine Way, Gloucester GL4 4AE, tel. +44 (0) 1452 411491, fax +44 (0) 1492 425609, email: chairman@lrta.org

(LRTA Publications), 13a The Precinct, Broxbourne EN10 7HT, tel. +44 (0) 20 7918 3116, fax +44 (0) 20 7799 1846

Living Streets (new name in August 2001 for the Pedestrians' Association), 3rd Floor, 31–33 Bondway, London SW8 1SJ, tel. +44 (0) 20 7820 1010, fax +44 (0) 20 7820 8208

Local Authorities Road Safety Officers Association (Wendy Broome), Hertfordshire County Council, Road Safety Unit, 96 Victoria Street, St Albans AL1 3TG, tel. +44 (0) 1727 816960

London Cycling Campaign, Unit 228, Great Guildford Business Centre, 30 Great Guildford Street, London SE1 0HS, tel. +44 (0) 20 7928 7220, fax +44 (0) 20 7928 2318

London School of Cycling, tel. +44 (0) 20 7249-3779

London Walking Forum (Jim Walker, Director), Suite 425, The London Fruit Exchange, Brushfield Street, London EC4 4XP, tel. +44 (0) 20 7239 4034, fax +44 (0) 20 7248 2583, email: info@londonwalking.com

Motorcycle Industry Association, tel. +44 (0) 1203 227427

National Assembly for Wales Transport Directorate, 2nd Floor, Cathays Park, Cardiff CF10 3NQ, tel. +44 (0) 29 2082 6502

National Bicycle and Pedestrian Clearinghouse (US), email: bfareports@aol.com

National Center for Bicycling and Walking, 1506 21st St NW, Suite 200, Washington DC 20036, voicemail: +1 (0) 202 463 6622, fax +1 (0) 202 463 6625, email:ncbw@bikefed.org

National Children's Bureau (including Young TransNet which uses IT and the Internet to assist children and young people in transport research and action), 8 Wakley Street, London EC1V 7QE, tel. +44 (0) 20 7843 6325, fax: +44 (0) 20 7278 9512, email: youngtransnet@ncb.org.uk

National Cycling Archive, The Modern Records Centre, University of Warwick Library, University of Warwick, Coventry CV4 7AL

National Cycling Forum Secretariat (c/o Department for Transport, CLT Division, Zone 3/23, Great Minster House, 76 Marsham St, London SW1P 4DR, tel. +44 (0) 20 7944 2977 (NCF publications via DFT Orderline, tel. +44 (0) 20 7944 2979)

National Institute for the Blind, tel. +44 (0) 20 7387 1266

National Society for Clean Air and Environmental Protection (Mary Stevens), 44 Grand Parade, Brighton BN2 2QA, tel. +44 (0) 1273 878770, fax +44 (0) 1273 606626, email: cleanair@mistral.co.uk

National TravelWise Association (John Sykes), Environment Department, Hertfordshire County Council, County Hall, Pegs Lane, Hertford SG13 8DN, tel. +44 (0) 1992 556117

NEMO (Netzwerk für Mobilitätsmanagement Öesterreich) (Austrian Mobility Management Network), tel. +43 (0) 316 810451-16, fax +43 (0) 316 810451-75, email: nemo@ne-mo.at

New Zealand Sustainable Transport Network (Contact: Elizabeth Yeaman), Transport and Local Government Executive, Energy Efficiency and Conservation Authority, PO Box 388, Wellington, New Zealand, email: Elizabeth.Yeaman@moc.govt.nz

Northern Ireland Roads Service, Belfast, tel. +44 (0) 28 9054 0473

Nottingham Commuter Planners Club (Co-ordinators: Sue Flack and Jeremy Prince), Transport Strategy Team, Nottingham City Council, Development and Environmental Services Department, Exchange Buildings North, Smithy Row, Nottingham NG1 2BS, tel. +44 (0) 115 915 5218, fax +44 (0) 115 915 5483

Oeko-Institut e.V, Postfach 6226, D-79038 Freiburg, fax +49 (0) 761 475437

Office of the Rail Regulator (OPRAF), 1 Waterhouse Square, 138–142 Holborn, London EC1N 2TQ

Organisation for Economic Co-operation and Development (OECD), Paris, fax +33 (0) 1 4524 7876

Oxford University Transport Studies Unit, 1 Bevington Road, Oxford OX2 6NB, tel. +44 (0) 1865 274715

Parliamentary Advisory Council for Transport Safety (PACTS), St Thomas' Hospital, Lambeth Palace Road, London SE1 7EH, tel. +44 (0) 20 7922 8112, fax +44 (0) 20 7401 8740

Planners' Press, American Planning Association, Chicago, tel. +1 (0) 312 786 6344, fax +1 (0) 312 431 9985, email: BookService@planning.org

Planning, Transport, Research and Computing (PTRC), 1 Vernon Mews, Vernon St, London W14 0RL, tel. +44 (0) 20 8741 1516, fax +44 (0) 20 8741 5993, e-mail: ptrc@cityscape.co.uk

RAC Foundation, 14 Cockspur St, London SW1Y 5BH

Railfuture (former Railway Development Society), Roman House, Room 207, The Colour-works, 2 Abbott Street, London E8 3DP, tel. +44 (0) 20 7249 5533, fax +44 (0) 20 7254 6777, email: alix@theRailCampaign.org.uk

Rail Passengers' Council (former Central Rail Users' Consultative Committee), Clements House, 14–18 Gresham Street, London EC2V 7NL, tel. +44 (0) 20 7505 9090, email: rpc@gtnet.gov.uk

Rails to Trails Conservancy (US), 1100 17th Street NW, 10th Floor, Washington DC 20036, tel. +1 (0) 202 331 9696

Re-cycle (sends second-hand bikes to less developed countries), (Merlin Matthews), 60 High Street, West Mersea, Essex CO5 8JE, tel. +44 (0) 1206 382207, email: info@re-cycle.org

Research Unit of Integrated Transport Studies, The Hague, fax +31 (0) 70 351 6558

Road Danger Reduction Forum (RDRF):
   (a) c/o Ken Spence, City of York Road Safety Unit, 9 St Leonard's Place, York YO1 2ET, tel. +44 (0) 1904 551331
   (b) Mike Baugh (RDRF 'New Agenda' newsletter editor), Senior Road Safety Officer, Bath and North East Somerset Council, Riverside, Temple Street, Keynsham, Bristol BS18 1LA, email: Mike_Baugh@BathNES.gov.uk

RoadPeace, PO Box 2579, London NW10 3PW, tel. +44 (0) 20 8838 5102 (office) or +44 (0) 20 8964 1021 (helpline), email: info@roadpeace.org.uk

Royal Association for Disability and Rehabilitation (RADAR), 12 City Forum, 250 City Road, London EC1V 8AF, tel. +44 (0) 20 7250 3222

Royal National Institute for Deaf People, tel. +44 (0) 20 7296 8000

Royal Society for Mentally Handicapped People, tel. +44 (0) 20 7454 0454

Royal Society for the Prevention of Accidents (RoSPA), Edgbaston Park, 353 Bristol Road, Birmingham B5 7ST, tel. +44 (0) 121 248 2000, fax +44 (0) 121 248 2001

Royal Town Planning Institute (RTPI), 41 Botolph Lane, London EC3R 8DL, tel. +44 (0) 20 7929 9494, fax +44 (0) 20 7929 9490, email: online@rtpi.org

Scottish Cycling Development Project contacts:
   (a) David Marsh, Glasgow, tel. +44 (0) 141 287 9374
   (b) Gareth George, Edinburgh, tel. +44 (0) 131 445 7485, email: info@scottishcycling.co.uk

Scottish Executive, Development Department, Transport Division, Victoria Quay, Edinburgh EH6 6QQ

Scottish Executive (publications), Edinburgh, tel. +44 (0) 131 244 7243

Scottish Forum for Transport and the Environment (SFTE), Edinburgh,
fax +44 (0) 131 455 5141

Skills Development Program for Sustainable Transportation (Lucie Maillette, Co-ordinator), 761 Queen Street West, Suite 101, Toronto, Ontario M6G 1JI, Canada

Slower Speeds Initiative, 14 Hopton Road, Hereford HR1 1BE, tel. +44 (0) 1432 277857, email: Paige@speed-campaign-info.fsnet.co.uk

Stationery Office, PO Box 276, London SW8 5DT, tel. orders +44 (0) 20 7873 0011, fax orders +44 (0) 20 7873 8200, email: books.orders@theSO.co.uk

Strategic Rail Authority, 55 Victoria Street, London SW1H 0EU, tel. (020) 7659 6000

Streets for People Network c/o Transport 2000, tel. +44 (0) 20 7388 8366

Sund Bytrafik (Danish Association for Sustainable Transportation), Svend Gønges Vej 33, DK-2700 Brønshøj, Denmark, tel. +45 (0) 31284468, fax +45 (0) 33 916244, email: jean-paul.bardou@dkb.dk

Surface Transportation Policy Project, 1100 17th St NW, 10th Floor, Washington DC 20036, USA, tel. +1 (0) 202 939 3470, email: stpp@transact.org

Sustrans, 35 King Street, Bristol BS1 4DZ, tel. +44 (0) 117 926 8893,
email: sustrans@sustrans.org.uk

Sustrans Scotland, 3 Coates Place, Edinburgh EH3 7AA, tel. +44 (0) 131 623 7600

Swedish National Road Administration/Vagverket, S-781 87 Borlange,
tel. +46 (0) 243 75170, fax +46 (0) 243 75939, email: bert.svensson@vv.se

SWOV (Dutch Institute for Road Safety Research), PO Box 1090, NL-2260 BB Leidschendam, The Netherlands, tel. +31 (0) 70 320 9323, fax +31 (0) 70 320 1261, email: swov@swov.nl

Tandem Club (c/o Peter Hallowell), 25 Hendred Way, Abingdon, Oxfordshire OX14 2AN

Technical Advisers Group (of Local Authorities) (TAG) via tel. +44 (0) 20 7645 9153

TRANSform Scotland (sustainable transport campaign group), 72 Newhaven Road, Edinburgh EH6 5QG, tel. +44 (0) 131 467 7714, fax +44 (0) 131 554 8656,
email: colin@transformscotland.org.uk

TransPlan (Transport and Planning Research Network), (Director: Chris Wood), 45 Beatrice Road, Norwich NR1 4BB, tel./fax +44 (0) 1603 667314,
email: transplan@the-cutting-edge.freeserve.co.uk

TransportAction (Energy Saving Trust/DTLR) Hotline on 0845 602 1425

Transport and Health Study Group, c/o Department of Public Health Medicine, Stockport Health Authority, Springwood House, Poplar Grove, Hazel Grove, Stockport, Cheshire SK7 5BY, tel. +44 (0) 161 419 5467

Transport and Planning Research Network (TransPlan), 45 Beatrice Road, Norwich NR1 4BB, tel. (01603) 667314, email: transplan@the-cutting-edge.freeserve.co.uk

Transport for London (TfL), Windsor House, 42–50 Victoria Street, London SW1H 0TL, tel. +44 (0) 20 7960 6537, fax +44 (0) 20 7960 5128

Transport Directorate, National Assembly for Wales, Cardiff, tel. +44 (0) 29 2082 6502

Transport Management Solutions (Dave Holladay), PO Box 15174, Glasgow G3 6WB, tel. (0141) 332 4733, fax (0141) 354 0076, email: tramsol@aol.com

Transport Planning Society, 1 Great George Street, London SW1P 3AA, tel. +44 (0) 20 7665 2229, fax +44 (0) 20 7799 1325, email: Davies_k@ice.org.uk

Transport Research Laboratory publications (TRL), PO Box 304, Crowthorne, Berkshire RG45 6YU, tel. +44 (0) 1344 770203/770783/84, email: enquiries@trl.co.uk

Transport Stats.com, 108 Kingston Road, London SW10 1LX, tel. +44 (0) 20 8296 9661, fax +44 (0) 20 8542 4515, email: enquiry@transportstats.com

Transport Statistics Users' Group (TSUH), c/o IRN, Davis House, 129 Wilton Road, London SW1V 1LD, tel. +44 (0) 20 7416 8107, fax +44 (0) 20 7828 2030, email: TSUG@irn-research.com or c/o Peter Norgate (Chairman) Mott MacDonald, St Anne House, Wellesley Road, Croydon CR9 2UL, tel. +44 (0) 20 8774 2888, fax +44 (0) 20 8681 5706, email: pjn@mm-croy.mottmac.com

Transport 2000 (T2000), The Impact Centre, 12–18 Hoxton Street, London N1 6NG, tel. +44 (0) 20 7613 0743, fax +44 (0) 20 7613 5280, email: transport2000@transport2000.demon.co.uk

Transport Visions, c/o Dr Glenn Lyons, Transportation Research Group, Department of Civil and Environmental Engineering, University of Southampton, Highfield, Southampton SO17 1BJ, tel. +44 (0) 23 8059 4657, fax +44 (0) 23 8059 3152, email: g.lyons@soton.ac.uk

Transportation Association of Canada (TAC), tel. +1 (0) 613 736 1350

Transportation Options, 761 Queen St West, Suite 101, Toronto ON M6J 1G1, Canada, tel. +1 (0) 416 504 3934, email: detour@web.net

Transportation Research Board (TRB), 2101 Constitution Avenue NW, Washington DC 20418, tel. +1 (0) 202 334 3212

TravelWise Association via Lancashire County Council, tel. +44 (0) 1772 263649

Tricycle Association (David Heighway), 24 Manston Lane, Crossgates, Leeds LS15 8HZ, tel. +44 (0) 113 260 5290 (evenings)

Umwelt Prognose Institut (UPI), Landstrasse 118a, D-69121, Heidelberg, Germany

Union Cycliste Internationale (UCI), 37 route de Chavannes, case postale, CH-100 Lausanne 23, Switzerland, tel. +41 (0) 21 622 0580, fax +41 (0) 21 622 0588, email: admin@uci.ch

Union International des Chemins de Fers (UIC), 16 rue Jean-Rey, F75015 Paris, fax +33 (0) 1 4449 2029

Union Internationale des Transports Publics (UITP), Rue Sainte Marie 6, B-1080 Brussels, tel. +32 (0) 2 673 6100, fax +32 (0) 2 660 1072, email: publications@uitp.com

Vagverket (Swedish National Road Administration), tel. +46 (0) 24375710, fax +46 (0) 24375939, email: bert.svensson@vv.se

Verkehrsclub Deutschland (VCD) (German Environmental Traffic Club), Bonn, tel. +49 (0) 228 985 8510, email: service@vcd.org

Verkehrsclub der Schweiz (VCS) (Swiss Environmental Traffic Club), Bern, tel. +41 (0) 31 318 5411

Verkehrsclub Oesterreich (VCOe) (Austrian Environmental Traffic Club), Dingelstedt-gasse 15, A-1150, Wien (Vienna), tel. +43 (0) 1 893 2697, fax +43 (0) 1 893 2431, email: service@vcoe.at

Vervoermanagement Nederland, tel. +31 (0) 654645175, fax +31 (0) 302971497, email: bureau@vmnl.nl

Veteran Time Trials Association, 50 Briar, Armington, Tamworth, Staffordshire B77 4DY, tel. +44 (0) 1827 68885

Victoria Transport Policy Institute (VTPI), 1250 Rudlin Street, Victoria, BC, V8V 3R7, Canada, tel./fax +1 (0) 604 360 1560, email: litman@vtpi.org

Wheels for All (cycling for disabled people), Margaret Biggs, 1 Enterprise Park, Agecroft Road, Pendlebury, Manchester M27 8WA, tel. +44 (0) 161 745 9099/9088, email: cpnw@cycling.org.uk

Workbike Association of the UK, c/o Andrea Casalotti, Zero Emissions, Real Options Ltd (London), tel./fax +44 (0) 20 7723 2409, email: wa@workbike.org

World Wide Fund for Nature (WWF):
  WWF International, Avenue de Mont Blanc, CH-1196 Gland, Switzerland
  WWF-UK, Panda House, Godalming, Surrey, tel. +44 (0) 1483 426444

Wuppertal Institute (Transport Division), Secretariat, Doeppersberg 19, D-42103 Wuppertal, Germany, tel. +49 (0) 202 249298118, fax +49 (0) 202 2492 263, email: Andreas.Pastowski@wupperinst.org

Young TransNET project, c/o National Children's Bureau, 8 Wakley Street, London EC1V 7QE, tel. +44 (0) 20 7843 6325

Zero Emissions, Real Options Ltd (London), tel./fax +44 (0) 20 7723 2409, email: zero@workbike.org

Websites

Access Board (US) website: http://www.access-board.gov

Accessible Transportation database for travellers with disabilities and the travel industry developed in the USA by the KFH Group, Inc, in conjunction with James D Fleming and Associates: http://www.projectaction.org

Access Sustainable Transport Forum:
http://www.the-commons.org/access/eehome.html

Adventure Cycling Association (USA) (former BikeCentennial):
http://www.adv-cycling.org

Allgemeiner Deutscher Fahrradclub (ADFC) (German National Cycling Club):
http://www.adfc.de

ADFC Forschungsdienst Fahrrad (Cycle Research Service):
http://www-2.informatik.umu.se/adfc/index.html

All Wales Transport Forum: http://www.awtforum.org.uk

Alternative Travel in Towns (ALTER) project: http://www.alter-europe.co.uk

American Association of State Highway and Transportation Officials (AASHTO):
http://www.aashto.org

American Council of the Blind: http://www.acb.org/pedestrian

American Public Transit Association: http://www.apta.com

America Walks (coalition of walking advocacy groups) site: http://americawalks.org/

ARGUS (Arbeitsgemeinschaft Umweltfreundlicher Stadtverkehr) (Austria):
http://www.argus.or.at/argus/

Association for Commuter Transport (ACT): http://www.act-network.demon.co.uk

Association for Commuter Transportation: http://tmi.cob.fsu.edu/act/act.htm

Association for European Transport (AET): http://www.aetransport.co.uk/

Association of Bicycle-friendly Cities & Communities of North-Rhine Westphalia/ Nordrheinwestfalen (Germany): http://www.fahrradfreundlich.nrw.de

Association of British Drivers: http://www.abd.org.uk

Association of Cycle Traders (ACT): http://www.cyclesource.co.uk

Association of Pedestrian and Bicycle Professionals (APBP) (USA): http://www.apbp.org

Association of Transport Co-ordinating Officers (ATCO): http://www.atco.org.uk/

Association Vervoermanagement Nederland (Mobility Management Netherlands):
http://www.vmnl.nl

Associazione Italiana Citta Ciclabili (Italian Association of Cycle Cities): http://ancma.mall.it

A to B Magazine (folding bikes and links between cycling and public transport, including electric bikes): http://www.a2bmagazine.demon.co.uk

Australian Cycling Promotion Fund (a voluntary contribution by companies supplying bicycle products and related services in Australia): http://www.cycling-australia.com

Bicycle Federation of Australia: http://www.bfa.asn.au/

Bicycle Helmet Safety Institute (USA) http://www.bhsi.org/

Bike Bus'ters (Danish cycling and health promotion project): http://www.i4.auc.dk/TRG

Bike Magic (cycling clubs, merchandise outlets and routes): http://www.bikemagic.com

Bike Rail carriage information: http://www.atob.org.uk/Bike_Rail.html

Bikes and Public Transport Information: http://www.a2bmagazine.demon.co.uk

Bikes on buses (Seattle, Washington State, USA) information: http://www.swnw.com and http://www.swnw.com

Bike to Work home page: http://biketowork.itelcom.com/

Brake: http://www.brake.org.uk

British Cycling Federation (BCF): http://www.britishcycling.org.uk

British Horse Society (BHS): http://www.bhs.org.uk

British Medical Association (BMA): http://www.bma.org.uk

British Medical Journal on the web at http://www.bmj.com/

British Schools Cycling Association (BSCA): http://www.bsca.org.uk

British Waterways: http://www.britishwaterways.com or http://www.british-waterways.org

(BW in Scotland): http://www.scottishcanals.co.uk/

Bundesministerium für Verkehr (Austrian Federal Ministry for Transport, Innovation and Technology): http://www.bmv.at

Bundesministerium für Verkehr, Bau- und Wohnungswesen (Federal German Ministry of Traffic and Construction): http://www.bmvbw.de

Bureau of Transportation Statistics (US): http://www.bts.gov

BYPAD (European bicycle policy quality management audit project): http://www.bypad.org

Cambridge Travel to Work (Cycle-friendly employers) Project: http://www.cfe.org.uk

Car-Busters (magazine and resource centre): http://www.antenna.nl/eyfa/cb

Car-Free Cities site: http://www.carfree.com

Carfree.com: http://www.carfree.com

Car-Sharing: http://www.members.aol.com/CarSharing/

CarSharing Consortium (Ecoplan Casebook on):
http://www.ecoplan.org/carshare/cs_index.htm

CAST (The Centre for Alternative and Sustainable Transport, Staffordshire University):
http://www.staffs.ac.uk/geography/CAST

Center for a New American Dream (transport pages):
http://www.newdream.org/transport/

Center for Liveable Communities: http://www.lgc.org/clc

Center for Urban Transportation Research (US): http://www.arch.usf.edu/flctr/projects/

Centre for Education in the Built Environment (transport remit):
http://cebe.cf.ac.uk/resources/index.html?urlindex.html#planning

Centre for Independent Transport Research in London (CILT):
http://www.cilt.dial.pipex.com

CERTU (Centre d'études sur les reseaux, les transports, l'urbanisme et les constructions
publiques): http://www.certu.fr/transports

Children on the Move (EcoPlan, Paris): http://www.ecoplan.org/children

Children's Play Council Home Zones: http://www.homezonenews.org.uk

Cities for Cyclists Project (European Cyclists' Federation) c/o Dansk Cyklist Forbund
(Danish Cyclist Federation): http://www.ecf.com/cfc

City of Portland (Oregon, USA) Bicycling Program: http://www.trans.ci.portland.or.us

Clear Zones: http://www.foresight.gov.uk/transport

Commission for Integrated Transport (CfIT): http://www.cfit.gov.uk

Commission of the European Communities Environment Directorate:
http://europa.eu.int/comm/environment

Commission of the European Communities (CEC) Sustainable Urban Development EU
Framework for Action: http://www.inforegio-cec.eu.int/urban/foram

Community Car Share Network (CCSN) (UK): http://www.carshareclubs.org.uk

Commuter Choice Program: http://www.commuterchoice.com

'Commuter Financial Incentives' chapter of the VTPI (Victoria Transport Policy Institute, Canada) Online TDM Encyclopedia: http://www.vtpi.org/tdm/tdm8.htm

Confederation of Passenger Transport (CPT): http://www.cpt-uk.org/cpt

Congress for New Urbanism (US): http://www.cnu.org

Consortium of Bicycle Retailers (CoBR): http://www.cobr.co.uk

Countrygoer: http://www.countrygoer.org.uk

Countryside Agency: http://www.countryside.gov.uk

CROW (Dutch Traffic Research Organisation): http://www.crow.nl

CTC (Cyclists' Touring Club) http://www.ctc.org.uk

CTC CycleDigest (Acrobat PDF version) available at http://www.ctc.org.uk

Cutting your car use (Anna Semlyen): http://www.cuttingyourcaruse.co.uk

CyberCyclery: http://www.cyclery.com

Cycle Campaigns Network (CCN): http://www.cyclenetwork.org.uk

Cycle City Guides: http://www.cyclecityguides.co.uk

CycleSafe (CTC): http://www.cyclesafe.org.uk

Cycling International: http://www.cycling.org/

Cykelframjandet (Sweden): http://195.84.56.180/cykelfr/

Danish Road Directorate (Vejdirektoratet): http://www.vd.dk

Dansk Cyklist Forbund (DCF) (Danish Cycling Federation): http://webhotel.uni-c.dk/dcf/

David Engwicht (Queensland, Australia) traffic reduction site: http://www.lesstraffic.com/traffic_calming_2.doc

Defining Walkable Communities site (Bicycle Federation of America): http://www.prowalk.org

Department for Transport: http://www.dft.gov.uk

Department of Health (Health Promotions Division): http://www.doh.gov.uk

Detour (Toronto pro-bike anti-car): http://www.web.net/~detour/

Directory of Policy Journals in Transport, Urban Planning and Sustainability (Institute for Sustainability and Technology Policy (ISTP), Murdoch University, Australia) site: http://wwwistp.murdoch.edu.au/research/journal/

Directory of Transportation Libraries and Information Centers (Transportation Division of the Special Libraries Association, USA): http://ntl.bts.fov/tldir/

EcoPlan Sustainable Transport: http://www.ecoplan.org/

(Echte Nederlandse) Fietserbond (ENFB) (Dutch National Cyclists' Union): http://www.pz.nl/fbenfb

Energy Efficiency and Transport: http://www.energy-efficiency.gov.uk/transport

Environmental Protection Agency (EPA) (US) Commuter Choice Program: http://www.epa.gov/orcdizux/transp/comchoice

Environmental Transport Association (ETA): http://www.eta.co.uk

EPA Office of Mobile Sources (US): http://www.epa.gov/oms/transp.htm

EPA Transportation Partners (US): http://www.epa.gov/tp/

EPOMM (European Platform on Mobility Management): http://www.epommweb.org

EU Energy and Transport Directorate: http://www.europa.eu.int/en/comm/dg07/tif/

EU Transport in Figures: http://europa.eu.int/en/comm/dg07/tif/index/htm

Eurobike (travel): http://www.eurobike.com/index.html

European Car-Free Cities Project: http://www.edc.eu.int/second.html

European Car-Sharing: http://www.carsharing.org

European Conference of Ministers of Transport (ECMT): http://www.oecd.org/cem/index.htm

ECMT (traffic crash data): http://www.oecd.org/cem/index.htm

European Cyclists' Federation (ECF): http://www.ecf.com

ECF bicycle research reports index site: http://www.ecf.com/html/research.htm

ECF Cities for Cyclists project: http://www.ecf.com/cfc

ECF cycle helmets debates: http://webhotel.uni2.dk/dcf/helmet/

ECF European Cyclist: http://www.ecf.com/magazine

ECF Euro-velo routes http://www.ecf.com/eurovelo

ECF Velo-City conferences http://www.ecf.com/velocity

European Cyclist Magazine (European Cyclists' Federation): http://www.ecf.com/magazine

European Environment Agency (Copenhagen) Air Pollution in Europe: http://www.eea.eu.int

European Federation for Transport and Environment (T&E), Brussels: http://www.t-e.nu

European Federation of Road Traffic Victims (FEVR), Geneva: http://www.fevr.org

European Target Project (run by Yorkshire & Humber TravelWise in partnership with several European cities including Göteborg and Bremen): http://www.eu-target.net

European transport statistics site: http://www.europa.eu.int/comm/eurostat

European Youth for Action works to reduce the burden of traffic in Europe: http://antenna.nl/eyfa/

Eurostat (provides transportation and crash data for European countries): http://www.europa.eu.int

Euro-velo (proposed European cycle routes being promoted by the European Cyclists' Federation): http://pauli.uni2dk/dcf/htm/eurovelo.htm

FCC (Friends of City Cycling, Budapest, Hungary): http://www.vbb.hu/

Federal Highway Administration (US): http://www.fhwa.dot.gov

FHWA bicycling information: http://www.fhwa.dot.gov/environment/bikeped

FHWA pedestrian program: http://www.ota.fhwa.dot.gov/walk

FHWA traffic calming: http://www.fhwa.dot.gov/environment/tcalm/index.htm

Federal Transit Administration (USA): http:/www.fta.doc.gov.

Federal Transit Administration Commuter Choice: http://www.fta.dot.gov/library/policy/cc/cc.html

FGM (Forschungsgesellschaft Mobilität) / AMOR (Austrian Mobility Research), Graz: http://www.fgm-amor.at

FIAB (Italy) Cycling Group: http://www.arpnet.it/~bici/fiabeng.htm

Folding Society (folding bikes): http://www.whooper.demon.co.uk/foldsoc/

FUBICY (Fédération des Usagers de Bicyclette) (France): http://www.union-fin.fr/natcog/fubicy/

Full Costs of the Car: http://www.flora.org/ago/fullcostscar.html

Fussverkehr Schweiz (Swiss Pedestrians' Association): http://www.fussverkehr.ch/

Green Transport Plan website (DLTR): http://www.local-transport.dtlr.gov.uk/gtp/index.htm

Greenways (Countryside Agency project): http://www.hea.org.uk

Grona Bilister (Swedish Green Motorists): http://www.gronabilister.com

Guide Project (EU project on the concept and practice of 'interchange') (Group for Urban Interchanges Development and Evaluation): http://www.cordis.lu/transport/src/guide.htm

Guide to taking your bike by train (UK): http://www.railinfo.freeserve.co.uk/railtravel/bikes.html

Health Development Agency (successor to the Health Education Authority): http://www.hda-online.org.uk/pubhealth.htm

Healthy Schools Initiative (Department of Health/Department for Education and Employment): http://www.wiredforhealth.gov.uk

Highways Agency: http://www.highways.gov.uk

Highways, Traffic and Transportation Engineers' Resource Centre: http://www.RoadSource.com

Home Zones (1): http://www.ncb.org.uk/cpchxz.htm

Home Zones (2): http://www.homezones.org

Home Zones (Children's Play Council): http://www.homezonenews.org.uk

IG Velo (Interessengemeinschaft Velo – Swiss cycling organisation): http://www.igvelo.ch

Indicators of the Environmental Impact of Transportation: http://www.epa.gov/tp/

Institute for European Environmental Policy (IEEP), London: http://www.ieep.org.uk

Institute for Transportation & Development Policy: http://www.itdp.org

Institute of Highway Incorporated Engineers (IHIE): http://www.ihie.org.uk

Institute of Logistics and Transport (IoLT): http://www.iolt.org.uk/

Institute of Transportation Engineers (ITE) (US): http://www.ite.org

Institute of Transport Studies, University of Leeds (ITS): http://www.its.leeds.ac.uk

Institution of Civil Engineers (ICE): http://www.ice.org.uk

Institution of Highways and Transportation (IHT): http://www.iht.org

Insurance Institute for Highway Safety (US): http://www.highwayssafety.org

Intelligent Speed Adaptation (ISA) project (Aalborg University): http://www.infati.dk/uk/

Interbike: http://www.interbike.com

International Bicycle Fund (Seattle, USA): http://www.ibike.org

International Bicycle Fund bike to work, bicycling promotion and bicycle encouragement ideas: http://www.ibike.org/bikeday.htm

International Car Free Day Consortium Website: http://www.ecoplan.org/carfreeday/

International Car Share Consortium: http://www.ecoplan.org/carshare/

International Council for Local Environmental Initiatives (ICLEI) Car-Sharing: http://www.the-commons.org

International Human Powered Vehicles Association (IHPVA): http://www.ihpva.org

International Police Mountain Bike Association: http://www.ipmba.org/ International

International Road Federation (IRF): http://www.irfnet.org/irfnet

International Road Traffic and Accident Database: http://www.bast.de/htdocs/fachthemen/irtad//english/we2.html

International Union of Railways: http://www.uic/asso.fr

ITS Focus (The UK Association for the Promotion of Intelligent Transport Systems): http://www.its-focus.org.uk

Jane's Road Traffic Management: http://www.janes.com

Japan Bicycle Promotion Institute: http://www.jbpi.or.jp

John Franklin's cycling (including helmets) resources site: http://www.lesberries/co.uk/cycling/cydigest.htm

Journal of Transportation and Statistics: http://www.bts.gov/jts

Journal of World Transport Policy and Practice: http://www.ecoplan.org/wtpp

Laboratoire d'économie de transport (LET): http://www.ish-lyon.cnrs.fr/let

Langzaam Verkeer vzw (Belgium): http://www.langzaamverkeer.be

Le Monde à Bicyclette (Montreal, Quebec, cycling campaign): http://www.cam.org/~lemab/

Light Rail Transit Association (LRTA) (promotes better public transport through light rail, tramway and metro systems): http://www.lrta.org

Living Streets (new name in August 2001 for Pedestrians' Association): http://www.pedestrians.org.uk

London Accident Prevention Councils: http://www.lapc.org.uk

London Car-Share: http://www.london-share.co.uk

London Cycle Network: http://www.londoncyclenetwork.org

London Cycling Campaign (LCC): http://www.lcc.org.uk/

London Driver Improvement Service (LDIS): http://www.ldis.co.uk

London School of Cycling: http://www.londonschoolofcycling.co.uk

London Transport: http://www.londontransport.co.uk

London Walking Forum: http://www.londonwalking.com

Ludgate Public Affairs (free monitoring of government and parliament and analysis of transport and planning issues): http://www.ludgate.co.uk/public-affairs

LUTRAG (1000 Friends of Oregon Land Use Transportation and Air Quality): http://www.teleport.com/~friends/lutraq.htm

MASTER programme (Managing Speeds of Traffic on European Roads): http://www.vtt.fi/yki/yki6/master/master.htm

Mean Streets (US surface transportation policy review): http://www.transact.org

Mile End Park (East London linear ecology park and bridge): http://www.mileendpark.co.uk

Millennium Trails (USA): http://www.millenniumtrails.org

Mimic Project (EU project on the concept and practice of 'Interchange') (Mobility, Intermodality and Interchanges): http://www.cordis.lu/transport/src/mimic.htm

Ministry of Economic Affairs, Energy & Transport for North-Rhine-Westphalia (Nordrhein-Westfalen), Germany: http://www.mwmev.nrs.de

MKM (Slovenia) Cycling Group: http://kamen.uni-mb.si/other/mkm

Molasses road safety schemes database (TRL): http://www.trl.co.uk/molasses

Motorbike Action Group: http://dredd.megn.ucl.ac.uk/www/mag/

Moving the Economy (MTE) (Toronto) on-line database: http://www.city.toronto.on.ca/mte

National Bicycle and Pedestrian Clearinghouse (US): http://www.bikefed.org/clear.htm

National Bikes on Transit Guide (US): http://www.bikemap.com

National Car-Share: http://www.nationalcarshare.co.uk

National Center for Bicycling and Walking (US): http://www.bikewalk.org

National Center for Statistics and Analysis (USA) (collects and analyses traffic crash data): http://www.nhtsa.dot.gov/people/ncsa

National Children's Bureau (including Young TransNet which uses IT and the Internet to assist children and young people in transport research and action): http://www.youngtransnet.org.uk

National Cycle Network Information Service (Sustrans):
http://www.nationalcyclenetwork.org.uk

National Cycling Forum Secretariat: http://www.local-transport-dtlr.gov.uk/ncs/ncs.htm

National Cycling Strategy (Department of Transport) (1996):
http://www.dtlr.gov.uk/dot/ncs/ncs.htm

National Cycling Strategy (promotion site, launched in 2000):
http://www.nationalcyclingstrategy.org.uk

National Highway Safety Administration (US): http://www.nhtsa.dot.gov

National Society for Clean Air and Environmental Protection: http://www.nsca.org.uk

NSCA website for people to calculate their transport emissions:
http://www.travelcalculator.org

National Transportation Week Pedestrian Website (US):
http://www.ota.fhwa.dot.gov/ntw/bikeped.htm

National TravelWise Initiative: http://www.travelwise.org.uk

NEMO (Netzwerk für Mobilitätsmanagement Öesterreich) (Austrian Mobility Management Network): http://www.ne-mo.at

New Zealand Sustainable Transport Network: http://www.eeca.govt.nz

North Central Texas Council of Governments On-line Bike Commuter Guide:
http://nctog.dst.tx.us/envir/bikeped/goview/index.htm

North-Rhine-Westphalia (north-west Germany)/Nordrhein-Westfalen cycling information:
http://www.fahrradfreundlich.nrw.de

North Sea Cycle Route Project: http://www.northsea-cycle.com

Northern Ireland cycling website (Roads Service of Northern Ireland):
http://www.Drdni.gov.uk/roads/cycling

Northern Ireland Road Service (NIRS): http://www.nics.gov.uk

Northern Ireland transport statistics: http://www.doeni.gov.uk/statistics/transport.htm

Nottingham Travel Wise Centre: http://www.nottinghamtravelwise.org.uk

Nottinghamshire County Council traffic and travel information:
http://www.nottsscc/gov.uk/utc

Odense (Denmark) Europe's No. 1 Cycling City by 2002 Project (in Danish):
http://www.cyclecity.dk

Oeko-Institut e.V, Freiburg, Germany: http://www.oeko.de

Online sustainable urban travel resources: http://www.nottingham.ac.uk/sbe/planbiblios

Ordre des Cols Durs UK Cycloclimbing: http://www.ocd.org.uk

Oregon Bicycle and Pedestrian Planning site:
http://www.odot.state.or.us/techserv/bikewalk/

Organisation for Economic Co-operation and Development (OECD) transport pages:
http://www.oecd.org.env/transp

PACTS (Parliamentary Advisory Council for Transport Safety): http://www.pacts.org.uk

Partnership for a Walkable America: http://www.walkable.org

Pedestrian and Bicycle Information Center: http://www.bicyclinginfo.org

PIRATE Project (EU project on the concept and practice of 'Interchange') (Promoting
Interchange Rationale, Accessibility and Transfer Efficiency):
http://www.cordis.lu/transport/src/pirate.htm

Planners Web (Planning Commissioners' Journal): http://www.webcom.com/pcj

Planning for a Car-Free Day: http://www.ecoplan.org/carfreeday/

PTRC Education and Research Services Ltd:
http://www.cityscape.co.uk/users/cz57/index.htm

QueryNet (MassTransport.Com) (World Travel and Transportation Information):
http://www.MassTransport.com

Quiet Roads (Countryside Agency): http://www.quiet-roads.gov.uk

Radfahren (cycling magazine, Germany): http://www.radfahren.de

Railfuture (new campaigning name from March 2001 for the Railway Development
Society): http://www.railfuture.org.uk

Rails to Trails Conservancy (US): http://www.railstotrails.org

Railtrack: http://www.railtrack.co.uk

Re-cycle (sends second-hand bikes to less developed countries): http://www.re-cycle.org

Resource Centre for Urban Design Information (RUDI): http://rudi.herts.ac.uk/

Rides (alternative commuting in the San Francisco Bay areas): http://www.rides.org/

Right of Way (New York): http://www.rightofway.org

Road Busters: http://members.gn.apc.org/~roadbusters/

Road Danger Reduction Forum (RDRF): http://www.rdrf.org.uk

RoadPeace: http://www.roadpeace.org.uk/

Roadsource (US Highway, Traffic and Transportation Engineers' Resource Center):
http://www.roadsource.com/

Road User Exposure to Air Pollution: http://www.eta.co.uk/camp/press.htm

Royal National Institute for the Blind (RNIB): http://www.rnib.org.uk

Royal Society for the Prevention of Accidents (RoSPA): http://www.rospa.org.uk

Royal Town Planning Institute (RTPI): http://www.rtpi.org.uk

Rues pour tous (Swiss Traffic Club Tempo 30 campaign): http://www.rues-pour-tous.ch

Safe and Sound (Department of Health and DTLR School Travel Challenge): http://www.doh.gov.uk/safetrav

SBB/CFF (Swiss Federal Railways): http://www.sbb.ch

Scottish Cycling Database: http://www.scottishcycling.co.uk/

Scottish Cycling Development Project: http://www.scottishcycling.co.uk

Scottish Executive, Edinburgh: http://www.scotland.gov.uk

Scottish Executive (publications): http://www.scotland.gov.uk/publications

SLF (Norwegian cycling group): http://www.oslonett.no/home/slf-bike

Slower Speeds Initiative site: http://www.slower-speeds.org.uk

Smogbusters Australian national website: http://www.environment.gov.au/smogbusters

Spokes (East Lothian Cycling Campaign, Edinburgh, Scotland): http://www.spokes.org.uk

Sportswork bike racks for buses and quick load bike racks (US): http://www.swnw.com

Sprawl Watch Clearinghouse (USA): http://www.sprawlwatch.org

Stockholm Cycle Network Plan: http://www.gfk.stockholm.se/gata-park/cykelplan

Stockley Park Transport Plan: http://www.stockleypark.co.uk

Strasbourg City Council (transport policy background): http://www.sdv.fr/strasbourg

Strategic Rail Authority: http://www.opraf.gov.uk

Sund Bytrafik (Danish Association for Sustainable Transportation): http://sunsite.auc.dk/sound_transport/

Surface Transportation Policy Project (US): http://www.transact.org

Sustainable Access to Leisure Sites and Amenities (SALSA Project, London Borough of Ealing): http://www.ealing.gov.uk/salsa

Sustainable Transportation Monitor (Centre for Sustainable Transportation, Toronto, Canada): http:/www.web.net/~cstctd

Sustrans: http://www.sustrans.org.uk/

Sustrans National Cycle Network information service:
http://www.nationalcyclenetwork.org.uk

Svenska Cykelsallskapet (Sweden): http://www.svenska-cykelsallskapet.se

Swedish National Road Administration (Vaegverket): http://www.vv.se

Swiss Accidents Prevention Office: http://bpa.ch

Swiss Federal Railways (SBB/SBB/CFS): http://www.sbb.ch

Swiss Sustainability Criteria in Transport/Transport and Environment Programme:
http://www.snf.ch/nfp41/home.htm

SWOV (Dutch Institute for Road Safety Research): http://www.swov.nl

T & E (European Federation for Transport and Environment): http://www.t-e.nu

Tandem Club (UK): http://www.tandem-club.org.uk

Traffic calming information: http://www.its.leeds.ac.uk/primavera/p_calming.html

TrafficLinq (website directory for traffic and transportation experts with links to sites on
ITS, public transport, transportation planning, traffic engineering, road pricing, electronic
toll collection, traffic information, parking, traffic safety, bicycling, commuting, software,
organisations and traffic industry): http://www.trafficlinq.com

Trails and Greenways Clearing House (US): http://www.trailsandgreenways.org

Transact (Transportation Partners and the Surface Transportation Project, STPP)(US):
http://transact.org

TRANSform Scotland (sustainable transport campaign group):
http://www.transformscotland.org.uk

Transport and Development Politics directory:
http://www.preservenet.com/TransDev.html

Transport and Health Study Group (THSG): http://www.nhs.uk/transportandhealth

Transport and Logistics Links site (University of Huddersfield):
http://hud.ac.uk/schools/applied_sciences/transweb.htm

Transport and Social Exclusion: http://www.art.man.ac.uk/transres

Transport Canada: http://www.tc.gv.ca/

Transport Demand Management: http://env.fpm.wisc.edu/cerp/tdm.html

Transport for Liveable Communities Network: http://www.ticnetwork.org

Transport for London: http://www.transportforlondon.gov.uk

Transport Planning Society: http://www.tps.org.uk/

Transport Professional Organisations (The Transport Web): http://transportweb.com/bodies.html

Transport Research & Information Network (TR&IN): http://www.platform8.demon.co.uk

Transport Research Institute, Napier University, Edinburgh, Scotland: http://www.tri.napier.ac.uk/

Transport Research Laboratory (TRL): http://www.trl.co.uk

TRL Molasses database of road safety schemes: http://www.trl.co.uk/molasses

Transport statistics site (DTLR): http://www.transtat.dtlr.gov.uk

Transport Statistics Users Group (TSUH): http://www.irn-research.com/tsug.html/

Transport 2000: http://www.transport2000.org.uk

Transport Visions Network (via the Southampton University Transportation Research Group website): http://www.trg.soton.ac.uk/research/TVNetwork

Transport Web: http://transportweb.com

TransportAction (energy saving trust/DTLR): http://www.transportaction.org.uk

Transportation Action Network (US) (including Mean Streets annual report): http://www.transact.org/

Transportation and Community and System Preservation Pilot Program: http://tscp-fhwa.volpe.dot.gov/

Transportation Association of Canada (TAC), Ottawa: http://www.tac-atc.ca

Transportation Research Board (US): http://www.nas.edu/trb/reb.html

Transportation Research Information Services (TRIS): http://www.nas.edu/trb/about/tris.html

Transportation Resources: http://www.ldc.lu.se/~ttsak/nyborta.htm

TravelWise site: http://www.travelwise.org.uk

Travelwise Net: http://www.travelwisenet.com

Trento Cycling Pages (touring and off-road cycling in Europe and the Mediterranean): http://www-math.science.unitn.it/Bike

TRIS (Transportation Research Information Service) (the world's largest and most comprehensive bibliographic database on transportation): http://www.bts.gov/ntl/tris/ or http://tris.amti.com/

TRL (Transport Research Laboratory): http://www.trl.co.uk

TRL Molasses database of road safety schemes: http://www.trl.co.uk/molasses

UIC (Union Internationale des Chemins de Fer): http://www.uic.asso.fr

UITP (International Association of Public Transport): http://www.uitp.com

Umweltbundesamt (Federal German Environment Office), Berlin:
http://www.umweltbundesamt.de

Union Cycliste Internationale (UCI): http://www.uci.ch

Universities Transport Studies Group (UTSG): http://www.its.leeds.ac.uk/utsg/

University College London Transport Studies Unit:
http://www.ucl.ac.uk/transport-studies/

University of California Transportation Center: http://socrates.berkeley.edu/~uctc

University of Technology (Sydney) transport site: http://www.sinta.uts.edu.au/

US Department of Transportation: http://www.dot.gov

US DOT Bureau of Transportation Statistics (BTS): http://www.bts.gov

US Trails: http://www.traillink.com

Vagverket (Swedish National Road Administration): http://www.vv.se

Vauban car-free development (Freiburg, Germany): http://www.vauban.de

Vejdirektoratet (Danish Road Directorate): http://www.vd.dk

Velo-Journal (Switzerland): http://www.velojournal.ch

Velo-Mondial Secretariat: http://www.velomondial.net

Velonet: http://www.cycling.org

Verkerhrsclub der Schweiz (VCS-ATE) (Swiss Traffic Club): http://www.verkehrsclub.ch

Verkerhrsclub Deutschland (VCD) (German Traffic Club): http://www.vcd.org

Verkehrsclub Oesterreich (VCOe) (Austrian Traffic Club): http://www.vcoe.at

Vervoermanagement Nederland (Netherlands): http://www.vmnl.nl

Vias Verdes (Spanish Greenways): http://www.viasverdes.com

Victoria Transport Policy Institute (VTPI) (Canada): http://www.vtpi.org

Walkable Communities Inc (US): http://www.walkable.org

Washington State Bicycle Commuting Guide, on-line PDF version:
http://www.olywa.net/leveen/

Wheels for All (cycling for disabled people): http://www.cycling.org.uk

Workbike (USA): http://www.workbike.org/news

World Health Organisation (transport/environment pamphlets): http://www.who.dk/environment.pamphlets

World Transport Policy and Practice electronic edition: http://www.ecoplan.org/wtpp

Wuppertal Institute (Transport Division), Germany: http://www.wupperinst.org

Young TransNet (uses IT and the Internet to assist children and young people in transport research and action): http://www.youngtransnet.org.uk

Y-Tech (revived 'drag and drop' white bicycles in Amsterdam): http://www.dds.nl/~y-tech/

Zero Emissions, Real Options Ltd (London): http://www.workbike.org

Index